Food Production: Perspectives and Challenges

Food Production: Perspectives and Challenges

Edited by **Hugh Brennan**

New York

Published by Callisto Reference,
106 Park Avenue, Suite 200,
New York, NY 10016, USA
www.callistoreference.com

Food Production: Perspectives and Challenges
Edited by Hugh Brennan

International Standard Book Number: 978-1-63239-340-1 (Hardback)

This book contains information obtained from authentic and highly regarded sources. Copyright for all individual chapters remain with the respective authors as indicated. A wide variety of references are listed. Permission and sources are indicated; for detailed attributions, please refer to the permissions page. Reasonable efforts have been made to publish reliable data and information, but the authors, editors and publisher cannot assume any responsibility for the validity of all materials or the consequences of their use.

The publisher's policy is to use permanent paper from mills that operate a sustainable forestry policy. Furthermore, the publisher ensures that the text paper and cover boards used have met acceptable environmental accreditation standards.

Trademark Notice: Registered trademark of products or corporate names are used only for explanation and identification without intent to infringe.

Printed in the United States of America.

Contents

Preface   VII

Part 1   **Sustainable Food Production: Case Studies**   1

Chapter 1   **To Meet Future Food Demands We Need to Change from Annual Grain Legumes to Multipurpose Semi-Perennial Legumes**   3
Henning Høgh-Jensen

Chapter 2   **Issues in Caribbean Food Security: Building Capacity in Local Food Production Systems**   25
Clinton Beckford

Chapter 3   **Food Security and Challenges of Urban Agriculture in the Third World Countries**   41
R.A. Olawepo

Chapter 4   **Climate Change Implications for Crop Production in Pacific Islands Region**   53
Morgan Wairiu, Murari Lal and Viliamu Iese

Chapter 5   **Permanent Internal Migration as Response to Food Shortage: Implication to Ecosystem Services in Southern Burkina Faso**   73
Issa Ouedraogo, Korodjouma Ouattara, Séraphine Kaboré/Sawadogo, Souleymane Paré and Jennie Barron

Chapter 6   **Strengthening Endogenous Regional Development in Western Mexico**   87
Peter R.W. Gerritsen

Chapter 7   **Achieving Household Food Security: How Much Land is Required?**   103
P. Ralevic, S.G. Patil and G.W. vanLoon

Chapter 8   Enhanced Food Production by Applying
a Human Rights Approach – Does Brazil Serve
as a Model of Best Practice?   119
Hans Morten Haugen

Part 2   Scientific Methods for Improving
Food Safety and Quality   143

Chapter 9   Physical Factors for Plant Growth
Stimulation Improve Food Quality   145
Anna Aladjadjiyan

Chapter 10   Rapid Methods as Analytical Tools for Food and Feed
Contaminant Evaluation: Methodological
Implications for Mycotoxin Analysis in Cereals   169
Federica Cheli, Anna Campagnoli,
Luciano Pinotti and Vittorio Dell'Orto

Chapter 11   Natural Hormones in Food-Producing Animals:
Legal Measurements and Analytical Implications   189
Patricia Regal, Alberto Cepeda and Cristina A. Fente

Chapter 12   Milk Biodiversity: Future Perspectives of
Milk and Dairy Products from Autochthonous
Dairy Cows Reared in Northern Italy   215
Ricardo Communod, Massimo Faustini,
Luca Maria Chiesa, Maria Luisa Torre,
Mario Lazzati and Daniele Vigo

Chapter 13   Aluminium in Acid Soils: Chemistry,
Toxicity and Impact on Maize Plants   231
Dragana Krstic, Ivica Djalovic,
Dragoslav Nikezic and Dragana Bjelic

Chapter 14   Genetic Characterization of Global Rice
Germplasm for Sustainable Agriculture   243
Wengui Yan

Permissions

List of Contributors

Preface

The perspectives and challenges related to the field of food production are covered in this book. It talks about food production, manufacturing and certain issues related with its demand and supply in various countries of the world. Due to the prevailing food crisis, it is very important that a proper technique of food production is adopted and quality and safety of the food produced is also checked. This book discusses social issues like food shortage which emerge from insufficient food production in the third world countries. There is special focus on sustainable food production case studies in countries like semi-arid Jamaica, Africa, Caribbean, Nigeria, Pacific Island, Mexico and Brazil. This book also highlights the problems like quality check and safety and basically deals with scientific methods for improving food quality and safety. It discusses, in detail, the alternate method of control on deterioration of food, which is physical, rapid and analytical. Other problems related to animal husbandry, dairy production and hormones in food producing animals; and approaches and tasks in maize and rice production, have been discussed in the book.

The researches compiled throughout the book are authentic and of high quality, combining several disciplines and from very diverse regions from around the world. Drawing on the contributions of many researchers from diverse countries, the book's objective is to provide the readers with the latest achievements in the area of research. This book will surely be a source of knowledge to all interested and researching the field.

In the end, I would like to express my deep sense of gratitude to all the authors for meeting the set deadlines in completing and submitting their research chapters. I would also like to thank the publisher for the support offered to us throughout the course of the book. Finally, I extend my sincere thanks to my family for being a constant source of inspiration and encouragement.

Editor

Part 1

Sustainable Food Production: Case Studies

To Meet Future Food Demands We Need to Change from Annual Grain Legumes to Multipurpose Semi-Perennial Legumes

Henning Høgh-Jensen
Department of Environmental Science,
Aarhus University,
Denmark

1. Introduction

The last meal of an Iron Age man buried in a Danish bog included at least 60 plant species, including barley, linseed and species we now consider weeds. A modern man relies in contrast on a remarkable small number of crop plants, mostly cereal stables like wheat, rice and maize (Evans, 1998). Both land and people in Sub-Saharan Africa are suffering. Natural resource management is in distress and most rural Africans remain poor and food insecure despite widespread macroeconomic, political and sectorial reforms. Most predictions are that these Africans will remain food insecure in the foreseeable future (Pinstrup-Andersen & Pandya-Lorch, 2001). Innovations are, however, changing this landscape much faster than we could expect.

A market-oriented agriculture has been promoted by many agents of change. And change is happening. The last 10 years a renewed optimism has taking root in the fact that a number of African countries are demonstrating high economic growth rates (Radelet, 2010). We do not know the winners and the losers yet – just that they are there. Not all farmers will have the capacity to join the market orientations by high-value commodities. They are simply not able to innovate.

Nitrogen is a major limiting nutrient for food production but the growing demand for food is met in two ways. One through fossil fuel driven fixation of nitrogen, Haber-Bosch nitrogen (Erisman et al., 2008), is one way and symbiotic fixation of nitrogen, leguminous nitrogen (Giller, 2001), is the other. Feeding approximately half of humanity is made possible by Haber-Bosch nitrogen, the other half by leguminous nitrogen. With the current focus on reducing emissions of greenhouse gasses while simultaneously increasing the biomass production for food, fibre, feed and fuel, the use efficiency of the leguminous nitrogen must be improved.

Annual grain legumes basically satisfy their own need for nitrogen via their capability for fixing atmospheric nitrogen (see e.g. Unkovich et al., 2010). However, they seldom contribute much to soil fertility or to subsequent crops. Further, due to their annual structure they must be reseeded every season with consequences for investing resources and potential susceptibility for unfavourable growth conditions during the renewed crop establishment phase.

Legume seeds hold a carbon-nitrogen ratio of approximately 10 compared to values up to 30 for cereals. Thus from a diet point of view, grain legumes are very valuable protein sources. This importance has been recognized since ancient history (Cohen, 1977). In addition to the nitrogen located in the grain, some nitrogen pools are located in the residues, which can be utilized for fodder or returned to the soil. Another important leguminous nitrogen pool is in the roots and rhizodeposits (Wichern et al., 2008).

2. Innovation

Innovation is a buzz-word and there are a multitude of definitions. Within the business management literature, innovation is mostly seen as a tool used by entrepreneurs to create a resource that will give them an advantage over their competitors (Drucker, 1985, p. 27). Or more broadly, some see an innovation in an idea, practice or object that is new to the individual; a newness that gives a value to the individual when implemented (Rogers & Shoemakers, 1971, p. 19; Urabe, 1988, p. 3). So we can say that innovation is linked to entrepreneurs and it represents *newness*, it has a relation to *invention* or to its process of *adoption* and is as such both a *process* and an *outcome*, where the final feature of involving change or a *discontinuity* with the prevailing product/service or market paradigm may be the most important.

Innovation can be triggered by many factors. It may be a farmer that explores new possibilities to solve an irritating problem. Or it may be a social way of responding and adapting to changes in access to natural resources, assets or markets. The photo in Figure 1 illustrates an innovation developed by the entrepreneur AMFRI Farm, a private company in Uganda exporting organic fruits, species and pulps overseas. The plastic bag contains just two different chillies, a ginger tuber and lemongrass. Based on this simple combination, the net profit per unit of specie is larger than if each species is sold in bulk amounts.

The innovation is that one bag fits into a busy dinner-shopping westerner who like fresh spices but do not want to buy a whole lot of chillies. The innovation is that it gives an important value to the customer; a value which makes the customer very little sensitive to how many cents that she is paying per individual chilli. And it is all packed in one little bag – ready to be shopped and go! Other examples of such market-oriented value chain innovations are testing of consumers preferences. Jones et al. (2002) describes how green pods of pigeonpea were presented to a UK market segment and how they responded to colour differences in the pods.

A cropping-oriented form of innovations is a more local oriented innovation than market-oriented innovation often are, simply because it has to take into account the spatial variability in soil fertility, precipitation, crop growth preferences, etc. The innovations are often found in the management of the crops. Examples of such documented and published innovation include the development of MBILI systems (Woomer et al., 2004) in Kenya where double rows of legumes and cereals are planted to decrease the competitive pressure on the legumes. Another is the push-pull plant protection technologies developed by the international research institute ICIPE (Khan et al., 2011), which by planting designs try to manipulate the pests.

Fig. 1. Illustration of innovations at AMFRI FARMS LTD.'s packing site in Kampala, Uganda. Photo by Henning Høgh-Jensen.

Other forms of innovations take place in research. This includes the researcher developed but farmer-centred models of plant breeding (Bänzinger et al., 2006) or the introduction of modern varieties of pigeonpea in eastern and southern Africa (Myaka et al., 2006); the latter we will return to later. Innovations also include the attempts to let nutrient additions solve the lack of yields by a micro-dosing approach (Twomlow et al., 2010) that basically reflects the denial of scientists to accept that inorganic fertilizers are not used by farmers for stable cereals, thereby also the failure of the Green Revolution technology package in Africa.

Such approaches generally assume that response curves are favourable, e.g. generating approximately 5 kg of additional grain for each kg of applied inorganic nitrogen. The example of Twomlow et al. (2010), however, demonstrates that the scientists apparently never checked the socioeconomic conditions for the farmers but only tested the geophysical-ecological conditions by trials across multiple farms. They did not ask the simple question to the farmer: "are you able to pay for this investment?"

Three long-term experiments in Africa, one each from Kenya, Nigeria and Burkina Faso, outline the basic concept that we operate under this sub-continent. The extracted outcome from Kapkiyai et al. (1999) presented in Figure 2 illustrates the benefits on crop yields of feeding the soil various carbon sources. Franke et al. (2008) and Bostick et al. (2007) largely support the learnings from these empirical effects of various technologies on yield and soil quality. Table 1 illustrates a practical finding from such trials, namely that the total soil carbon content only change little and slowly but it is in the intermediate carbon pools that the management differences are detectable. Similar findings under Asian conditions have been presented by Wen (1984), by Khan et al. (2007 for North American conditions, and under North Scandinavian conditions by Ågren & Bosatta (1996).

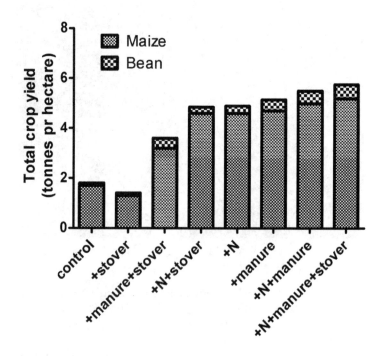

Fig. 2. Long-term effect of residue/manure management in Kabete, Kenya, after 18 yr on a Humic Nitisol. Manure signifies 10 tonnes farm yard manure per hectare and N signifies 120 kg nitrogen and 52 kg phosphorus per hectare in inorganic fertilizers (modified from Kapkiyai et al., 1999).

Soil feeding strategy	Particulate organic carbon (mg C kg^{-1} soil)	Total soil organic carbon (tonnes C ha^{-1})
No addition (control)	713	24.5
Maize stover retention	748	25.5
FYM	1459	26.0
No organic material + inorganic fertilizers	695	24.0
Maize stover retention + inorganic fertilizers	942	25.0
FYM + inorganic fertilizers	1514	26.0
Maize stover retention + FYM + inorganic fertilizers	1613	28.0

Table 1. Long-term effects on soil organic carbon pools of different soil feeding strategies (after Kapkiyai et al., 1999).

3. Crop nutrient limitations for feeding mankind

The use of agrochemicals by many small-scale farmers is reaching a low of 9 kg in Sub-Saharan Africa (FAO, 2003). The reason for this is often sought in the variability of returns to their use in drought-prone climates (e.g. Benson, 1998). However, this simplistic cause-effect relation is inadequate. An important aspect is that blanket recommendations disseminated through the extension systems have been based on the aim of maximizing crop yields (see e.g. Khan et al., 2007). The farmer and the farming families may, however, have many other issues on their agenda that cause such recommendations to fail (Barrett et al., 2002a). Furthermore, the extension systems often target males despite the fact that many of the farmers are female (Gilbert et al., 2002). In the most dramatic manifestations, the non-use of fertilizers can be observed in for example Southern Malawi among smallholders, who nevertheless still attempt to crop soils that are depleted beyond the situations where they can sustain a maize crop without additional nitrogen inputs. A major reason why these small-holders use very little, or no, fertilizer is that they cannot afford to buy it. However, simple cost-benefit analyses have also shown that it is not profitable to apply fertilizer on maize in southern Malawi under present prize conditions (Whiteside, 1998) and that, as a rule of thumb, farmers need an additional 5 kg of grain for every kg of N applied (Twomlow et al., 2010).

Under temperate conditions, leguminous nitrogen can be controlled and transferred to subsequent crops, even when the nitrogen providing crops are incorporated (Eriksen et al., 2004; Høgh-Jensen and Schjoerring, 1997). That makes the use of annual leguminous crops a sustainable approach. Under tropical conditions, however, leguminous nitrogen cannot be conserved and controlled in the same manner and thus transfer to subsequent crops when using annual crops are small (Laberge et al., 2011). That may not make the use of annual leguminous crops an unsustainable approach but clearly the nitrogen pools left by the crops are lost and not available to the subsequent crop. This may be due to leaching following heavy rain showers before the subsequent crops' roots can assess the nitrogen and/or there may be gaseous losses involved following denitrification.

Approaches to quantify the amounts of atmospheric nitrogen fixed by grain legumes erroneously base their estimates on the nitrogen located in harvested biomass above-ground (Herridge et al., 2008; Unkovich et al., 2010). This approach ignores the depositions below-ground associated with crop roots (Gregory, 2006); deposits can benefit subsequent crops (Laberge et al., 2011).

The initial transfers of nitrogen from annual grain legumes to subsequent crops are relatively small (Høgh-Jensen et al., 2005; Laberge et al., 2011). However, Høgh-Jensen et al. (2006) concluded that N-rich leaf litter together with root residues from pigeonpea grown as annual crops well into the dry season is able to enhance a sustainable maize yield that is approximately twice the yield of maize grown in sole stand. The potential of the crops when cultivated as semi-perennials are yet not well-understood but has be partly tested in agroforestry systems (Daniel & Ong, 1990; Odeny, 2007).

4. Semi-perennial legumes in Africa

The current section will investigate the use of semi-perennial legumes in Africa in particular as this continent to a very limited degree has taken up the use of Haber-Bosch nitrogen. Traditional African agriculture is often pictured as inefficient and unproductive. The prejudice in the statement of a Rhodesian administrator in 1926 that intercropping is nothing more than "hit and miss planting in mixtures" (Juggens, 1989, as cited in Barrett et al., 2002a) derogates a view that persists until today. The challenges remain of inducing a sustainable intensification by improving productivity and natural resource management. But much progress thas been made – both in understanding the socio-biophysical complexity and decision making rationality of the farmer as well as in feasible and attractive options.

Scholars are often quite pessimistic when viewing the potential of legumes to contribute further to feeding other crops (Breman & van Reuler, 2002; Palm et al., 1997). But their point of departure is often annual legumes like beans, soybean, cowpea, and groundnuts. Legumes have a C_3 photosynthesis that is characterized by relatively low optimum growth temperatures and water use efficiencies and further, often they do not have an erect growth. Thus, these annual grain legumes are all sensitive to competition for phosphorus, water and light. Further - and maybe even more important - is that evaluations may be based on the false premise that semi-subsistence agriculture cannot support widespread improved natural resource management in contemporary Africa (Barrett et al., 2002b). In other words, improved natural resource management must re-pay the investment on a rather short term basis and thus it is linked to high-value cash crops!

Intercropping of maize and grain legumes is a common practice in many areas in Africa although the rationale behind is not always clear. Pigeonpea is a multipurpose leguminous shrub, which thrives on relatively poor soils (Daniel & Ong, 1990; Odeny, 2007). It is grown with the aim of increasing household cash income and for food, fodder, firewood, and soil fertility improvement. There is an increasing international market for pigeonpea grain. Thus, maize intercropped with long-duration pigeonpea has emerged as a highly productive system with multiple beneficial effects on the farming systems (Kumar Rao et al., 1983; Nene & Sheila, 1990). Consequently, these cropping systems are widespread in some areas of eastern and southern Africa. Intriguingly, however, few kilometres away pigeonpea may not be found and this without apparent socio-economic, cultural or biophysical causes to explain the change. Previous research has shown that integration of pigeonpea into the maize-dominated cropping systems may be a significant low-cost, low-technological step towards increased food production, improved household diet and alleviation of soil degradation and poverty among small holders in semi-arid Africa (Jones et al., 2002; Mergeai et al., 2001; Versteeg and Koudokpon, 1993).

5. Socio-economic aspects of pigeonpea

Africa's declining per capita food production is paradoxical as most African economies are agriculturally based and approx. 75 % of the population lives in rural areas. As a multiple purpose (Table 2), the drought-resistant crop pigeonpea (Figure 3) provides many benefits to the resource-poor farming families. The integration of pigeonpea into the maize-dominated cropping systems of southern and eastern semi-arid Africa has been shown to lead to multiple benefits.

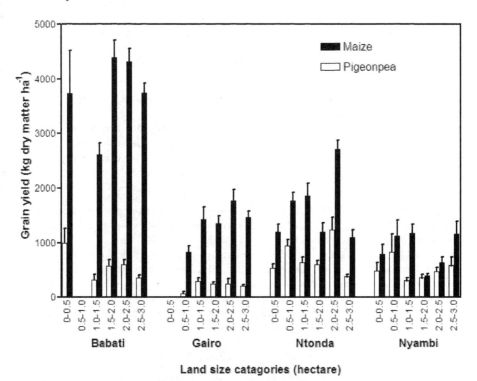

Fig. 3. Relation between land size and mean grain yields from four sites of maize and pigeonpea over two consecutive cropping seasons. Bars represent means±standard error; n varies depending on category (Høgh-Jensen & Odgaard, unpublished data).

Consequently, leguminous nitrogen may be added to the system in substantial quantities if the right uses of the crop are possible. The crop traits to look for may be similar to pigeonpea, i.e. semi-perennial and with use for food, fuel, fodder and of course cash. This example indicates that even under tropical conditions, leguminous nitrogen can efficiently be used to boost production.

As the global economies try to reduce their fossil fuel dependencies, the request of bio-based resources will increase as, in principle, all organic chemicals can be obtained from biomass. Physical and economic constraints must be expected for satisfying this demand in a similar manner as for food. In this light, the potential higher nitrogen use efficiency that semi-perennial legumes offer together with the multipurpose use that is exemplified by pigeonpea will be required to obtain a sustainable development. Furthermore, pigeonpea is drought tolerant and well-adapted to diverse environments. However, one thing is the potential benefits in terms of increased productivity and improved soil fertility, which we have documented above. A different question is how farmers perceive the benefits of integration of pigeonpea.

The main reason for women being less productive than men in relation to maize yields appears to be that women have less access to chemical inputs and technical know-how than men and it is stressed that when women have equal access to such inputs their productivity matches that of men (Gilbert et al., 2002; Gladwin, 2002). Further, the households with most land obtained the highest maize yields. However, pigeonpea showed exactly the opposite effect; favoured by the households with small land access (Odgaard R & Høgh-Jensen H, unpublished). Pigeonpea grain yields were further unaffected (P=0.69; data not shown) by gender of household head.

Due to its multipurpose nature, grain yield of pigeonpea may not meaningfully be taken as a reliable productivity indicator alone. In addition to the grain yield, the importance of green pods in the diet and food security, firewood, soil fertility boasting and livestock benefit (Table 2) must be taken into consideration. Finally cash diversification is a very important issue as possible surpluses of maize often are sold at harvest. As pigeonpea can be sold both as green pods, as dry grain and as seed it is a potential source of income (and food) during a large part of the year.

The high rating of green pods that Tanzania farmers attribute to pigeonpea crops (Table 2) and the very low mature grain yield that was harvested in Malawi (partly shown in Figure 2) indicate that pigeonpea plays an important role in the diet of farmers. As one out of five malnourished children reside in Sub-Saharan Africa (Rosegrant et al., 2001), pigeonpea may play an undervalued role in health and food security through its contribution to a balanced diet. Furthermore, as surplus maize is sold in May-June to cover school fees and debts, among others, the crops of pigeonpea that can be harvested in August-November gives the farmer a small income for which new farming inputs can be acquired. This diversification is obtained with very minor additional labour demands, which is an important issue for female farmers as the stressed by the female participants (Høgh-Jensen & Odgaard, unpublished data), which agrees with Snapp & Silim (2002).

In Malawi, the two sites do not differ in relation to the question of cash and food as being major purposes of growing pigeonpea: 68 % of the farmers at both sites attribute pigeonpea to cash, 95 % to food. However, at Nyambi (a very poor soil site) 100 % attribute pigeonpea to firewood whereas only 68 % do so at Ntonda (a better soil site). The main explanation for the difference in relation to firewood is the question of distance to other sources of firewood. In Tanzania, all farmers at both sites see pigeonpea as a food as well as a cash crop, but also firewood is seen as an important product of the crop. However, as indicated by Table 2, pigeonpea is attributed a number of other functions by the respondents.

	Gairo (% of respondent)	Babati (% of respondent)
Cash	100	100
Food	96	94
Eat dry	32	50
Eat green	60	88
Firewood	89	94
Soil fertility	58	75
Medicinal uses for mainly human	31	13
Feed for livestock	15	94

Table 2. The purpose of growing pigeonpea by Tanzanian farmers at two sites (Høgh-Jensen, 2011).

Although all respondents at the two sites in Tanzania state that pigeonpea is grown both for food and cash purposes, the cash amounts obtained from pigeonpea sales differs in the two areas. While there is a fairly well established market for pigeonpea in Babati and a reasonably well-organized trade in the crop out of the district, the pigeonpea trade in Gairo is different. Here pigeonpea is mainly sold at local markets or to local traders, who ferry the crop mainly to the market in Dar es Salaam, a 4 hours drive at tarmac roads. It appears that it is mainly the men who are involved in pigeonpea trade in Babati, whereas both men and women are involved in the various forms of trade taking place in Gairo.

In Babati, the crop has mainly been introduced as a cash crop and consequently the male sex seems to be the main manager of the income from the crop, and also the main decision maker in relation to use of the crop and the land on which it is grown. Considering the heavy labour input provided by women in farming activities of all kinds, a voice of caution as to the potential role of pigeonpea in improving gender relations needs to be raised. The strong emphasis on the commercial aspects of growing pigeonpea in Babati, combined with the male control of cash income and other resources, may, as discussed, lead to increasing gender inequality instead of reducing it. For similar effects of rapid commercialization see for example Mbilinyi (1991). Moreover, the commercialization of land and other resources in Babati has contributed to constraints for the young generation in relation to getting access to arable land. The growing of pigeonpea has enhanced these constraints.

In Gairo, on the other hand, very few men claim to be main managers of the income. The majority of the respondents irrespective of sex claimed that both husband and wife are involved in selling the crop, controlling the income and making decisions.

The situation in Malawi, especially in Ntonda where more than 50 % of the respondents are women, seems to be similar to the one in Gairo, apart from the fact that women seem to be more involved in the local pigeonpea trade than men.

In many African societies, men and women do have separate income streams and different rights to cash crops grown on the families land. In some cases, it is the females who cultivate the majority of the food crops. Pigeonpea appeared as one of the few examples where the products cut across these patterns as both men and women in various ways and to varying degrees enjoy the multiple benefits, which can be derived from it. Some research has shown, though, that different services associated with the pigeonpea crop, are valued differently by men and women. It has been found for example that women do not value soil fertility services but value food consumption (Snapp & Silim, 2002). Compared to our findings this is much too general a statement. While men seem to be more concerned about soil fertility than women in for example Babati, where the majority of the respondents are male, all the female respondents in the traditionally pigeonpea-growing area of Ntonda maintained that pigeonpea improves soil fertility.

There is effective market demand for both whole grain pigeonpea as well as processed pigeonpea products from eastern and southern Africa in several global markets (Jones et al., 2002). In Malawi, pigeonpea is sold and bought on local markets and the local markets are functioning as outlets to the larger absorbing world market. The Indian market has a huge absorbing capacity in August-September where the home supply is not able to meet the demand. Mostly the harvest in Malawi will be too late for that market, using medium-to-long duration varieties, and thus the crops can not fetch premium prices. No farmers have apparently showed sufficient entrepreneurship to venture into production of short-duration varieties to obtain a high-value cash and export crop. And it is anticipated that a significant backup must be created to make this a realistic strategy for resource-poor farmers.

6. Nutrient limitations to the technology

In the case of cereals, like maize, the relation between harvest grain and total biomass (so-called harvest index) is relatively linear. Figure 4 illustrates this point quite nicely with grain yields ranging from below 1 to as high as 9 tonnes maize grain per hectare. It is also seen that the grain yield was about 54 % of total biomass at all yield levels at Babati and Gairo, the two Tanzanian sites. Hence the conditions for grain development were optimal in Tanzania. Drought conditions can, however, lower the harvest index significantly as grain development may be hindered as demonstrated in the two Malavian sites of Ntonda and Nyambi (Figure 5).

The same data can due to the range in yields be used for investigating the effect of soil fertility on grain yields. Using an upper boundary line analytical approach (Anderson & Nelson, 1974; Walworth & Summer 1988) the following figure can be generated (Figure 6), demonstrating that soil nitrogen availability was determining biomass yields. The variability in soil-N fertility and maize yield from farm to farm were associated with the year span and of intercropping maize and pigeonpea in the past. Hence, it appears that the accumulation during the years of the added effect of pigeonpea in the cropping system can increase soil-N fertility from 1 tonne maize per ha to 9 tonnes maize per ha. A threshold of around 0.10 % soil nitrogen was identified (Figure 6). Consequently farming practice is very important and variability in soil nitrogen availability can be created by the year span of maize and pigeonpea intercropping.

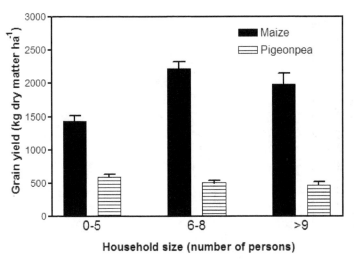

Fig. 4. Relation between household size and mean grain yields of maize and pigeonpea over two consecutive cropping seasons. Bars represent means±standard error; n=480 for pigeonpea and 640 for maize (Høgh-Jensen & Odgaard, unpublished data).

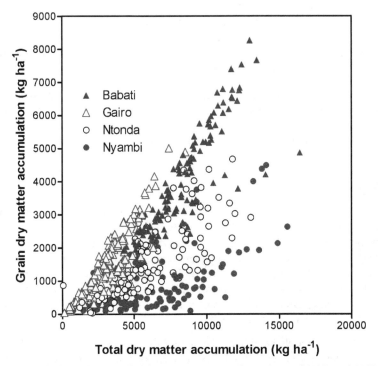

Fig. 5. Maize yields from farmers fields in Tanzania in the season of 2002 and 2002 from four different sites.

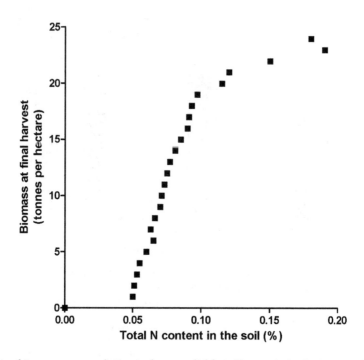

Fig. 6. Maize biomass accumulation in farmers fields in Tanzania in the season of 2002 in relation to soil nitrogen (Nielsen & Høgh-Jensen, 2006).

No relation were seen between the concentration of nitrogen in the ear leaf tissue sampled at the tasseling stage and the final harvest or the level of soil-N because the N uptake at that time still was limited by the capability of the plants to absorb N into new tissue. The main reason is the accumulation of nitrate due to the N-flush from the onset of the rainy season (Høgh-Jensen et al., 2009).

7. Modelling

The flux of nitrogen through the grain legume in mixtures with non-legumes may be a major driving force for sustaining productivity in smallholder systems. This flux is enhanced through biological N_2 fixation and comprises three major flows of nitrogen. First, legume nitrogen is present in the living plant material, that can be harvested and the residues are incorporated in the soil and utilised by subsequent crops. Secondly, the turnover of legume tissue can lead to nitrogen transfer to associated species. Thirdly, the turnover of maize tissue can lead to deposits under conditions where nitrogen is not rapidly mineralised and made available to the plant.

Dynamic simulations models have been developed that are capable of simulating different aspects of the nitrogen cycle and dynamics, as well as biomass production in wheat and other cereals. Furthermore, models have been developed for leguminous grain crops like soyabean (Lim et al., 1990; Thies et al., 1995), faba bean (Stützel, 1995) and groundnut

(Hammer et al. 1995). Nevertheless, none of these approaches take N_2-fixation into account in a mechanistic manner in spite of its importance for nitrogen dynamics of legume-based cropping systems in the short as well as in the longer term. Simulation models can compliment traditional field experiments in researching alternative management options (Carberry et al., 2002).

7.1 The model in brief

DAISY (a soil-plant atmosphere system model) is a one-dimensional open-source (http://www.dina.kvl.dk/~daisy/) agroecosystems model that, in brief, simulates crop growth, water and heat balances, organic matter balance, and the dynamics of N in agricultural soils. This simulation is based on information on management practices and weather data. The simulation of the organic matter balance and the N dynamics is strongly interconnected; hence the organic matter model is considered an integral part of the overall N balance model. Weather data are used as driving variables (input to DAISY). These variables can be viewed as connections to the surrounding environment. The minimum data requirement are daily values of global radiation, air temperature and precipitation (Hansen et al., 1991).

The model allows simultaneous growth of multiple species, which is utilised in the present study. In the crop model, the crop is divided into shoots and roots. The shoot is characterised by dry matter content and leaf area index, while the root system is characterised by dry matter content, rooting depth and root density. Gross canopy photosynthesis may be limited by water and/or N deficiency. The information of water supplied in the driving variables and N status is calculated using information on the actual soil profile.

Nitrogen fixation is simulated as described by Høgh-Jensen (Høgh-Jensen, 1996) and can be understood as a negative feedback regulation of N_2-fixation when external inorganic N is available (Soussana et al., 2002). N_2-fixation is governed by two factors. First, pigeonpea compete with associated crops for inorganic N. This competition is determined by growth and associated N uptake kinetics as described by Michaels Mentons' V_{max}. Secondly, an arbitrary fixation rate, set at 0.95, describes the nitrogen uptake by N_2 fixation relative to the crops optimal N demand at the present development stage according to Ryle et al. (1979).

7.2 Scenario management

The greater complexity of intercropping systems compared to mono crops poses an enormous difficulty for modelling them. However, one way to tackle this problem of complexity is to start by modelling crops in pure stands and to identify key processes that are of importance for the research question.

In the present study, the return of residues and senescence rate of plant tissue was identified as determining the long-term impact of incorporating pigeonpea into the maize cropping system. Following this approach, two crop modules were developed for maize and pigeonpea that fitted to the development and yields obtained in field trials in southern Malawi under farmers conditions. This includes concentrations of N in different organs of the crops as well as pigeonpea leaf litter.

All simulations were conducted by DAISY version 2.47. The temperature function that influence soil organic matter turnover was modified to decrease mineralization in dry soil. The driving weather data were obtained from Chileka airport, 5 km north of Blantyre, Malawi, and supplemented, in particular on global radiation, with the use of the weather generator MarkSim® (www.ciat.cgiar.org).

All scenarios assumed a 100 % return of all residues from the maize to the system. Similarly, pigeonpea residues were assumed incorporated or recycled. However, as stems normally are utilised for firewood only 20 % of stems were recycled. No fertiliser was added in the simulations. These management options agree with real life situations of southern Malawi as the stocking density of ruminants is very low and maize residues are being incorporated into the soils. This often takes place right after harvesting the maize grain. As the crops are grown on ridges, incorporation is easy.

However, maize residues are not incorporated in the simulations before preparing for seeding a new crop. Further, pigeonpea shed their leaves during the dry season. Thus, the amount of green leaves is low at the time of the last harvest of the pods. Due to severe resource constraints, most resource-poor smallholders only apply minimal or no mineral fertilisers to their maize crops. Data on soil profiles were obtained from soil samples and supplemented with literature values (see Mwanga 2004).

7.3 Simulations outcomes

Technologies to improve productivity of nutrient-poor soils have historically depended on high levels of inputs. However, smallholder farmers rarely have sufficient cash to invest in fertilisers. Nor do they have labour or land to invest in the production of green manures or compost. Therefore, biologically based interventions to improve soils and farmers food security must necessarily have their starting point in the cropping already taking place; that is in the cropping of maize. In contrast to the use of agrochemicals like fertilisers, it is possible for even the poorest farmers to intercrop the maize with pigeonpea varieties. This may be why this technology dominates the whole of southern Malawi and parts of southern Tanzania.

The development of crop modules is based on a combination of literature values and experimental data on crop yields and crop quality. Even so, there is a need for further refinement and test of these modules. However, a realistic development - relative and in time - of crop organs was obtained (Figure 7) when simulated in pure stands and intercropped.

In the field situation, dead material does not accumulate to any significant extent on the soil surface as long as the soil is moist. Nevertheless, as pigeonpea shed most of its leaves during the dry season, these leaves accumulate on the dry soil surface although soil fauna still incorporate it into organic matter. Furthermore, the importance of root senescence of a short duration crop, like maize, may not be significant whereas the long duration pigeonpea may have substantial impact.

In an intercropping situation, the input from the pigeonpea to the systems in one cropping season exceeded the input from maize 3-fold of dead organic carbon and 50-fold of dead organic nitrogen. This contrasts to the common view that residues are normally of a low

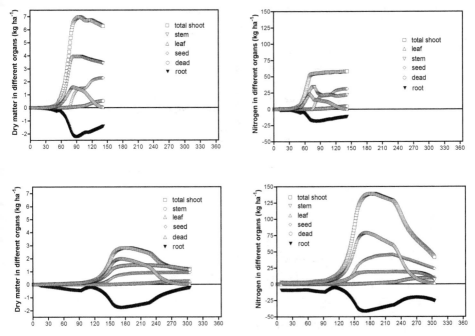

Fig. 7. Simulated dry matter (right) and nitrogen (left) simulations in different pools over one cropping season (days) of maize (top) and pigeonpea grown in mixture (bottom).

quality that is not able to sustain the nutrient requirement of subsequent crops (Palm et al., 1998). However, a high quality of approx. 20 g kg⁻¹ nitrogen of pigeonpea leaves is well documented (see references in Sakala et al., 2003). In a simulation study of maize using the model CERES-Maize, Harrington and Grace (1998) set residue addition to 500 kg dry matter pr hectare without considering a death rate on the roots.

Simulation of pure stand maize demonstrates how small the productivity in these systems is without fertiliser applications (Figure 8). The accumulation of N is very small as the flux of N in the systems are small due to the fact that these fluxed deviates from humus and the small amounts of nitrogen located in previous crop residues. This results in the grain yields of less than 1 tonnes per hectare that are so common for low-input Sub-Saharan Africa (Figure 8).

Adding 70 kg N ha⁻¹ yr⁻¹ improve the grain yields to approx. 2 tonnes per hectare (Table 3). However, similar maize yields can be obtained by intercropping maize with pigeonpea without fertiliser applications (Figure 8; Table 3). In addition, the intercropped systems have the additional beneficial grain yields of pigeonpea (Table 3). Furthermore, firewood, (see stem in Figure 7), which is a scarce resource in many of these areas, is also a pigeonpea by-product that is highly valued. Thus, in terms of productivity of the systems, the intercropped systems come out very convincingly.

The balance between maize and pigeonpea yields fluctuates but at the end of the second season the effect of pigeonpea was very clear (Figure 8; Table 3). However, these balances are strongly affected by the simulation of organic matter turnover in the soil and although DAISY is successful under temperate and subtropical conditions (Jensen et al., 1997), there is

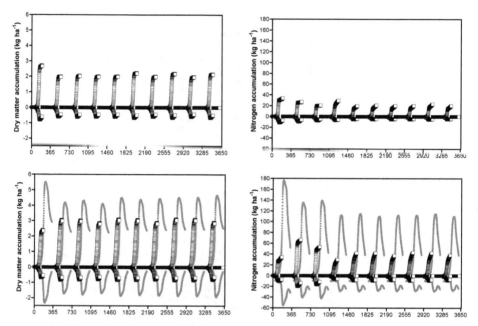

Fig. 8. Simulated dry matter (left) and nitrogen (right) over 10 years of maize (square) in pure stand (top) and maize-pigeonpea (line) intercropped (bottom).

Cropping system	Grain yield (kg DM ha⁻¹)									
	Yr 1	Yr 2	Yr 3	Yr 4	Yr 5	Yr 6	Yr 7	Yr 8	Yr 9	Yr 10
Pure stand maize without fertilizer	1382	816	880	838	835	993	843	969	816	947
Pure stand maize with 70 kg N ha⁻¹ yr⁻¹	2469	2013	1743	2010	2037	1842	1935	2087	1807	1906
Maize in mixture without fertilizer	1459	2094	2091	1904	2138	2104	1943	2147	2121	1943
Pigeonpea in mixture without fertilizer	945	963	931	943	943	963	931	943	943	943

Table 3. Simulated dry matter (DM) grain yields of maize in pure stand when receiving nil or 70 kg nitrogen ha⁻¹ annually or unfertilised maize-pigeonpea over a 10-years period.

a need for testing the model on well-defined long-term trials under tropical conditions. The current dataset contains yield data from maize in Malawi, which means that these maize trials contain two seasons residue effect. When comparing all maize in mixture with all maize in pure stand from this third cropping season the mixture yielded 1910 kg grain dry matter pr hectare compared to 1560 kg grain dry matter pr hectare from the pure stands.

When comparing the nitrate losses from the two different systems, i.e. maize in pure stand and maize-pigeonpea intercropped, the intercropped system is associated with a substantial larger loss of nitrate (data not shown); from less than 20 kg N ha^{-1} yr^{-1} in the pure maize stand to more than 70 kg N in the intercropped stand. The size of this loss is detectable during the second season that the crops are intercropped. This simulated loss is in agreement with the nitrate concentrations found at the end of the dry season (Borgen, 2004).

Total seasonal rainfall and crop yield is not well correlated under precipitation patterns like Malawi. Maize is especially susceptible to periods of droughts during the flowering phase of crop development (see Figure 5). Compared to the cropping season of 2001-2002, February 2002 had nearly half the rainfall of the thirty-year's mean in Malawi, which is critical as February is the normal month for flowering and crop grain yields may be affected.

8. Conclusions

Maize is a major stable food in Sub-Saharan Africa but low soil fertility, limited resources and terminal drought are the main determining factors affecting the productivity of maize in the developing world. Consequently, new technologies and approaches must be innovated as those currently on the shelves do not seem to solve this major challenge (Schlecht et al., 2006).

An obvious approach is to seek among the legume-based technologies for answers to some of the problems. This is not necessarily a new approach (Giller, 2001; Graham & Vance, 2003; Lupwayi et al., 2011; Vesterager et al., 2007) but the scientific world divides in two camps when promoting solutions. One focuses on a reversal of the apparent decline in soil fertility (see e.g. discussions in Kumwenda, 1998; Snapp, 1998; and Schlecht et al., 2006) and another focuses on strategies that utilize crops and crop management to get the best out of the existing biophysical environment. The current chapter acknowledge that legume N in crop residues easily exceed the 50 kg fertilizer-nutrient per hectare recommended by African heads of states at the Fertilizer Summit held in 2006 in Abuja, Nigeria. The chapter, however, also acknowledge that this legume nitrogen is difficult to manage. But most importantly it acknowledge that many of the regions most starving communities are in locations that cannot sustain crops that produce the above-mentioned equivalents of 50 kg fertilizer-nitrogen (Sumberg, 2002). Under such conditions, perennial legumes are one possible important answer to an erratic and harsh environment.

Pigeonpea is one of the few leguminous crops with a high potential to enhance soil fertility due to its complementarity with maize (McCown et al., 1992). Furthermore, associated labour inputs are minimal and seed costs are low compared to other green manure or agroforestry species (Sakala et al., 2003). In risky environments farmers are reluctant to invest in fertilisers (Mwanga, 2004) because they have limited access to cash and they are not sure of the returns.

Simulations confirm empirical data that N-rich leaf litter together with root residues from pigeonpea are able to enhance a sustainable maize grain yield that is approximately twice the yield of maize grown in pure stand. Simultaneously, nitrate leaching is predicted to increase substantially following the inclusion of the pigeonpea in the maize-based cropping systems. The simulated data (Figure 8) agrees partly with the measured data in the sense that the residue effect is measurable after two cropping season.

Maize is frequently intercropped with pigeonpea in eastern and southern Africa (Odeny, 2007; Myaka et al., 2006; Snapp & Silim, 2002). However, the poor seed quality and low plant stand of the pigeonpea crop often result in low residual effects of nutrients -

particularly nitrogen and phosphorus - to the subsequent maize crop. Consequently, such systems must enhance the legume population to make a significant impact. This requires availability of modern seed varieties.

Further, alternatives to intercropping must be developed. In some areas, the rainfall is not sufficient to sustain two intercropped crops (Postel, 2000). Under such conditions, rationing of the pigeonpea should be considered, which would maintain the seeding capacity of the trees but eliminate the hassle and risks by re-establishing the crop.

The market orientation is the predominate paradigm in Agricultural Research for Development (see Høgh-Jensen et al., 2010, for further discussion) but approaches like suggested above with semi-perennial legumes is not in conflict with this paradigm. In contrast it holds the potential for being included in both a market-driven development (Jones et al., 2002) and in poverty alleviation or food security oriented development efforts due to its adaptation to harsh environments.

9. References

Ågren, G.I. & Bosatta, E. (1996). *Theoretical Ecosystem Ecology. Understanding Elements Cycles.* Cambridge University Press, Cambridge.

Anderson, R.L. & Nelson L.A. (1974). A family of models involving straight lines and concomitant experiential designs useful in evaluating response to fertilizer nutrients. *Biometrics,* 31, pp. 308-318.

Barrett, C.B., Place, F., Aboud, A. & Brown, D.R. (2002a). The challenge of stimulating adoption of improved natural resource management practices in African Agriculture. In: *Natural Resource Management in African Agriculture* (eds.: C.B. Barrrett, F. Place & A.A. Aboud), pp. 1-22, CABI Publishing, Wallingford.

Barrett, C.B., Place, F., Aboud, A. & Brown, D.R. (2002b). Towards improved natural resource management in African agriculture. In: *Natural Resource Management in African Agriculture* (eds.: C.B. Barrrett, F. Place & A.A. ABoud), pp. 287-296, CABI Publishing, Wallingford.

Bänziger, M., Setimela, P. S., Hodson, D. & Vivek, B. (2006). Breeding for improved abiotic stress tolerance in maize adapted to southern Africa. *Agriculture and Water Management,* 80, pp. 212-224.

Benson, T.S. (1998). Developing flexible fertiliser recommendations for smallholder maize production in Malawi. In: *Soil Fertility Research for Maize-Based Farming Systems in Malawi and Zimbabwe* (eds.: S.R. Waddington, H.K. Murwira, J.D.T. Kumwenda, D. Hikwa and F. Tagwira), pp. 275-285. Harara, Zimbabwe: Soil Fert Net and CIMMIT-Zimbabwe.

Borgen, S. (2004). *Mineralization potential of tropical soils after cultivation with pure stand maize or maize/pigeonpea intercrop mixture.* BSc thesis. Copenhagen, Department of Agricultural Sciences, KVL. 39 p.

Bostick, W.M., Bado, V.B., Bationo, A., Soler, C.T., Hoogenboom, G. & Jones, J.W. (2007). Soil carbon dynamics and crop residue yields of cropping systems in Northern Guinea Savanna of Burkina Faso. *Soil & Tillage Research,* 93, pp. 138-151.

Breman, H. & van Reuler, H. (2002). Legumes: When and where an option? (No panacea for poor tropical West African soils and expensive fertilizers). In: *Natural Resource Management in African Agriculture* (eds.: C.B. Barrrett, F. Place & A.A. ABoud), pp. 285-298, CABI Publishing, Wallingford.

Carberry, P.S., Probert, M.E., Dimes, J.P., Keating, B.A. & McCown, R.L. (2002). Role of modelling in improving nutrient efficiency in cropping systems. *Plant and Soil*, 245, pp. 193-203.

Cohen, M.N. (1977). *The Food Crisis in Prehistory: Overpopulation and the Origins of Agriculture*. Yale University Press, New Haven, CT.

Daniel, J.N. & Ong, C.K. (1990). Perennial pigeonpea: a multi-purpose species for agroforestry systems. *Agroforestry Systems*, 10, pp. 113-129.

Drucker, P. (1985). *Innovation and Entrepreneurship*. Heinemann, London.

Eriksen, J., Askegaard, M. & Kristensen, K. (2004). Nitrate leaching from an organic dairy crop rotation: the effects of manure type, nitrogen input and improved crop rotation. *Soil Use and Management*, 20, pp. 48-54.

Erisman, J.W., Sutton, M.A., Galloway, J., Klimont, Z. & Winiwarter, W. (2008). How a century of ammonia synthesis changed the world. *Nature Geoscience*, 1, pp. 636-639.

Evans, L.T. (1998). *Feeding the Ten Billion. Plants and Population Growth*. Cambridge University Press, Cambridge.

FAO (2003). *FAOSTAT*. FAO, Rome.

Franke, A.C., Laberge, G., Oyewole, B.D. & Schulz, S. (2008). A comparison between legume technologies and fallow, and their effects on maize and soil traits, in two distinct environments of the West African savannah. *Nutrient Cycling in Agroecosystems*, 82, pp. 117-135.

Gilbert, R.A., Sakala, W.D. & Benson T.D. (2002). Gender analysis of a nationwide cropping system trial survey in Malawi. *African Studies Quarterly*, 6(1), retrieved on http://web.africa.ufl.edu/asq/v6/v6i1a9.htm.

Giller, K.E. (2001). *Nitrogen fixation in tropical cropping systems*. CABI Publishing, Wallingford.

Gladwin, C.H. (2002). Gender and Soil Fertility in Africa: Introduction. *African Studies Quarterly*, 6(1), retrieved on http://web.africa.ufl.edu/asq/v6/v6i1a9.htm.

Graham, P.H. & Vance, C.P. (2003) Legumes: Importance and constraints to greater use. *Plant Physiology*, 131, pp. 872-877.

Gregory, P.J. (2006). Roots, rhizosphere and soil: the route to a better understanding of soil science. *European Journal of Soil Science*, 57, pp. 2-12.

Hammer, G.L., Sinclair, T.R., Boote, K.J., Wright, G.C., Meinke, H. & Bell, M.J. (1995). A peanut simulation model: I. Model development and testing. *Agronomy Journal*, 87, pp. 1085-1093.

Hansen, S., Jensen, H.E, Nielsen, N.E. & Svendsen, H. (1991). Simulation of nitrogen dynamics and biomass production in winter wheat using the Danish simulation model DAISY. *Fertilizer Research*, 27, pp. 245-259.

Harrington, L. & Grace, P. (1998). Research on soil fertility in southern Africa: Ten awkward questions. In: *Soil Fertility Research for Maize-based Farming Systems in Malawi and Zimbabwe* (eds.: S.R. Waddington, H.K. Murwira, J.D.T. Kumwenda, D. Hikwa & F. Tagwira), pp. 3-11, CIMMIT.

Herridge, D.F., Peoples, M. & Boddey, R.M. (2008). Global inputs of biological nitrogen fixation in agricultural systems. *Plant and Soil*, 31, pp. 1-18.

Høgh-Jensen, H. (1996). Simulation of nitrogen dynamics and biomass production in a low-nitrogen input, water limited clover/grass mixture. In: *Dynamics of Roots and Nitrogen in Cropping Systems of the Semi-Arid Tropics* (eds.: O. Ito, C. Johansen, J.J. Adu-Gyamfi, K. Katayama, J.V.D.K. Kumar Rao & T.J. Rego), pp 585-601, Japan International Research Center for Agriculture, Ibaraki.

Høgh-Jensen, H. (2011). Legumes' rhizodeposition feeds a quarter of the world. In: *Clovers: Properties, Medicinal Uses and Health Benefits* (eds.....), pp. xx-xx, Nova Publishers. *Forthcoming*

Høgh-Jensen, H. & Schjoerring, J.K. (1997). Residual nitrogen effect of clover-ryegrass swards on yield and N uptake of a subsequent winter wheat crop as studied by use of ^{15}N methodology and mathematical modelling. *European Journal of Agronomy*, 6, pp. 235-243.

Høgh-Jensen, H., Sakala, W.D., Adu-Gyamfi, J.J., Kamalongo, D. & Myaka, F.A. (2006). Can modelling fill the gap in our knowledge of the nitrogen dynamics of the maize-pigeonpea intercrop? In: *Pigeonpea-based cropping systems for smallholder farmer's livelihood* (ed.: H. Høgh-Jensen), pp. 43-53, DSR Publishing, Copenhagen.

Høgh-Jensen, H., Kamalongo, D., Myaka, F.A. & Adu-Gyamfi, J.J. (2009). Multiple nutrient imbalances in ear-leaves of on-farm cultivated maize in Eastern Africa. *African Journal of Agricultural Research*, 4, pp. 117-112.

Høgh-Jensen,. H., Oelofse, M. & Egelyng, H. (2010). New challenges in underprivileged regions calls for people-centred research for development. *Society and Natural Resources*, 23, pp. 908-915.

Jensen, L.S., Mueller, T., Nielsen, N.E., Hansen, S., Crocker, G.J., Grace, P.R., Klír, J., Körschens, M. & Poulton, P.R. (1997). Simulating trends in soil organic carbon in long-term experiments using the soil-plant-atmosphere model DAISY. *Geoderma*, 81, pp. 5-28.

Jones, R., Freeman, H.A. & Monaco, G.L. (2002). *Improving the access of small farmers in eastern and southern Africa to global pigeonpea markets*. AGREN Network Paper No. 120. January 2002.

Kapkiyai, J.J., Karanja, N.K., Qureshi, J.N., Smitson, P.C. & Woomer, P.L. (1999) Soil organic matter and nutrient dynamics in a Kenyan nitisol under long-term fertilizere and organic matter input management. *Soil Biology & Biochemistry*, 31, pp. 1773-1782.

Khan, Z., Midega, C., Pittchar, J., Pickett, J. & Bruce, T. (2011). Push–pull technology: a conservation agriculture approach for integrated management of insect pests, weeds and soil health in Africa. UK government's Foresight Food and Farming Futures Project. *International Journal of Agricultural Sustainability*, 9, pp. 162-170.

Khan, S.A., Mulvaney, R.L., Ellsworth, T.R. & Boast, C.W. (2007). The myth of nitrogen fertilization for soil carbon sequestration. *Journal of Environmental Quality*, 36, pp. 1821-1832.

Kumar Rao, J.V.D.K., Dart, P.J. & Sastry, P.V.S.S. (1983). Residual effect of pigeonpea (*Cajanus cajan*) on yield and nitronge response of maize. *Experimental Agriculture*, 19, pp. 131-141.

Kumwenda, A.S. (1998). Soil Fertility Research for Maize-based Farming Systems in Malawi and Zimbabwe. In: *Soil Fertility Research for Maize-based Farming Systems in Malawi and Zimbabwe* (eds.: S.R. Waddington, H.K. Murwira, J.D.T. Kumwenda & F. Tagwira), pp. 263-269, Proceedings of the Soil Fert Net Results and Planning Workshop, 7-11 July 1997, Mutara, Zimbabwe. CIMMIT.

Laberge, G., Haussmann, B., Ambus, P. & Høgh-Jensen, H. (2011). Cowpea rhizodeposition and belowground transfer of nitrogen to millet in intercrop and rotation on a sandy soil of the Sudano-Sahelian eco-zone of Niger. *Plant and Soil*, 340, pp. 360-382.

Lim, J.T., Wilkerson, G.G., Raper, C.D. & Gold, H.J. (1990). A dynamic growht model of vegetative soya bean plants: Model structure and behaviour under varying root temperature and nitrogen concentration. *Journal of Experimental Botany*, 41, pp. 229-241.

Lupwayi, N.Z., Kennedy, A.C. & Chirwa, R.M. (2011). Grain legume impact on soil biological process in sub-Saharan Africa. *African Journal of Plant Science*, 5, pp. 1-7.

Mbilinyi, M. (1991). *Big Slavery. Agribusiness and the Crisis in women's employment in Tanzania*. Dar es Salaam University Press. Dar es Salaam.

McCown, R.L., Keating, B.A., Probert, M.E. & Jones, R.K. (1992). Strategies for sustainable crop production in semi-arid Africa. *Outlook on Agriculture*, 21, pp. 21-31.

Mergeai, M., Kimani, P., Mwang'ombe, A., Olubayo, F., Smith, C., Audi, P., Baudoin, J.-P. and A. L. Roi (2001). Survey of pigeonpea production systems, utilization and marketing in semi-arid lands of Kenya. *Biotechnology, Agronomy, Society and Environment*, 5(3), pp. 145-153.

Myaka, F.A., Sakala, W.D., Adu-Gyamfi, J.J., Kamalongo, D., Ngwira, A., Odgaard, R., Nielsen, N.E. & Høgh-Jensen, H. (2006). Yields and accumulations of N and P in farmer-managed maize-pigeonpea intercrops in semi-arid Africa. *Plant and Soil*, 285, pp. 207-220.

Mwanga W 2004 Low use of fertilizers and low productivity in sub-Saharan Africa. Nutr. Cycl Agroecosys. 45, 135-147.

Nene, Y.L. & Sheila, V.K. (1990). Pigeonpea: geography and importance. In: *The Pigeonpea* (eds.: Y.L. Nene, S.D. Hall & V.K. Sheila, pp. 1-14, CAB International and ICRISAT.

Nielsen, N.E. & Høgh-Jensen, H. (2006). Soil fertility evaluation of farmers' fields in semi-arid Tanzania and Malaiwi. In: *Pigeonpea-based cropping systems for smallholder farmer's livelihood* (ed.: H. Høgh-Jensen), pp. 43-53, DSR Publishing, Copenhagen.

Odeny, D.A. (2007). The potential of pigeonpea (*Cajanus cajan* (L.) Millsp.) in Africa. *Natural Resources Forum*, 31, pp. 297-305.

Palm, C.A., Myers, R.J.K. & Nandwa, S.M. (1997). Combined use of organic and inorganic nutrient sources for soil fertility maintenance and replenishment. In: *Replenishing Soil Fertility in Africa* (Eds.: R.J. Buresh, P.A. Sanchez & F. Calhoun), pp. 193-217, SSSA Special Publication Number 51, Madison.

Palm, C.A., Murwira, H.K. & Carter, S.E. (1998). Organic matter management: from science to practice. In: *Soil Fertility Research for Maize-based Farming Systems in Malawi and Zimbabwe* (eds.: S. Waddington, H.K. Murwira, A.S. Kumwenda, D. Hikwa & F. Tagwira), pp. 21-27, Proceedings of the Soil Fert Net Results and Planning Workshop, 7-11 July 1997, Murara, Zimbabwe. CIMMIT.

Pinstrup-Andersen, P. & Pandya-Lorch, R. (2001). Meeting food needs in the 21st century. How many and who will be a risk? In: *Who will be fed in the 21st century? Challenges for Science and Policy* (eds.: K. Weibe, N. Ballenger & P. Pinstrup-Andersen), pp 3-14, IFPRI, Washington.

Postel, S.L. (2000) Entering an era of water scarcity: the challenge ahead. *Ecological Applications*, 10, pp. 941-948.

Radelet, S. (2010). *Emerging Africa. How 17 Countries are Leading the Way*. Center for Global Development, Washington.

Rogers, E. & Shoemaker, F. (1971). *Communication of Innovation*. Free Press, New York.

Rosegrant, M.W., Paisner, M.S., Meijer, S. & J. Witcover (2001). *Global Food Projections to 2002. Emerging Trends and Alternative Futures*. IFPRI, Washington, DC.

Ryle, G.J.A., Powell, C.E. & Gordon, A.J. (1979). The respiratory costs of nitrogen fixation in soyabean, cowpea, and white clover. I. Nitrogen fixation and the respiration of the nodulated root. *Journal of Experimental Botany*, 30, pp. 135-144.

Sakala, W.D., Kumwenda, J.D.T. & Saka, A.R. (2003). The potential of green manures to increase soil fertility and maize yields in Malawi. *Biological Agriculture and Horticulture*, 21, pp. 121-130.

Schlecht, E., Buerkert, A., Tielkes, E. & Bationo, A. (2006). A critical analysis of challenges and opportunities for soil fertility restoration in Sudano-Sahelian West Africa. *Nutrient Cycling in Agroecosystems*, 76, pp. 109-136.

Snapp, S.S. (1998). Soil nutrient status of smallholder farms in Malawi. *Communications in Soil Science and Plant Analysis*, 29, pp. 2571-2588.

Snapp, S.S. & Silim, S.N. (2002). Farmers preferences and legume intensification for low nutrient environments. *Plant and Soil*, 245, pp. 181-192.

Soussana, J.-F., Minchin, F.C., Macduff, J.H., Raistrick, N., Abberton, M.T. & Michaelson-Yeates, T.P.T. (2002). A simple model of feedback regulation for nitrate uptake and N_2 fixation in contrasting phenotypes of white clover. *Annals of Botany*, 90, pp. 139-147.

Stützel, H. (1995). A simple model for simulation of growth and development in faba beans (*Vicia faba* L.) 1. Model description. *European Journal of Agronomy*, 4, pp. 175-185.

Sumberg, J. (2002). The logic of fodder legumes in Africa. *Food Policy*, 27, pp. 285-300.

Thies, J.E., Singleton, P.W. & Bohlool, B.B. (1995). Phenology, growht, and yield of field-grown soybean and bush bean as a function of varying modes of N nutrition. *Soil Biology & Biochemistry*, 27, pp. 575-583.

Twomlow, S., Rohrbach, D., Dimes, J., Rusiki, J., Mupangwa, W., Scube, B., Hove, L., Moyo, M., Masingaidze, N. & Mahposa, P. (2010). Micro-dosing as a pathway to Africa's green revolution: evidence from broad-scale on-farm trials. *Nutrient Cycling in Agroecosystems*, 88, pp. 3-15.

Unkovich, M.J., Baldock, J. & Peoples, M.B. (2010). Prospects and problems of simple linear models for estimating symbiotic N_2 fixation by crop and pasture plants. *Plant and Soil*, 329, pp. 75-89.

Urabe, K. (1988). Innovation and the Japanese management style. In: *Innovation and Management. International Comparisons* (eds.: K. Urabe, J. Child & T. Kagono), pp. 3-25, de Gruyter, Berlin.

Vesterager, J.M., Nielsen, N.E. & Høgh-Jensen, H. (2007). Nitrogen budgets in crop sequences with or without phosphorus-fertilised cowpea in the maize-based cropping systems of semi-arid eastern Africa. *African Journal of Agricultural Research*, 2 (6), pp. 261-268.

Versteeg, M.N. & Koudokpon, V. (1993). Participatory farmer testing of four low external input technologies, to address soil fertility decline in Mono Province (Benin). *Agricultural Systems*, 42, pp. 265-276.

Whiteside, M. (1998). *Living Farms. Encouraging Sustainable Smallholders in Southern Africa.* Earthscan Publications, London.

Wichern, F., Eberhardt, E., Mayer, J., Joergensen, R.G. & Müller, T. (2008). Nitrogen rhizodeposition in agricultural crops: Methods, estimates and future prospects. *Soil Biology & Biochemistry*, 40, pp. 30-48.

Walwort, J.L. & Summer, M.E. (1988). Foliar diagnosis: A review. In: *Advances in Plant Nutrition* (eds.: B. Tinker & A. Läuchli), pp. 193-241, Praeger Publishers.

Wen, Q.-X. (1984) Utilization of organic materials in rice production in China. In: *Organic Matter and Rice*, pp. 45-56. IRRI.

Woomer, P.L., Lan'gat M & Tungani J.O. (2004). Innovative maize-legume intercropping results in above- and below-ground competitive advantages for understorey legumes. *West African Journal of Applied Ecology*, 6, pp. 85-94.

2

Issues in Caribbean Food Security: Building Capacity in Local Food Production Systems

Clinton Beckford
Faculty of Education,
University of Windsor, Ontario,
Canada

1. Introduction

1.1 The concept of food security

In this section I will discuss some key concepts of food security which will frame the discussion of the issue in the Caribbean context. The Food and Agricultural Organization (FAO) defines food security as a condition where "... all people at all times, have physical and economic access to sufficient, safe and nutritious food to meet their dietary needs and food preferences for an active and healthy life" (FAO, 2002). Four broad dimensions of food security are usually identified- availability-the supply of food in an area, access-the physical and economic ability of people to obtain food, utilization- the proper consumption of food and stability- the sustainability of food supplies (World Food Program, 2009). *Availability* speaks to the supply of food and is influenced by factors such as food production, stockpiled food reserves and trade (EC-FAO Food Security Programme, 2008). Aspects of food availability include the agro-climatic essentials of crop and animal production and the socio-cultural and economic milieu in which farmers operate (Schmidhuber and Tubiello, 2007). The second dimension *access* addresses the ability of individuals and households to purchase food. It takes into consideration the availability of financial resources to acquire adequate food both in terms of quantity and quality. Concerns about access take cognizance of the fact that availability of adequate food at the national or international level does not guarantee individual or household food security (EC-FAO Food Security Programme, 2008; Schmidhuber and Tubiello, 2007). The issue of *entitlements* is therefore critical (Sen, 1981). Entitlements maybe defined as "the set of those commodity bundles over which a person can establish command given the legal political, economic and social arrangements of the community of which he or she is a member" (Schmidhuber and Tubiello, 2008 p.19703). The dimension of *utilization* is closely related to consumption patterns and behaviour which impact nutritional status and hence health and productivity. It is also related to food safety, preparation, and diversity in diets (EC-FAO Food Security Programme, 2008; Schmidhuber and Tubiello, 2007). The fourth dimension *stability* refers to long term consistency in the other three dimensions. It accounts for the reality of individual or households losing access and becoming food insecure periodically, seasonally, temporarily or permanently (EC-FAO Food Security Programme, 2008; Schmidhuber and Tubiello, 2007). Food security objectives cannot be genuinely met unless these four dimensions are concurrently fulfilled.

Food insecurity may be described as *chronic, transitory* or *seasonal* (Schmidhuber and Tubiello, 2007). Chronic food insecurity which occurs when individuals cannot meet minimum food needs is long term and sustained and results from persistent poverty and lack of assets and resources. In contrast transitory food insecurity is short term and temporary occurring suddenly when the ability of people to produce and access food diminishes as a result of factors such as variations in food production and supply, increases in food prices and reduction in income. Seasonal food insecurity is generally predictable given that it results from a regular and cyclical pattern of inadequate availability and access to food which may be caused by seasonal climatic variations, cropping patterns and unavailability of work (Schmidhuber and Tubiello, 2007).

The final term which will be central to this discussion is *vulnerability* defined here to mean conditions which create susceptibility to food insecurity. In other words the concept refers to possible and future threats to food security even though food security may prevail at the present time. This concept underscores the fact that food security is not a static condition but is dynamic and is influenced by risk factors, the ability of people to manage risks, and vulnerability to certain events (Schmidhuber and Tubiello, 2007).

The FAO suggests that the world produces enough food to provide all its 6 billion plus inhabitants with sufficient daily nutrition (FAO, 2008). Despite this nearly a billion people qualify as being hungry and between 2007 and 2008 some 115 million people were added to the global figures of the chronically hungry (Josette Sheeran, Executive Director, World Food Program, 2009). The world's poorest people continue to face an uphill battle in food security (Chen and Ravillion, 2004). In this context the global food crisis and the prevalence of hunger is indeed a paradox: the paradox of hunger in the midst of plenty. Food insecurity is the absence of food security implying that hunger exists as a result of problems with availability, access and utilization or that there is susceptibility to hunger in the future (World Food Program, 2009).

2. The Caribbean and Jamaican context

Food production in Jamaica is mainly the purview of thousands of small-scale farmers who cultivate small holdings on mostly marginal land in the hilly interior, certain river valleys and flood plains and the dry southern coastal plains (Barker and Beckford, 2008). Most of this food is sold, processed, resold and consumed locally, thus providing the foundation of people's nutrition, incomes and livelihoods and contributing to rural and national development (Beckford and Bailey, 2009). This is achieved despite enormous documented challenges facing small-scale food producers (McGregor, Barker and Campbell, 2009; Campbell and Beckford, 2009; Beckford and Bailey, 2009; Beckford, 2009; Barker and Beckford, 2008; Beckford, Barker and Bailey, 2007). The significance of agriculture in Jamaica is historical and goes beyond satisfying household needs. It makes an indispensable contribution to national, community and household food security. Agricultural policy in the 1970s shifted to focus on food self-sufficiency in which domestic food production and eating locally grown foods were prioritized. Agriculture which had historically been the backbone of the economy and the small-scale domestic food sector which had always been the driver of food security became even more important. Domestic food farming in Jamaica is still dominated by traditional farming techniques.

Fig. 1. A farmer transport farm inputs to a farm plot on a mule.

Despite the importance of agriculture and the small-scale food production sector in particular, Jamaica like almost every country in the CARICOM region except perhaps Guyana has seen dramatic reduction in food output and has become a net importer of food. In his welcoming remarks to a forum on agriculture, food production and food security in the Caribbean and Pacific regions in 2005, CARICOM Secretary General Edwin Carrington pointed out that up to the mid-1980s the CARICOM region was a net exporter of food but had since become a net importer of food. Food insecurity in Jamaica and the wider CARICOM (with the exception of Haiti) tends to be under stated. It is often construed simply in terms of availability and complacency exists because again with the exception of Haiti, the dramatic and sensationalized incidents of hunger often seen in parts of Sub-Saharan Africa and Asia are largely unknown in the region. However, it is now acknowledged that the region including Jamaica, faces urgent and significant food security challenges. Jamaica, like the rest of the CARICOM has been experiencing declining agricultural productivity, decreasing earnings from traditional export crops, a high and growing dependence on imported food, increasing levels of poverty and increases in diet-related diseases like diabetes, hypertension and obesity. The World Food Summit (WFS) set a goal of reducing global hunger by 50 percent by 2015. To this end the FAO established a Trust Fund for Food Security and Food Safety to be used to strengthen and sustain projects within the FAO Special Program for Food Security (SPFS). The increasing concerns about food security in the Caribbean prompted the CARIFORUM to ask the FAO to prepare a CARIFORUM Regional Special Programme for Food Security (CRSPFS) under the SPFS. The original deadline of 2007 was later extended to 2010. In 2002 the FAO and the CARICOM

Secretariat collaborated and launched the US$26 Million Food Security Project. In April, 2003 a $5million a joint food security project was launched under the aegis of the CARICOM, the CARIFORUM, the FAO and the Government of Italy. Then in December, 2003 the Caribbean Food Security, Health and Rural Poverty Program was launched by the CARIFORUM and the FAO aimed at ensuring food security, reducing poverty, and improving nutrition and health in the region (Caribbean Food Emporium, 2003) In July 2009, the University of the West Indies, Mona hosted academics, agriculturalists and leaders in a conference on Food Security and Agricultural Development in the Americas. These events clearly indicate a certain level of urgency no doubt spurred by the recognition that food security is now a national and regional priority which cannot be ignored.

3. Factors affecting food security in the Caribbean

Food security and insecurity in the Caribbean is affected by several major factors. i) declines in productivity of land, labour and management in the agricultural sector resulting in a weakening capacity to supply food competitively; ii) decline in earnings from traditional export crops resulting in a reduced ability to purchase food; iii) the erosion and threatened loss of trade preferences for traditional export crops, the earnings of which are used to buy imported food; iv) the very high dependence on imported food and the uncertainty of food arrival associated with external shocks; v) the increasing incidents of pockets of poverty which affects peoples access to food; vi) concerns over the association of the high use of imported foods and growing incidents of diet-related diseases as people become estranged from local traditional foods and environment and adopt North American foods and lifestyles. These issues are all manifested in Jamaica where domestic food production has plummeted from the halcyon period of the mid-1990s when food production peaked over 650,000 tons. Since then a number of factors have combined to decrease food production. Significant among these were a series of devastating hazards including hurricanes, droughts and floods. It is estimated that agricultural losses just from hurricanes in 2007 was around US$285 million (McGregor, Barker and Campbell, 2009). Small-scale food producers are also facing daunting competition from cheap foreign imports. With their low resource base, high price of inputs, unsophisticated marketing and distribution, general lack of access to financial resources, and inability to engage in scale economies many have succumbed to this competition mainly from the USA and have been forced out of farming (Beckford and Bailey, 2009). An entrenched structural dualism in Jamaican agriculture has resulted in certain resource allocation biases against the domestic food production sector as the lion's share of resources goes to the traditional export crop sector including, sugar, coffee, citrus, and bananas. This dualism has influenced agricultural policy, creating asymmetrical relationships between small-scale food farmers and centers of economic and political power (Beckford, Barker and Bailey, 2007). These problems are exacerbated by limited size of the domestic market for the range of products offered by local farmers and limited farmland (FAO, 2007).

Assessed in the context of the various dimensions of food security, the situation in the Caribbean becomes clearer. The famine and hunger which characterize much of Sub-Saharan Africa and parts of Asia are typically not associated with the Caribbean – with the notable exception of Haiti. However, in light of declining food production, great reliance on imported food, growing poverty, and the growing incidence of diet-related diseases, food security in Jamaica may be described as precarious or, to use a technical food security term *vulnerable* (Beckford and Bailey, 2009). In the case of availability it might be argued that the Caribbean region is safe. Food availability is determined by local production, agro processing, food aid, food trade and food reserves or stockpiles. We have already seen that local production has declined significantly over the last two to three decades and CARICOM countries as a whole have moved from net exporters to net importers of food. The region is now very heavily dependent on food imports to meet its food needs (Beckford and Bailey, 2009). In 2006 for example, Jamaica imported some US$1.64 billion worth of food which was half the country's total import bill (Beckford and Bailey, 2009).

The situation is similar in many other CARICOM states and the developing world where markets have been opened up through trade liberalization (Short, 2000; Spitz, 2002; Walelign, 2002). I would argue that this dependence on food imports constitutes a major threat to Caribbean food security. First of all, purely from a livelihood perspective it does immeasurable damage to local producers and rural development. Faced with unfair competition and the dumping of cheap, heavily subsidized food mainly from the USA, many farm families experience difficulty providing a satisfactory livelihood for themselves (Beckford and Bailey, 2009; CIOEC, 2003; Via Campasina, 1996, 2003; UNDP, 2005). Most of the imported food to the Caribbean comes from the USA where heavily subsidized production enables farmers to sell for less than the cost of production (Windfuhr, 2002, 2003; Windfuhr and Jonsen, 2005). Local farmers are therefore forced into unfavourable, often times insurmountable competitive situations and in Jamaica for example, many have succumbed to this dumping of cheap exports and gone out of business (Beckford and Bailey, 2009).

It might be argued that opening up local markets to international competition is beneficial to consumers through lower prices and that this competition should stimulate more efficient local production thereby providing even greater access to affordable food. The problem with this argument is that unfettered competition from heavily subsidized foreign food producers has coincided with the removal of subsidies from local producers creating an uneven playing field. The over-reliance on imported food raises other obvious dangers as well. For one thing the structure of the world economy means that external shocks often reverberate throughout the system with devastating consequences for the most vulnerable nations and people. Market stability is a major concern here causing uncertainty of supplies and raising prices which could both result in food shortages in the region. Given the dependence of the Caribbean on food from the United States, terrorist attacks on the food system in America could have serious implications for Caribbean food security. The US$26 Million Food Security Project implemented in 2002 was partly in response to the near food crisis in some Caribbean countries in the aftermath of the September 11 terrorist attacks on the United States. Food safety is an important part of food security which perhaps does not get enough attention in the literature on the subject. In the context of the Caribbean I would suggest that this should be a real concern. Apart from the threat to safe food posed by terrorist attacks, the long distance traveled by food imported from distant places significantly increases the risk of food being contaminated (Halweil, 2005). Beckford and Bailey (2009) pointed out that the longer food travels the more it changes hands thus increasing the risks for contamination. Halweil, (2005) argues that the centralized nature of American food production and processing increases the risks of contaminated food while the sheer size and uniformity of farm operations creates ideal conditions for the rapid spread of diseases. The dependence on foreign food and especially American food is therefore cause for concern.

In terms of the food security dimension of access, the rising and persistent poverty among pockets of the population in some Caribbean countries has been a concern which has featured in the regional food security strategy. While it is no doubt true that the poverty, hunger and starvation common in some parts of the developing world is largely unknown in the Caribbean, it is also true that many households and individuals in the region experience hunger from time to time with rising use of food stamps and other food aid programs being observed. It could be argued that there is incomplete and perhaps distorted knowledge about the extent of hunger in Caribbean populations.

4. Building capacity and empowering local food producers

The food security challenges of the developing world cannot be solved by food aid or dependence upon food imports. As we have seen, despite the claims that globally there is enough food to feed everyone, world hunger is at its worse with dire prospects. This paper reflects the view that to achieve real food security developing countries must become more food self-sufficient by increasing productivity, diversifying and expanding the range of crops with a focus on maximizing the use of traditional foods, reducing post harvest losses, improving the marketing and distribution of farm produce and increasing women's participation in the food security endeavour. The paper is also framed within the general principles of *food sovereignty* (McMichael, 2009b), or as some prefer, *food democracy* (Lang, 2009a). Food sovereignty speaks to the right of local farmers and peoples to define their own food and agriculture in contrast to having food largely subject to international market forces (Beckford and Bailey, 2009). Windfuhr and Jonsen (2005) described food sovereignty as a platform for rural revitalization at the global level based on equitable distribution of resources, farmers having control over planting stocks and productive small farms supplying consumers with healthy, locally grown food. A food sovereignty approach advocates the right of people to be able to protect and regulate domestic agriculture and trade in order to achieve sustainable development goals: to determine the extent to which they want to be self reliant; and to restrict dumping of products in their markets (Beckford and Bailey, 2009). This incorporates into the discussion the issue of *agency*, with the empowerment of farmers and rural peoples to solve their own problems. Such approaches would specifically draw on local agro-ecological knowledge and wisdom of elders.

According to Dr Kayano Nwanze, President of the United Nations International Fund for Agricultural Development, "Smallholder farmers supply 90 percent of the food for developing countries and feed one-third of the world." This means that for small developing island states like Jamaica and the rest of the CARICOM Region any serious effort at enhancing food security must start with increasing local production and improving self sufficiency. In this regard building the capacity of small-scale food producers to increase agricultural output is fundamental. Of critical importance is the reimagining of the role of women in the production and marketing of food. This is important to national food security but it is critical to community and household food security and nutrition. To be effective and sustainable food security strategy in Jamaica and CARICOM must create the conditions for females to improve their own food security and their families, improve nutrition, and achieve greater economic independence.

The empowerment of local food producers in the Caribbean should focus on activities aimed at strengthening local food production and distribution systems, increasing the capacity of farmers to increase food production through sustainable systems and practices and increasing income and improving livelihoods. As part of the strategy to accomplish this, governments must address the role of women in the production and marketing of farm produce and nutrition. This paper suggests the following thematic priority areas:

1. Contributing to the development of gender responsive technologies and innovations to increase agricultural productivity, improve nutrition and reduce post harvest losses. The focus should be initiatives which will increase women's participation in food production and marketing and improve their food security, nutrition and economic livelihoods.
2. Supporting on-farm research informed by sound social and gender analysis to identify technological adoption benefits and economic and ecological viability of small-scale farming. Farmers should be engaged in on-farm adaptive research through field trials, demonstrations, crop experimentation under normal farm field conditions, field schools and other techniques.
3. Developing underutilized species for the achievement of food, nutrition, and income security. This can be addressed through explorations of local/traditional knowledge about wild edible plants and their food security, nutritional and medicinal uses.

The overall goal should be the development and implementation of an integrated and comprehensive strategy aimed at building the capacity of small-scale food producers in the region to increase productivity and improve livelihoods and income through gender responsive sustainable agricultural technologies and practices. The main objectives should be to:

1. Increase food security and nutrition and enhance the role of women through on-farm adaptive applied research and education of farmers and food distributors aimed at strengthening local food production and marketing systems;
2. Utilize the local/traditional knowledge and epistemologies within a framework of collaborative and participatory research to help small-scale farmers explore solutions to some of their most pressing problems.

There are a number a number of urgent specific areas to be addressed if the capacity of local farmers to produce more, more diverse, and better quality and affordable food.

Enhancing Farming Expertise. This paper takes the position that increasing farmers' knowledge and understanding and ability to apply this knowledge and understanding is a fundamental issue in capacity building and empowerment of local food producers. Research among small-scale farmers in the Caribbean consistently point to a wealth of traditional or local knowledge based on intergenerational knowledge and experience. This has served farmers well over the years and is largely responsible for the success and survival of many small-scale food producers who have been forced to survive without any significant institutional support. The education of farmers being conceived of here is a structured and systematic program of practical information dissemination based on evidence-based identification of the gaps in farmers' knowledge and priorities for action. This dissemination should be done through *agricultural extension services, farmer field schools* and *on-farm adaptive research* based on participatory and collaborative principles. Based on recent research from

Jamaica, priority areas identified by farmers include post harvest storage of crops, irrigation, pest management and control, marketing and distribution of crops, organic farming, record keeping, grading of fresh produce, soil management and adaptation and coping strategies with regard to meteorological hazards which are ubiquitous to the Caribbean (Campbell and Beckford, 2009; McGregor, Barker and Campbell, 2009; Beckford and Bailey, 2009; Rhiney, 2009).

A major problem within the small-scale food production sector in general is the lack of knowledge about proper post harvest food storage, absence of proper storage facilities for farmers, and knowledge and experience of general advanced post harvest management of farm produce. Farmers do not keep good records and so the volume of post harvest crop loss is hard to quantify but evidence gathered from field research suggest they are significant. Crops most affected include perishable fruits, vegetables, condiments, and peas and beans which are susceptible to weevil. There is little or no capacity to store produce during periods of over-supply and releasing food to market at different times thereby sustaining income over longer periods. The economic viability of domestic food cultivation can be significantly enhanced by reduction in food loss due to damage and decay. This would allow farmers to increase production knowing that they would have a much longer window on marketing their produce while regulating market supply (Beckford, Campbell and Barker, 2011, Farr, 2010). Examples from places with similar experiences suggest that solutions can be simple and inexpensive. In Tamil Nadu, India for example, post harvest losses of potatoes was significant as they were stored in mounds and the ones at in the middle and bottom rotted quickly due to trapped moisture. This was mitigated by inserting plastic pipes with holes drilled in them into the piles which facilitated air exchange and circulation and slowed down the decay process (Bechard, 2010). Also in Tamil Nadu, post harvest losses of chilli peppers have been reduced by 95% using simple solar tunnel dryers (Bechard, 2010).

Education and training in the proper cleaning, sorting and grading of fresh produce - especially for the export market-is also needed. This would allow farmers to market produce differently based on quality. Training in sustainable pest management is also very important. There needs to be more efforts at promoting non-chemical pest control protocols among small-scale food producers. Just as important is the promotion of organic fruit and vegetable production among small-scale producers. This would cut the use of chemical fertilizers and raise farm incomes through higher prices while supplying healthy and nutritious produce to the market and protecting the environment.

Education about natural hazard mitigation is also needed. The region is prone to a host of such hazards with hurricanes getting most of the attention, but droughts, floods, and landslides are all features of Caribbean life. Farmers need to be aware of what they can do before and after extreme events to mitigate losses. For example, how can plants already in the ground be safely removed until after a hurricane? What crops can be harvested early and how should they be stored?

Small-scale domestic food farmers are notorious for their poor record keeping and the absence of a business approach to farming. This makes it difficult to accurately analyze their operations and identify reliable solutions to their problems. Their operations are typically very informal with little application of principles of business. Even for the multitude of farmers for whom crop cultivation is their only source of income, farming appears to be more of a way of life than a business. Record keeping would ensure accurate applications of inputs and facilitate better planning. Farmers also need education about the marketing and distribution of produce.

5. Rediscovery of local foods

This paper submits that the problem of food security in the Caribbean requires local solutions which should revolve around the discovery and rediscovery of local or traditional foods. Food security in the Caribbean is being undermined by changes in tastes and diet to North American influences as people become estranged from their local foods and consumption habits. This is not unique to the Caribbean. In Lebanon for example, it has been found that food security is compromised as many Lebanese transition from a traditional, diverse Mediterranean diet to Western style diets which are deficient in micro-nutrients and heavy on white flour, corn, sugar and vegetable oils which are not as nutritious as local olive oil (Boothroyd, 2010). This has resulted in a high incidence of high blood pressure and high cholesterol among people 40-60 years old.

In Jamaica, many traditional foods, wild and edible plants have lost their place in local diets. There is a need for reintroducing some of these plants and foods back into the local diets. Research and education is needed about local edible wild plants. These should be documented highlighting their uses, preparation and nutritional and health benefits. A study in Lebanon funded by Canada's International Development Research Center (IDRC) found that villagers who regularly used wild edible plants and kept gardens enjoyed greater food security and better health than those who did not. Researchers studied the nutritional value of over 40 edible plants, how they were used, and identified local nutritious and healthy affordable dishes which were then widely promoted (Boothroyd, 2010).

6. Increasing women's participation in food production

Sustainable food security in the Caribbean requires the effective participation of women in food production. This is significant in the context of the dimensions of availability, access and nutritious foods and the implications for overall household food security. There are many commercial female farmers in the Caribbean but using the Jamaican context as an example, women are mainly involved in the marketing and distribution of food as they make up a disproportional amount of sellers in local produce markets across the country. The strategic participation of women in food production could be an effective strategy for addressing food security at the household level. This can be done through a Kitchen Garden Project or Backyard Garden Project in which women receive training in growing organic foods especially fruits and vegetables mainly for home consumption. The aim would be to increase supplies of safe and nutritious foods for their households. Women should also receive training in food handling and preparation to maximise the nutritional value of their families' meals.

An interesting aspect of the IDRC study discussed earlier which holds lessons for the Caribbean is the development of a communal "Healthy Kitchen" by women in three villages. The project centered on the preparation of traditional dishes using wild plants and other produce. The women obtained training in commercial food preparation and marketing and became nutritional ambassadors selling their produce in local markets, catering at weddings and other events and operating an eco-lodge which show-cased their Healthy Kitchen cuisine. This is the kind of approach which may be necessary to bring Caribbean populations back to local and traditional foods. It is different from conventional eat local campaigns conducted through the media which have been largely unsuccessful due to the top down approach and lack of grassroots community and household engagement.

Fig. 2. A female farmer prepares land for cultivation

7. Community and household agro-processing

Most of the food which is produced in the Caribbean is sold as fresh produce. An important component of food security and women's participation should be initiatives to promote increases in agro-processing at the household or community levels. Agro-processing would drastically reduce post-harvest losses, preserve food, and add value thus increasing farm incomes. Cottage industries based on locally produced fresh farm produce should be promoted, encouraged and supported. In this regard there should be efforts to establish properly constituted cooperatives but household level industries should also be pursued. Women could also play an instrumental role here as the history of cottage industries in the region suggests that they have always taken a leadership role. Again there will be a need for training and ongoing learning in areas like food processing, business management, marketing and distribution and accounting.

8. Distribution and marketing of fresh produce

This is an area requiring urgent attention. The marketing and distribution of domestic food crops is done through various informal commercial activities. The primary strategy is where farmers sell their produce to people mainly women who sell in produce markets across the country. Traditionally these women who are called higglers would go around to various farms and purchase different kinds of farm produce which would then be transported in a truck to the market place. Some farmers now take their produce to these markets themselves where they are sold in bulk to higglers or retailed to shoppers. The marketing and distribution of domestic foods have been identified as a major obstacle to production in Jamaica. There is no regulated system in place and small-scale farmers are basically left to their own devices in marketing farm produce locally.

Fig. 3. A farmer prepares produce for sale at a roadside

A major irony of Jamaican and Caribbean agriculture more generally is what might aptly be described as the estranged relationship between the region's world famous tourist industry and its local agriculture. Local tourism is booming while local agriculture stagnates and declines (Thomas-Hope and Jardine-Comrie, 2007; Dodman and Rhiney, 2008). This is not a new phenomenon as indicated by research from the 1970s and 1980s lamenting the limited benefits the small-scale domestic food sector enjoyed from the tourism industry (Momsen, 1972; Belisle, 1983; Belisle, 1984). In more recent work it has been suggested that several changes including more openness towards serving local cuisine in resort facilities and globalization of food consumption habits and the desire of tourist to eat local foods, could serve to strengthen the link between local agriculture and tourism (Momsen, 1998; Torres, 2003; Conway, 2004; Rhiney, 2009). However, recent research into the role of tourism in local food supply chain suggests that there are still considerable problems for farmers [Rhiney, 2009]. This paper argues that the tourism sector is an area where the potential for creative use of more local foods in the cuisine could be harnessed and successfully promoted (Beckford, Campbell and Barker, 2011). The extent to which locally grown foods and traditional foods are used in the hotel kitchens requires research but indications are that use is limited.

Generally, a more proactive State role in the marketing and distribution of domestic food crops could enhance viability by providing stable markets and fair prices. Farmers in Jamaica have consistently identified the collapse of the government agency the Agricultural Marketing Cooperation (AMC) as a watershed event in their declining fortunes. The AMC was a government run marketing board which bought domestic fresh foods from farmers and ensured a reliable distribution outlet. Farmers should also be encouraged and educated in the establishment of local marketing cooperatives. Examples of successful marketing by small-scale farming cooperatives can be found in Jamaica and used as models (Rhiney, 2009; Timms, 2006). Recent experiments with Farmers Markets in Jamaica are encouraging. Organized by the Ministry of Agriculture these provide a space for farmers to sell fresh produce. They cut out the middle man allowing the farmer to sell retail directly to consumers increasing their profits and providing more affordable food to consumers.

The distribution and marketing of fresh foods is important to food security as lack of markets and profitability is a major hindrance to increasing the participation of people in commercial farming and raising food production.

9. Conclusion

This discussion has demonstrated that food security is considered to be an area requiring urgency in regional development. The region has made some strides in addressing food security concerns but there is still a great deal of work to be done from the standpoint of policy but also at the grassroots level. Erwin Larocque, CARICOM's Secretariat's Assistant Secretary-General for Regional trade and Economic Integration underscored this in highlighting two pressing regional food security issues. First he made reference to the impact of international developments on the Caribbean's ability to be self sufficient in food production and internationally competitive to afford necessary imports. Secondly, he stressed the need to address food security issues in the context of the Millennium Development Goals. He stressed that the region's dependence on imported food made it vulnerable and noted the climb in diet related diseases.

Improving food self-sufficiency and reducing dependence on imported food are thus big priorities for local agriculture and food security. Based on previous research with small-scale farmers in Jamaica this paper identified a number of specific suggestions for addressing food security in the region. These may be summarized as: (i) increasing farmer expertise to in areas such as post harvest storage of crops, irrigation, pest management and control, marketing and distribution of crops, organic farming, record keeping, grading of fresh produce, soil management and adaptation and coping strategies with regard to meteorological hazards. (ii) The discovery or rediscovery of local or traditional foods including wild and edible plants. (iii) Increasing the participation of women in local agriculture through for example, kitchen gardens, local kitchens, and cottage industries. (iv)

Improve the marketing and distribution of fresh foods in general and more specifically improve linkages between local agriculture and tourism. Together these strategies can help to increase food production and hence food self-sufficiency, reduce the need for food imports, improve availability of nutritious foods, increase value added for farmers thereby improving economic viability and rural livelihoods, and improve the sustainability of local small-scale food systems.

Policies and strategies to improve food security in Jamaica and the wider Caribbean should be informed by high quality research. Individual Caribbean governments and CARICOM should therefore draw on the expertise of the local research community and involve the University of the West Indies as a partner. In the past twenty five years or so there has been significant research in the region on the topic of food and agriculture including research about renewable and sustainable agriculture, food security, traditional knowledge and agriculture, hazards and local agriculture among other topics. This existing research provides an ideal starting point for dissecting the issue of regional food security. However, there is need for ongoing research in a number of vital areas including: ongoing analyses of the extent of food insecurity and hunger in the Caribbean; enhancement of the role, place and fortunes of women in agriculture to serve the goals of improved food security and nutrition; ecological and economic sustainability of small-scale farming systems in the region; the role and potential of local knowledge and epistemologies in enhancing food security and nutrition in the region; identifying the main obstacles faced by farmers in increasing the production of affordable, nutritious food; and the marketing and distribution of locally grown fresh foods with a focus on strategies for strengthening linkages between the local agriculture and tourism sectors.

The key to enhancing food security in the region is improving the capacity of local food producers to significantly increase the production and supply of affordable nutritious food produced using environmentally and economically sustainable production systems. To this end, this paper argues that this can be done by adopting elements of a food sovereignty approach (Beckford and Bailey, 2009; Holt-Gemenez, 2006; Schwind, 2005;). Food sovereignty has been described as a basis for the revitalization of rural spaces with equitable distribution of resources and small scale producers having the ability to supply locally grown healthy foods (Windfuhr and Jonsen, 2005). This concept also speaks to the rights of people to determine the source of their foods, to be able to protect and regulate domestic agriculture and trade and to restrict unfair competition from cheap foreign food imports (Via Campesina, 1996; Institute for Agriculture and Trade Policy, 2003). Food sovereignty philosophy does not eschew international trade but rather advocates the formulation of trade regulations which serve the interests of local peoples and farmers. A food sovereignty approach in the Caribbean would prioritize local agriculture by providing farmers with the capacity to produce affordable healthy and safe foods while protecting them from unfair competition which place their livelihoods and regional food security at risk (Schwind, 2005; Kent, 2001). It would enhance food self-sufficiency and reduce the dependence on food imports. Food sovereignty strategy should also increase the participation of farmers and local peoples in agricultural planning and decision-making (Beckford and Bailey, 2009; Stamoulis and Zezza, 2003).

10. References

Barker, D. and Beckford, C.L. (2008) Agricultural Intensification in Jamaican small-scale farming systems: vulnerability, sustainability and global change. *Caribbean Geography*, 15, 160-170.

Bechard, M. (2010) Feeding the world: scaling up projects will tackle food security through an interdisciplinary approach. *UniWorld*   March 2010, 5-7. Association of Universities and Colleges in Canada.

Beckford, C. L. (2009) Sustainable agriculture and innovation adoption in a tropical small scale food production system: the case of yam minisetts in Jamaica. *Sustainability,* 1, 81-96.

Beckford, C.L., Barker, D. and Bailey, S.W. (2007). Adaptation, innovation and domestic food production in Jamaica: Some examples of survival strategies of small-scale farmers. *Singapore Journal of Tropical Agriculture,* 28, 273-286.

Beckford, C.L., Campbell, D. and Barker, D. (2011) Sustainable food production systems and food security: Economic and environmental imperatives in yam cultivation in Tralawny, Jamaica. *Sustainability,* (3), 541-561.

Beckford, C.L and Barker, D. (2007) The role and value of local knowledge in Jamaican agriculture: Adaptation and change in small-scale farming. *Geographic Journal.* 173, 118-128.

Beckford, C.L. and Bailey, S.W. (2009) Vulnerability, constraints and survival on small-scale food farms in St Elizabeth, Jamaica: Strengthening local food production systems. In *Global Change and Caribbean Vunerability: Environment, economy and society at risk:* McGregor, D.F.M., Dodman, D., and Barker D., Eds.; The University of the West Indies Press: Kingston, Jamaica, 218-236.

Belisle, F. J. (1983) Tourism and food production in the Caribbean. *Annals of Tourism Research.* 10 (4), 497-513.

Belisle, F. J. (1984) The significance and structure of hotel food supply in Jamaica. *Caribbean Geography,* 1 (4), 219-233.

Boothroyd, J. (2010) Healthy food for the picking: Kitchens and wild plants could be key to Lebanon's food security. *UniWorld*   March 2010, 14-15. Association of Universities and Colleges in Canada.

Campbell, D. and Beckford C.L. (2009) Negotiating uncertainty: Jamaican small scale farmers' adaptation and coping strategies before and after hurricanes. A case study of Hurricane Dean. *Sustainability,* 1, 1366-1387.

Caribbean Food emporium (2003) Caribbean Food Security. Retrieved from http://www.caribbeanfoodemporium.co.uk/foodsecurity.htm.

Chen, S. and ravillion, M. (2004). How have the world's poorest fared since the early 1980s? Washington, D.C,: World Bank Development Research Group. www.worldbank.org/research/povmonitor/MartinPapers/How_have_the_poorest_fared_since_the_early_1980s.pdf

Conway, D. Tourism, environmental conservation and management and local agriculture in the Eastern Caribbean: Is there an appropriate, sustainable future for them? In *Tourism in the Caribbean: Trends, development and prospects:* Duval, D. T. Ed.; Routledge: London and New York, 2004, pp.187-204.

Coordinadora de Integracion Economicas Campesinas de Bolivia (2003) Towards a world convention on food sovereignty and trade. Retrieved from http://www.cioecbolivia.org/wgt/food_sovereignty.htm.

Dodman D. and Rhiney, K. (2008) We nyammin: Food authenticity and the tourist experience in Negril, Jamaica. In *New perspectives in Caribbean tourism:* Daye, M., Chambers, D., and Roberts, S., Eds.; Routledge: New York, 115-132.

EC-FAO Food Security Programme (2008) An introduction to the Basic Concepts of Food Security. *Food Security Information for Action Practical Guidelines.*

Farr, M. (2010) Targeted aid: Three UPCD projects tackle CIDA's priorities for effective development. *UniWorld* March 2010, 3-4. Association of Universities and Colleges in Canada.

Food and Agricultural Organization (2008) Crop prospects and food situation. FAO: Rome.

Food and Agricultural Organization (2007) Food security and trade in the Caribbean: FAO/Italy project helps CARICOM profit from trade liberalization. FAO Newsroom. Retrieved from http://www.fao.org/newsroom/en/news/2007/1000691/index.html.

Food and Agricultural Organization (2002) The State of food Insecurity in the World, 200. FAO: Rome.

Halweil, B. (2005) Farmland defense: How the food system can ward off future threats. In *New perspectives on food security* Glynwood Center Conference Proceedings, Cold Spring, New Yprk: Glynwood Center, 25-30.

Holt-Gimenez, E. (2006) Movimiento Campesino a Campesino: Linking sustainable agriculture and social change. Food first Backgrounder 12 (1) Winter/Spring. Institute for Food and Development Policy, Oakland, California.

Institute for Agriculture and Trade Policy (2003) Towards Food sovereignty: Constructing an alternative to the World Trade Organization's agreement on agriculture, farmers, food and trade. International Workshop on the Review of the AoA. Geneva. Retrieved from http://www.tradeobservatory.org/library.cfm?RefID=25961.

Kent, G. (2001) Food and trade rights. UN Chronicle, Issue 3. Retrieved from http://www.un.org/Pubs/chronicle/2002/issue1/010p7.html.

Lang, T. (2009a) How new is the world food crisis? Thoughts on the long-term dynamics of food democracy, food control and food policy in the 21st century. Presented to the Visible Warnings: The world food Crisis in Perspective Conference April 3-4, 2009 Cornell University, Ithaca, NY.

McMichael, P. (2009b) A food regime genealogy. *Journal of Peasant Studies*, (35), 139-169.

McGregor, D.F.M., Barker, D., and Campbell, D. (2009) Environmental change and Caribbean Food security: Recent hazard impacts and domestic food production in Jamaica. In *Global Change and Caribbean Vulnerability: Environment, economy and society at risk:* McGregor, D.F.M., Dodman, D., and Barker D., Eds.; The University of the West Indies Press: Kingston, Jamaica, 197-217.

Momsen, J.H. (1972) Report on vegetable production and the tourist industry in St Lucia. Calgary: Department of Geography, University of Calgary.

Momsen, J. H. (1998) Caribbean tourism and agriculture: New linkages in the global era? In Globalization and neoliberalism: The Caribbean context: Klak, T. Ed.; Rowman and Littlefield: Lanhan MD, 267-272.

Rhiney, K 2009 Globalization, tourism and the Jamaican Food Supply Network. In *Global Change and Caribbean Vulnerability: Environment, economy and society at risk:* McGregor, D.F.M., Dodman, D., and Barker D., Eds.; The University of the West Indies Press: Kingston, Jamaica, 237-258.

Schmidhuber, J. and Tubiello, F.N. (2007) Global food security under climate change. *Proceedings of the National Academy of Science,* 104: (50), 19703-19708.

Schwind, K. (2005) Going local on a global scale: Rethinking food trade in an era of climate change, dumping and rural poverty. *Food First Backgrounder* 11, (2), 1-4.

Sen, A. K. (1981) Poverty and famines: An essay on entitlements and deprivation. Oxford. Clarendon Press.

Short, C. (2000) Sustainable food security for all by 2020; food insecurity: A symptom of poverty. London: department for International Development (DFID). Retrieved from http://www.ifpri.org/2020conference/PDF/summary_short.pdf.

Spitz, P. (2002) Food security, the right to food and the FAO. *FIAN-Magazine* no. 2.

Stamoulis, K. and Zezza, A. (2003). A conceptual framework for national agricultural, rural development, and food security strategies and policies. ESA Working Paper, No. 03-17, November, 2003. Agriculture and Economic Development Division, FAO, Rome. Available at www.fao.org/documents/show_cdr.asp?url_file=/docrep/007/ae050e/ac050e00. htm

Thomas-Hope, E., and Jardine-Comrie, A. (2007) Caribbean agriculture in the new global environment. In *No Island is an Island: The impact of globalization on the Commonwealth Caribbean:* Baker, G. Ed.; Chatam House: London, 19-43.

Timms, B. (2006) Caribbean agriculture-tourism linkages in a neoliberal world: Problems and prospects for St Lucia. *International Development Planning Review.* 28, 35-56.

Torres R Linkages between tourism and agriculture in Mexico. *Annals of Tourism Research.* 30 (3): 546-566.

United Nations Development Program (2005) The Millenium Project 2005. Halving hunger: It can be done. Final report on the Task Force on Hunger. New York: Earth Institute at Columbia University. Available at http://www.unmilleniumproject.org/who/tf2docs.htm.

Via Campesina (1996) Food sovereignty: A future without hunger. Retrieved from http://www.viacampesina.org/imprimer.php3?id_article=38.pdf.

Via Campesina (2003) What is food sovereignty? Retrieved from http://www.viacampesina.org/IMG/_article_PDF/article_216.pdf

Walelign, T. (2002) The fifth P7 summit: food sovereignty and democracy- Let the world feed itself. GREEM/EFA International Relations Newsletter no. 6 (December).

Windfuhr, M. (2002) Food security, food sovereignty, right to food: Competing or complimentary approaches to fight hunger and malnutrition? Hungry for what is right. FIAN-Magazine no. 1.

Windfuhr, M. (2003) Food sovereignty and the right to adequate food. Heidelburg, Germany: FIAN-International.

Windfuhr, M. and Jonsen, J. (2005) Food sovereignty: Towards democracy in localized food systems. Rugby, UK: ITDG Publishing.

World Food Program (2009) Emergency Food Security Handbook. Second Edition, January, 2009. Http://www.wfp.org/operations/emergency_needs/EFSA_section1.pdf

3

Food Security and Challenges of Urban Agriculture in the Third World Countries

R.A. Olawepo

Department of Geography and Environmental Management,
University of Ilorin, Ilorin,
Nigeria

1. Introduction

Interest in food security has been very strong most especially, since the world food crisis of 1972-1974 (Ajibola, 2000, Muhammad-Lawal and Omotesho, 2006).The issue of food and nutritional development in the Third World countries over the years has also generated a lot of concerns and interests among the social scientists, researchers and both governmental and Non Governmental Organizations. This came out of identified incessant problems within the agricultural sector coupled with the dwindling resources and poverty levels among these countries. In many African countries, food insecurity is on the increase with the share that purchased food takes of the household budgets especially in the urban centers.This has also led to increase in the proportion of urban farmers. This increase in the number of urban farmers is in a way affecting positively food security in our urban environment.

Sawio (1993) indicated that urban populations worldwide are growing fast as a result of natural growth and rapid migration to the cities as people escape rural poverty, land degradation, famine, war, and landlessness. Feeding urban population adequately is a major problem in developing countries. Rural areas could no longer produce enough food to feed both rural and urban people and food importation is constrained by lack of sufficient foreign exchange. Countries like Zimbabwe, Kenya, Nigeria, Malaysia, Sri Lanka, Pakistan and Columbia spent a large proportion of their resources to develop agriculture as it is a very important contributor to their Gross Domestic Product, foreign exchange earner and a major employer of labour. The aim of this chapter is two folds. Firstly, to examine the issue of food security as it affects the rapidly growing third world countries, and secondly, to examine the issue of urban agriculture in these countries as a panacea to solving the emerging food crises and to proffer appropriate solutions to the challenges accruing from this development.

Food Security: Meaning and development. Food security has been recognized as an important goal the world over. This is in view of the resolution of the various world food conference and the establishment of the World Food Council among others (Muhammad-Lawal and Omotesho, 2006:71) The persistent hunger and famine in the developing world means ensuring adequate and nutritious food for the population will continue to be the principal challenge facing policy makers in many developing countries in the 21st century. As part of the Millennium Development Goals, the world leaders have in different occasions pledged to reduce poverty, hunger and improve accessibility to public goods and services.

Food security has been described as a widely debated and much publicized issue over the years with different authors giving meanings to reflect different purposes and objectives. According to Olayemi, (1998, and Ajibola 2000: 58) food security has individual, household, national as well as international perspectives. Food security is defined as "access by all people at all times to the food required for a healthy life" It addresses the risks of not having access to needed quantities and quality of food (Von Braun et,al 1993). This involves food availability, food accessibility, food utilization and the ability to acquire it. In the same vein, Demery, et al (1993) defined food security as access by all people at all times to enough food for an active, healthy life. Its essential elements are the availability of food and ability to acquire it. Thus, Tunde (2011) opined that food security is an objective of every family and household in the developing world, whether in urban or rural areas. A household is food secured if it can reliably gain access to food of a sufficient quality and quantities that allow all its members to enjoy an active and healthy life.

On the other hand, food security exists when all people at all times have physical and economic access to enough safe nutritious food to meet their dietary needs and food preferences for an active and healthy life style (World Food Summit, 1996). More than this, the availability and the quality of food available as well as its utilization are very essential. Food availability and accessibility are often associated with food production and supply, while utilization has to do with the nutritional aspects of food intake. Furthermore, Chung, et al (1997) adopted a conceptual framework to explain the relationship between the various forms of food security. These include food availability, accessibility and utilization. This framework according to them would help us to identify which food security indicators to tackle in order to bring in, efficient food security in a nation. From the framework, it is evidenced that while food availability is a function of resources utilization and production process, food accessibility relates positively to income and consumption while nutritional development arises from the process of food utilization .In the same vein, the work of Demery, et al, (1993) also showed a link that exists between agricultural Food Policy and food security within a National framework. This involves wholly, a Macro Economic Policy trend that would eventually increase household Income, food consumption and then Nutritional Status of the citizenry. Such policies may include trade, fiscal, monetary and employment policies among others .For the Third world to be food secure, a series of events and strategies would have to be put in place. Such issues include:

a.   residents must have food all the year round and in every part of the country,
b.   people must have access to a large supply of food and food products either being produced internally or being imported without stress,
c.   a large proportion of food production should come from the local content, and,
d.   the level of nutritional development would have to be on the increase.

Current evidence in many of the developing countries of the world reveals that apart from being food deficit nations, countries are characterized by escalating food prices, food scarcity, famine and post harvest losses. Many countries indigenes are having poor nutritional development especially those countries that experience incessant famine and drought. They include countries like Sudan, Ethiopia, Somali and Niger to mention a few from the African continent. This is because, apart from the natural catastrophes they experience, their national aggregate production of food is not balanced since demand can not meet the supply flow. According to Bergman and Renwick (1999) about 14% of the World population is chronically hungry today, with sub Saharan Africa as the most troubled region. Problems encountered in the process of increasing food production include inefficient application of fertilizers, lack of incentive for many farmers, and problems of land ownership among others. Since Malthus published his theory, the human population has increased from 1 billion in 1804 to almost 7 billion in 2011. The mass starvation he predicted however has not occurred, but a near occurrence is being experienced in several parts of the world, thus making food insecurity a threat that deserves attention.

Food insecurity on the other hand is when livelihood systems (capabilities, assets, quality of life) change or fail to adapt to the challenges and shocks of their external environment. These shocks include sudden price increases and unavailability in food, emanating from environmental, socio-economic and political problems among others. This also means that accessibility to food as well as its availability is affected widely, leading to an extensive low nutritional development and eventually starvation.

2. The emerging food crisis in the third world countries

Over the years, there has been short falls in food productivity and increase in food importation into many of the third world countries. For example in Nigeria (like many other developing countries) agriculture has always been the mainstay and livelihood for millions of people. Before independence in 1960, agriculture was the most important sector of the economy accounting for more than one-half of the annual GDP and for more than three-quarters of export earnings. Following the discovery and production of petroleum, the agriculture sector suffered severe decline. Between the mid 1960s and mid 1980s, Nigeria moved from position of self-sufficiency in basic food stuff to one of heavy dependency on imports.

In most of the developing countries of the world (like Nigeria, Ghana, Zimbabwe, Venezuela, Kenya, and some far east Asian countries) where agriculture has remained a mainstay of the economy, various efforts have been deduced to effect and ensure food security in the last five decades. Extensive areas that were scarcely utilized for farming in the past ages have been opened to productive agriculture by many nations. The United Nations report that between 1980 and 1993 for example, the World cropland area increased by another 3%, this has drastically increased to over 10% by 2009, and not less than 60% of these are found in the developing world. Most of these lands were opened by irrigation. Similarly, many food crops and improved seedlings have been transported to new areas by donor Agencies and Non Governmental Organizations to farming locations in the Third world countries. Apart from these, transportation and storage facilities development are on the increase while a large proportion of these countries have been developing improved land management techniques that would increase food productivities over the years. Large scale commercial farming through Agricultural Development Projects were also established to improve agricultural productivities coasting Billion of Dollars, and agricultural extension services are on the increase. Bergman and Renwick (1999:291) indicated that from the 1950s, an extensive effort was launched to develop new grain varieties and associated agronomic systems and to establish them in developing countries. It focused on certain crops (rice, wheat) and certain techniques (breeding for response to fertilizer inputs), this focused effort was known as green revolution. These and others were also adopted by various governments of the developing nations. There were other scientific revolutions which are on going in many of the developing nations that would make them self sufficient over the years to come, all things being equal. However, not all these developments have taken place equally everywhere.

Despite all these efforts to improve food productivities by developing countries, a large proportion of them still depend on food importation. For example up to 70% of the world's exportation of rice in 2010 is diverted to the developing Nations. Similarly, as at 1994, out of the ten greatest importers of Rice seven of them are from these developing countries of the world as shown on Table 1. The situation now however is more than doubled.

Exporters		Importers	
Thailand	4,859	Japan	2,536
USA	2,822	Brazil	987
Vietnam	1,970	Indonesia	630
China	1,630	China	517
Pakistan	984	Iran	475
India	891	Saudi Arabia	434
Myanmar	643	South Africa	431
Italy	619	Hong Kong	358
Australia	585	Nigeria	350
Uruguay	408	United Arab Emirates	350

Source : Adapted from Bergman and Renwick (1999)

Table 1. Top Ten Exporters and Importers of Rice (In Thousands of Metric Tons).

Apart from huge deficit between the exportation and importation of food in most of the developing world, there is a wave of emerging food crises around the corner. The amount of nutrients needed per capita in each country varies greatly, despite this, there has been a short fall in food production. The sub Saharan African countries clearly face the greatest problems of over all national food supplies. The United Nations in August 2011 indicated that over 605 of Southern Somali is hungry due to famine and ravaging wars over the years. In rich countries of the developing world (like Nigeria, India, China, Venezuela and Libya) many people are well fed but a large proportion of the people are poor. Most countries here are both importers and exporters of food and a few countries are net exporters of food despite the fact that portions of their own populations are under nourished. This may be because of injustice or strive and political instability as in the case of Sudan, Ethiopia and Somali.

Many reasons have been suggested as being the cause of the short falls and emerging food crises in these countries, they include among others:

i.   reduction in food production as a result of rural-urban migration;
ii.   the spread of food scarcity as a result of drought in the Sudano-Sahelian locations;
iii. reduction of food production as a result of logistic and transportation problems;
iv.   poor accessibility to land and crude occupancy/tenure systems;
v.   limited accessibility to capital, finance fertilizer and modern chemicals;
vi.   competition from other form of production especially cash crops for export markets;
vii. lack of agricultural extension services, storage facilities, improved seedlings and modernization;
viii. fragmentation of land due to poverty, poor input and subsistence productions;
ix.   effects of climate change causing excessive dryness, late onset of rain and flooding in some locations;
x.   increase growth rate ,urbanization and rising debt servicing components; and
xi.   poor and crude land management systems that progressively reduce fertility and farm output over the years;
xii. incessant famine in and around the desert areas of the world and dry areas especially in Somali and the surrounding vast land areas; and,
xiii. social injustice, poverty, unemployment, strife, wars and political instability.

With these and other problems facing food production in the developing countries, various measures are being taken by various levels of government. This ranges from expanded food policies to various planning options. The issues of women in agriculture and urban agriculture are thus on the increase. Urban agriculture is thus seen as one of the ways of creating food security in some of the developing nations.

3. Urban agriculture and food production

Urban agriculture has been defined by various scholars but the work of Axumite, et,al (1994) indicated that it refers not merely to the growing of food crops and fruit trees but that it also encompasses the raising of animals, poultry, fish, snails, bees, rabbits, guinea pigs, or other stock considered edible locally. In the same vein, Mougeot (1994) stressed that urban agriculture involves the production of food and animal husbandry, both within (intra) and fringing (peri) built up areas. Mougeot (1994...p18) expressed further that informal urban agriculture is one livelihood strategy that the urban poor use in combination with other strategies.

In order to meet a part of the food needs of poor urban dwellers, urban agriculture came into being, especially among the poor nations. Urban Agriculture, defined here as "crop growing and livestock keeping in both intra-urban open spaces and peri-urban areas" is becoming a common phenomena in urban areas in the developing world.(see, for example: Sanyal 1984,Wade 1986, Sawio 1993 and Tunde,2011).Urban agriculture has recently become familiar, almost permanent feature all over tropical Africa and in many developing countries, however, research on this social pattern is limited. Mougeot (1994...p18) expressed further that informal urban agriculture is one livelihood strategy that the urban poor use in combination with other strategies. A review of definitions commissioned by International Development Research Centre (IDRC) led Mougeot (2000) to propose the following:

"Urban agriculture is an industry located within (intra-urban) or on the fringe (peri-urban) of a town, a cit or a metropolis ,which growsand raises processes and distributes a diversity of food and non food products, (re-) using largely human and material resources, products and services found in and around that urban area, and in turn supplyinghuman material resources, products and services largely to that urban area".

Cai, et al (2004) remarked that the concept of urban Agriculture originated from the United States of America (USA) in 1950s, (although un-documented facts indicated that this form of agriculture originated from Africa) . It refers to agricultural activities in urban and peri urban areas by making use of the land, natural ecology and environmental resources. This is the growing ,processing, and distribution of food and other food products through intensive plant cultivation and animal husbandry in and around the cities.

For the purpose of this review and with our experience in Nigeria, Urban agriculture is any form of economic activities involving food production, farming, marketing and animal husbandry being practiced by the urban residents, within the city, around the city; and on rural land areas surrounding the city, using both human and non human resources that have affiliation with the urban set ups. These include free ranged poultry and animal productions found within and along urban roads as well as small scale and commercial productions in and around the city.

4. Types and structure of urban agriculture

In the course of this investigation, five major types were identified on the basis of location within and around the city areas. These are:

a. Market Gardening

This is practiced for the production of staple foods and perishables. They are found near homes, riverbanks, dumping site and other locations at the outskirt of the town. In Nigeria for example, these are financed through self sponsored irrigation projects in the wet areas and fadama regions along the city ways. These are commonly seen at the low density regions and urban fringes in places like Lagos, Ilorin and Ibadan among others in Nigeria. Some of these examples are also common in Yaoundé, Nairobi and Kampala. The farmers here are usually non indigenes and they produce vegetables like Lettuce, Spinach, Orchards and wine tapping especially in Southern Nigeria. Over 20% of the vegetables produced in urban areas in Nigeria are on these farms, especially along irrigation ponds.

b. Compound and Yard farming

This is commonly found within fenced houses especially in the core areas, newly developed locations, and within residential quarters scattered all over the town. These kinds of farms are not usually large scale and they are often fenced. Farmers here produce mainly perennial crops and grains for local consumption.

c. Subsistence farming on open lands

Urban cultivators and local farmers on surrounding villages land mostly own them. Landowners who have not developed their lands are also involved in these types of farming where most productions are for home consumptions and to supplement income. It was revealed that about 16% of urban dwellers are involved in Nigeria. This form of farming is mainly found among farmers who produce crops for consumption and for commercial purposes. This is the commonest of agriculture on urban landforms. The major crops involved here are grains, vegetables, tubers, orchards, and fadama farming on irrigated lands.

d. Expanded Commercial farming

These are found at surrounding villages and owned mostly by urban land owners, 'big time farmers' retirees, itinerant farmers and migrants among others. Large hectares of land are managed for fish farming, piggery, cattle rearing grain crops and in some cases mechanization is added. Modern fish farming are done in ponds, corporation fish farms, fishing on main rivers, lakes and pond around the city locations. 4.5% of urban farmers are involved in one form of fish production or the other.

e. Constricted surrounding land farming

These are farmlands in surrounding villages where farmlands have been engulfed by the growing city. The villagers owned and tilled the land for commercial productions. Live stock production is also commonly done here. Livestock production includes commercial poultry farming, local poultry farming, piggery, local animal husbandry, cattle rearing and domestic animal keeping and in addition to urban crop farming, 22% of urban farmers in Nigeria keep poultry for commercial purposes..

From the survey here in Nigeria, about 66% of urban families can survive with self produced food. 32% of them sell some of their products within the urban environment while about 15% women food vendors grow their own vegetables as supplement to other productions. The main urban crop production is done on multiple cropping in which a farmer combines more than one crop at a time. Rain fed cultivation of maize, corn, and tubers has also been found to be common especially on the upland locations. Presently, a farmer may cultivate as large as 1.5 hectare especially on farms found at the outskirt of the town while smaller sized farms are found within the town. Similarly, Fadama cultivation is common throughout the year on locations beside rivers.

5. Significance of urban food production

In a recent work by Tunde (2011:132) on motives of urban women farmers in Kwara State, Nigeria, three main motives of women in urban agriculture were isolated. These are food security, income supplement and accessibility to land. This means that women are involved in urban agriculture in order to boost food security, income generation and as a result of accessibility to land. An earlier work on urban farming in Ilorin, Nigeria (Olawepo, 2008) has also identified the issues of income, employment, food security, leisure and poverty alleviation as the main significant factors in urban agriculture. One of the goals of the study was to understand who was practicing urban agriculture. Table 2 shows the distribution of the occupational structure of the respondents in Ilorin. Three occupational categories dominate farming occupation among the 240 urban farmers. They include the urban permanent cultivators who are full time farmers (39.1%), these people are found mainly at the outskirt of the town and in the new developed areas of the city. The second group is the

Occupation	Total
Fulltime farming	94(39.1%)
Professionals	32 (13.3%)
Trading	28 (11.7%)
Civil Servants	54 (22.5%)
Un-employed	14 (5.8%)
Others	18 (7.5%)
Total	240 (100%)

Source- Local Fieldwork, 2007.

Table 2. Occupational Structures of Urban Farmers in Ilorin.

middle income civil servants and other public officers, about 22.5% falls into this category. These are people who majorly farm to supplement their income, and they are mostly found on open lands, and other un-used lands majorly at the outskirt of the town and the surrounding villages.

The professionals who are usually the artisans ranked third with about 13.3% of them in this category while about 11.7% of them are traders. About 5.8%of the un-employed are also involved in urban agriculture. The occupational categories referred to as others include casual workers and migrants who are just settling down in the last one year and they represent a large number of urban poor. This distribution shows that a large proportion of people add urban farming to their fulltime jobs to supplement income or a source of food security to feed immediate family.

The scarcity of food and unemployment has forced many urban poor into farming-at least to feed themselves and extending sale to the community. It is therefore a source of urban employment. This is true for a large portion of rural residents who migrate to the urban areas in search of employment. Many itinerant farmers are also engaged in farming related jobs in the urban areas. It is also clear that urban agriculture in many locations in developing countries makes a significant contribution to food self-reliance in major cities especially in Africa. Food self-reliance is not self-sufficiency but it can go a long way towards reducing food insecurity of vulnerable groups of people. Urban agriculture also supplements a significant share of cities needs and the quality of food they depend upon.

There are also indications that urban agriculture contributes to producers' well being in several ways, such as nutrition, health, cash saving and income generations Mougeot (1994:8) indicated that self-produced food accounted for as much as 18% of total household consumption in East Jakarta, while Olawepo (2008:294) indicated that this accounted for about 22% of urban local consumption in Ilorin, Nigeria. The percentages are much higher in some African countries. Thus in poorer countries and among the lower income groups, self produce food can cover considerable share of household's total food intake and can save or release larger share of household cash incomes. More and more people in our cities are trying to grow some of the food they need, even if it is not much nationally. Urban food production worldwide is for consumption and it can increase household income. Large proportion of urban farmers is doing it for commercial purposes while local productions support family expenditures and food security.

Despite the criticisms levied against urban agriculture in various urban centres in Nigeria, there has been a growing awareness on the need to recognize its relevance in contemporary Nigerian environment. It is not a small object that has generated attention especially in Nigeria, where the urge and need to feed 'more mouths' is on the increase. The increase in the area coverage of agricultural lands is a pointer to the fact that more people and migrants to urban centers are probably involved in farming practices as a result of its contribution to food production and income generation in time and space. It is a noticeable fact in Nigeria today that the successive Governments in the last three decades have been encouraging mass involvement in agricultural productions both in Nigeria's rural and urban centers. The introduction of government programmes such as the Green Rrevolution, Operation Feed the Nation (OFN) and better life for rural women among others has further boost the morale of men and women in agricultural practices in different locations, urban centres inclusive. This is noticeable in places like Lagos, Ibadan, Oshogbo, Oyo,Jos, Uyo, Sokoto, Maiduguri and Enugu to mention a few.

6. The challenges of urban agriculture

The problems of urban agriculture are numerous and they vary according to the types of farming or the locations where they are found. A recent study of Urban food production in Ilorin, Nigeria has afforded us the opportunity to share parts of the challenges of Urban Agriculture in developing World. Table 3 shows the list of urban agriculture problems from a recent study in Nigeria. For example, the expanded commercial farming at the surrounding villages face similar problems like those of other farmers in other locations in the state.85% or urban farmers indicated the problem of disturbance on farms from intruders and animals as the main problem. This affects majorly those in close locations to the city. This is so because most of the urban agriculture in developing world are not usually organized, except those owned by corporate organizations and governments. There are however variations from countries to countries. For example, commercial ranching is organized in countries like Kenya, Tanzania, Ethiopia and Zimbabwe. Whereas in a country like Nigeria, organized ranching are few while free ranged animal husbandry is common both within the city and at the outskirt of the city.

This problem was followed by unreliable supply of inputs such as fertilizers, agro-chemicals, and farm extension services. This problem is common among African urban farmers generally as a result of poor accessibility to capital which invariably is a constraint to increased agriculture and urban food production generally. Following closely is insecurity of tenure on farmland especially farmers within developed areas. This problem is on the increase as a result of sustained increase in the economy of most developing countries' urban set up. The rate of urban growth is creating increased demand in land use for both industrial and residential development. More land owners are demanding for their lands even before harvesting periods. Many at times, farmers will forfeit their crops with or without compensation from land owners.

Another important problem ranked high by 65% of farmers at restricting locations and ranked fifth by others is the issue of threat from Government officials especially those in charge of the city beautification. For example in Ilorin, Nigeria this includes the State Waste Management Corporation and Ministry of Land and Housing. This comes in the wake of Government's advocate for environmental protection within the city as 'defacing' the green acres and beauty of the town. Other problems faced in order of their importance include lack of financial support from the government, shortage of water for irrigation, poor yield due to late onset of rain, lack of storage facilities, dwindling price of agricultural products and insufficient time to work on the farm.

Problem	Rank	Remark/Causes
Disturbance from intruders & animals	1	Lack of fence and demarcation
Unreliable supply of inputs	2	Poor attention from government agencies
Insecurity of tenure on lands	3	Continuous development of land
Threat from government agencies	4	Misplaced priority
Poor financial support	5	Poor accessibility to fund
Shortage of water	6	Irrigation problem
Poor yield	7	late onset of rain the previous year
Dwindling price of farm products	8	Glut in the market
Insufficient time to work on the farm	9	Demand from primary occupation

Source: Local Fieldwork, 2007.

Table 3. Major Problems of Urban Agriculture

When asked to suggest the way out of these problems, the major consensus of the farmers is focused extensively on government recognition and provision of extensive farming land outside the main city location. This will boost production and improve food availability in the city. Food security will also be ensured not only in the urban setup, but in the surrounding localities. Various governments in the third world countries should lay emphasis on diversification of their economies. This would reduce dependency on single mode economy which relies mostly on a mineral resource. Apart from this, policies that would increase incomes and food productivity with high proportion of local content should be encouraged. In conclusion, in view of the importance of urban agriculture in the third world as indicated by these findings, urban agriculture should be retained as a part of the city's economy. This might entail zoning certain areas of the city for specifically agricultural uses (on the green belt model). We can alternatively alter existing bylaws to permit farming in certain parts of our urban cities-most notably in the residential suburbs and the more peri-urban areas. More must also be done to formulate planning policies that will directly increase the chances of the urban poor to enhance their livelihood by supporting urban agriculture, a promising but largely undeveloped sector.

7. References

Ajibola, O. (2000) Institutional Analysis of the National Food Storage Programme *Research Report*,:23 1-12 Development Policy Center, Ibadan

Axumite, G, Egziabher, D. B; Daniel G. M, Mougeot, L.J (1994) *An Example of Urban Agriculture in East Africa,* International Development Research Center, Ottawa Canada.,

Bergman, E.F. and Renwick, W.H. (1999) *Introduction Geography: People, Places and Environment* Prentice Hall, New Jersey.

Cai, J.Xie, L. and Yang, Z. (2004) Changing Role of Women in China for Urban Agriculture, *RUAF/Urban Harvest Women Feeding Cities Workshop,* Accra, Ghana.

Chung, K, Ramakrishna, R., and Riely, F. (1997) *Alternative Approaches to Locating the Food Insecure: Qualitative and Quantitative Evidence from South India Food Consumption and Nutrition Division,* Washington, D.C. International Food Policy Research Institute.

Demery,L. Ferrroni, M.,Grootaert,C. and D Wongovalle, J. (1993) *Understanding the Social Effects Policy Reform,* Washington D.C. World Bank.

Mougeot L.J (1994) *African City Farming from a World Perspectives,* International Development Research Center Ottawa, Canada.

Mougeot L.J (2000) Achieving urban Food and Nutrition security in Developing Countries: The Hidden Significance of Urban Agriculture, *IFPRI Brief Paper:* 6 1-15

Muhammad-Lawal, A. and Omotesho, O.A (2006) Farming Household Food Security in some Rural Areas of Kwara State, Nigeria, *Geo-Studies Forum,*3 (1&2):71-82

Olawepo, R.A. (2008) "The Household Logic of Urban Agriculture and Food Production in Ilorin, Nigeria", *European Journal of Social Sciences,* 6(2):288-296

Olayemi, J.K. (1998) Food Security in Nigeria, *Research Report,*:23 20-27,Development Policy Center, Ibadan

Sanyal. B. (1984) *Urban agriculture a strategy of survival in Zambia,* University of California at Los Angeles, Los Angeles, C. A. USA.

Sawio, C. J (1993) Feeding the urban masses? Towards an understanding of the dynamics of urban agricultures and land use change in Dar es Salaam, Tanzania, *Un-published Ph.D. Thesis,* Clark University, Worcester Mass, USA.

Tunde, A.M. (2011) Women Farmers and Poverty Alleviation in Small Towns of Kwara State, Nigeria, *Unpublished Ph.D. Thesis,* Department of Geography and Environmental Management, Unilorin, Nigeria.

Von Braun, J (1993) Improving Food Security of the Poor: Concept policy and Program, Washington D.C. , International Food Policy Research Institute.

Wade, I. (1986) *City food crop selection in third world cities Urban Resources Systems*, San Francisco, U.S.A.

World Food Summit (1996) *Food Security*, FAO World Food Summit, Rome

4

Climate Change Implications for Crop Production in Pacific Islands Region

Morgan Wairiu, Murari Lal and Viliamu Iese
Pacific Centre for Environment and Sustainable Development,
University of the South Pacific,
Fiji

1. Introduction

Climate plays an important role in crop production since plants require suitable temperature, rainfall and other environmental conditions for growth and development. Changes in temperature and rainfall would affect crop production, the degree of which varies with latitude, topography, and other geographic features of the location (Huey-Lin Lee, 2009).The IPCC 4th Assessment Report in 2007 predicted that the intensity of climate change especially temperature will increase in the future and stressed that many Pacific islands will be among the first to suffer its impacts. It further reported with high confidence that it is very likely that subsistence and commercial agriculture on small islands like those in the Pacific region will be adversely affected by climate change. The Asian Development Bank (2009) report also stressed that the Pacific Islands Countries and Territories (PICTs) are amongst some developing countries that are likely to face the highest reductions in agricultural potential in the world due to climate change. Furthermore, Secretariat of Pacific Regional Environment Programme (2009) emphasized that climate change impacts will be felt not only by current population but for many generations in the Pacific region because of the small island countries' high vulnerability to natural hazards and low adaptive capacity to climate change. The PICTs total land area is only 553 959 km² while they spread over almost 20 million km² of ocean. In other words, Pacific Islands land covers only 2 percent of the total Pacific region and Papua New Guinea (PNG) accounts for 83 percents of that total land area. The islands and their inhabitants are continuously exposed to a range of natural hazards, including cyclones, storm surges, floods, drought, earthquakes and tsunamis. The region's limited land area and vast ocean support the livelihood of approximately 9 million people (FAO, 2008).

The purpose of this chapter is to bring to the fore implications of climate change on the status of crop production in the Pacific Islands region. The Pacific Island people derive their livelihood or secure their food security from natural resources sectors including agriculture, forestry, fisheries and aquaculture; that is, their livelihood is depended on the environment. Any threat or impact on their environment will have profound impact on people's livelihoods. The PICTs limited land resources are under constant pressure from many factors including climate change. Agricultural crops contribute substantially to people's food security status.

2. Physical and natural environment of the Pacific region

The PICTs vary significantly in size, from PNG with a total land area of 460,330 km² to Nauru and Tuvalu that are smaller than 30 km² (FAO, 2008) The islands also have marked differences in geological resources, topographical features, soil types, mineral and water availability, diversity of terrestrial, freshwater and marine flora and fauna. Many Island countries especially the atolls have poorly developed infrastructure and limited natural, human and economic resources, and their populations are dependent on limited land and marine resources to meet their food requirements. The high islands support large tracts of intact forests including many unique species and communities of plants and animals.

The SPREP 2009 report described most of PICTs environment and development status in detail. Most of the PICTs economies are reliant on a limited resource base and are subject to external forces, such as changing terms of trade, economic liberalisation, and migration flows. The report further stated that demand by global market economies and increasing population of PICTs is resulting in significant commercial and subsistence harvesting of limited natural resources at unsustainable level. The activities include unsustainable logging, cultivation of steep and marginal lands, monocropping for commercial purposes, infrastructure development and mining. In the last 30 years, many terrestrial ecosystems have been heavily disturbed and degraded, increasing their vulnerability to global environmental changes including climate change. Further, the PICTs are located in the vast Pacific Ocean and are prone to natural hazards often of geological nature.

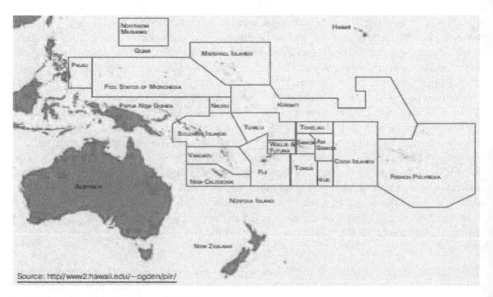

Source: http://www2.hawaii.edu/~ogden/piir/

Fig. 1. Map showing location of PICTs

The region hosts a population of approximately 9 million, a number expected to increase substantially by 2030 (FAO, 2008). The population densities vary from just over one person per kilometer for Pitcairn Island to almost 300 or more for Nauru and Tuvalu. The majorities of the population live in rural areas and rely heavily on agriculture, forestry and fisheries as a source of food security. However, urbanization is taking place very fast resulting in more than 40 percent of population residing in urban areas especially in small and atoll countries for example Kiribati and Tuvalu, putting pressure on fragile limited land and aquatic resources. Despite the strong geographical and cultural differences that characterize the region, many PICTs share common ecological and economic vulnerabilities especially to environmental and climate change.

3. Agriculture and crop production profile of Pacific region

The agriculture sector support the livelihoods of many Pacific islanders but it is one of the most vulnerable sectors to climate change. More than 70 percent of population in PICTs directly or indirectly relies on agriculture as a source of livelihood (ADB, 2009). Crop production practices in terms of size and production systems are just as diverse according the geographical diversity of the islands. For example, some diverse agricultural systems include the lowland sago management in PNG, systems of intensive dry cultivation of yams in Tonga, sunken fields dug to tap subsurface water for giant swamp taro cultivation on atolls in Kiribati and Tuvalu, and the remarkable landscapes of irrigated and bunded pond-fields for growing taro in New Caledonia and Fiji (Bellwood, 1989). McGregor (2006) classified the PICTs into three categories based on their diverse natural resource base and size. The larger island countries include Papua New Guinea (almost 90 percent of land area), Solomon Islands, Vanuatu and Fiji. These are mainly volcanic and generally rich in biological and physical resources. In marked contrast, the atoll countries (Federated States of Micronesia, Kiribati, Nauru, Niue, the Republic of the Marshall Islands, Tuvalu and Palau) are small, have limited natural resources and poor soils. The remaining countries (Cook Islands, Tonga and Samoa) fall in between the two categories above. The PICTs categories are shown in Table 1.

Almost all subsistence food, domestically marketed food, and export cash crops are grown by rural villagers on land that they access through customary land tenure arrangements or lease from traditional land owners. The mix of subsistence food production and small scale income generating activities can be broadly divided into:

- domestically marketed food (root crops and vegetables)
- export commodity crops (tree and root crops)
- minor cash crops (nuts and spices)
- livestock

		Land area (ha)	Arable land area (ha)	Population*	% Rural	Geographic	Importance of agricultural sector
Group 1 Relatively larger countries of Melanesia	Papua New Guinea	46,224,300	231,122	5,100,00 (2003)	85	High islands – a few small atolls	- Fundamental – overwhelming source of employment – provides a substantial proportion of net export earnings – subsistence a significant component of GDP

		Land area (ha)	Arable land area (ha)	Population*	% Rural	Geographic	Importance of agricultural sector
	Solomon Islands	2,853,000	17,118	515,870 (2009)	84	High islands – a few small atolls	Fundamental – overwhelming source of employment – provides a substantial proportion of net export earnings – subsistence a significant component of GDP
	Fiji Islands	1,827,200	168,102	837,271 (2007)	49	High islands, a few minor atolls	Fundamental – main employer and net foreign exchange earner, subsistence a significant proportion of GDP
	Vanuatu	1,219,000	207230	234,023 (2009)	76	High islands – a few small atolls	Fundamental – overwhelming source of employment – provides a substantial proportion of net export earnings – subsistence a significant component of GDP
	New Caledonia	1,910,300		220,000 (2000)		High islands	Important, particularly in the south
Group 2 Middle – sized countries of Polynesia	Samoa	293,500	25,828	178,200 (2003)	78	High islands	Fundamental – traditional agriculture the underlying strength of the economy
	Tonga	74,700		108,200 (2003)	57	High islands – a few small atolls	Fundamental – agricultural led economic growth in recent past

		Land area (ha)	Arable land area (ha)	Population*	% Rural	Geographic	Importance of agricultural sector
	Cook Islands	23,700		20,400 (2002)	30	High islands and atolls	Important – main export earner – subsistence a significant component of GDP
	French Polynesia	352100		233,500 (2000)		High islands and atolls	Some – small export earnings, domestic cash income, and subsistence
Group 3 Resource poor micro, predominantly atoll, states	Federated States of Micronesia	70,100		133,150 (2000)		High islands and atolls	Some – small export earnings, some domestic cash income, and some subsistence
	American Samoa	20,000		68,700 (2002)		High islands, with a few atolls	Minor – some subsistence and limited gardening
	Guam	54,100		163,941 (2003)		High island	Limited – some domestic market gardening
	Kiribati	81,100		98,600 (2003)	54	Predominantly atolls	Considerable – important for subsistence – copra important for out-island cash income and some foreign exchange
	Marshall Islands	72,000		73,600 (2002)	30	Atolls	Limited – some subsistence income earned from copra
	Nauru	2,100		12,329 (2001)		Raised coral island	Insignificant
	Niue	25,900		2,145 (2003)	68	Raised coral island	Significant – subsistence and some root crop exports
	Palau	48,800		19,000 (2001)	23	High islands and atolls	Some – market gardening
	Tokelau	1,000		1,400 (2003)		Atolls	Some subsistence

		Land area (ha)	Arable land area (ha)	Population*	% Rural	Geographic	Importance of agricultural sector
	Tuvalu	2,600		11,000 (2002)	58	Atolls	Some – subsistence and some cash income from copra
	Wallis and Futuna	25,500		14,900 (2003)		High islands and atolls	Some subsistence

Table 1. Pacific Island Countries and Territories Categories (adapted from McGregor 2006 and Secretariat of Pacific Community's Pacific Regional Information System, 2011)

Subsistence crop production represents a major strength of the PICTs economy because of the ability of people to feed themselves and support each other during periods of disasters, loss of cash income, and times of displacement. These traditional arrangements vary somewhat between the PICTs, and they are changing in response to modern economic development shifts and changing environment. Crop production involve cultivating, harvesting and managing food crops from different environments, the most important being shifting cultivation gardens, but also including fallow forests, primary forest, swamps and mangroves (Jansen et al., 2006). Soil fertility in gardens is maintained through a bush fallow in most cropping systems. However, subsistence crop production can sometimes fail, because of increasing population, diseases, pest and invasive species outbreaks, and extreme weather which interrupt with planting cycles. Climate change is now resulting in high frequency and severity of extreme weather events such as cyclones, drought, and excessive rainfall which impact on crop production. Many tropical crops such as yams (*Dioscorea spp.*), taro (*Colocasia esculenta*), cassava (*Manihot esculenta*) and sweet potatoes (*Ipomoea batata*) and other crops such as bananas (*Musa spp.*) and watermelon (*Citrullus lanatus*) form part of people's staple diet. For example, sweet potato is the most important subsistence crop in PNG, Solomon Islands and Vanuatu, while taro and cassava in Fiji, Samoa and Tonga. Sweet potato accounted for around 65 percent of the estimated 432 000 tonnes of staple food produced in 2004 in Solomon Islands (Bourke et al., 2006). For atoll islands, giant swamp taro (*Cyrtosperma chamissonis*) breadfruit (*Artocarpus altilis*), coconut (*Cocos nucifera*) are the main crops grown.

4. Observed changes in climate, trends and future projections

Historical climate data for the PICTs is limited, but there is some evidence of a trend towards warmer and drier conditions over the past 100 years. Despite limited climate data for the region, there is evidence that the climate is changing (FAO, 2008). The annual and seasonal ocean surface and island air temperatures increased from 0.6 to 1.0°C since 1910 throughout a large part of the South Pacific and decadal increases of 0.3 to 0.5°C in annual temperatures to the southwest of the South Pacific Convergence Zone (SPCZ) since 1970 (Folland et al., 2003). Hay et al. (2003) also reported that sea surface temperatures in the region have increased by about 0.4°C. At national level, the annual mean surface air temperature has increased by 1.2 °C since the reliable records began in Fiji, representing a rate of 0.25 °C per decade (Mataki et al., 2006). There was significant increase in the annual number of hot days and warm nights, and significant decrease in the annual number of cool days and cold nights, particularly in years after the onset of El Nino in the period 1961 to 2003 but extreme rainfall trends were generally less spatially coherent than extreme temperatures (Griffiths et al., 2003; Manton et al., 2001). Mataki et al. (2006) also examined the changes in the frequency of extreme temperature events and found that significant increases have taken place in the annual number of hot days and warm nights for both Suva and Nadi in Fiji, with decreases in the annual number of cool days and cold nights at both locations. The number of hot days (max temperature ≥ 32 °C) shows a significant increasing trend while the number of colder nights (min temperature < 18 °C) showed a decreasing trend at Suva. It is predicted that average temperatures are expected to rise by between 1.0 and 3.1°C. Air temperature could increase to 0.90°C -1.30°C by 2050 and 1.6°C -3.4°C by 2100 (World Bank, 2006).

The southern Pacific is now experiencing a significantly drier and warmer climate (by 15 percent and 0.8°C, respectively). The Central Equatorial Pacific, by contrast, is experiencing more intense rain (representing a change of about 30 percent) and a similarly hotter climate (0.6°C). There has been a small increase over ocean and small decrease in rainfall over land since 1970's. An analysis of monthly rainfall patterns at Goroka in Eastern Highlands Province of PNG from 1946 to 2002 found that there had been a shift to longer, but less pronounced, rainy seasons. Throughout the lowlands and highlands, villagers report similar changes in rainfall patterns. These changes are also linked in part to an increased frequency of El Nino events (Allen & Bourke, 2009). Observed rainfall at Nadi from 1941 to 2005 shows a large inter annual variability with no significant long term trend but there has been an increase in the frequency of extreme rainfall events over recent decades, a trend which is likely to continue into the future (GoF, 2011). The projected increases in surface air temperature and rainfall shown in table 2.

The global sea level gradually rose during the 20th century and continues to rise at increasing rates (Cruz et al., 2007). Small islands in the Pacific are particularly vulnerable to rising sea levels because of their proximity to the El Niño Southern Oscillation. Fifty-years or longer time-series data for sea-level rise from four stations in the Pacific reveal that the average rate of sea-level rise in this sub-region is 0.16 centimeters (cm) a year. Twenty-two stations with more than 25 years worth of data indicate an average rate of relative sea-level rise of 0.07 cm a year (Bindoff et al., 2007). In Asia and the Pacific, the sea level is expected to rise approximately 3–16 centimeters (cm) by 2030 and 7–50 cm by 2070 in conjunction with regional sea level variability (Preston et al., 2006). In Fiji over the period from October 1992 to December 2009, sea level increased by 5.5 mm per year, after taking into account the inverted barometric pressure effect and vertical movements in the observing platform. This is far greater than the estimated range of global sea-level rise over the past century, namely 1 to 2 mm per year.

Sea level is expected to rise between 9 and 90 cm by the end of the century, with the western Pacific experiencing the largest rise. Sea level rise is also likely to affect groundwater resources by altering recharge capacities in some areas, increasing demand for groundwater as a result of less surface water availability, and causing water contamination due to rising sea levels. Climate scenarios predict up to 14% loss of coastal land due to sea level rise and flooding by 2050 (Feresi et al., 2000), which are the prime coastal areas for economic activities including crop production.

Factor/Variable	Observation	Projections/Scenarios
Temperature	0.6 to 1.0 increase since 1910 0.3 to 0.5 decadal increase since 1970	Air temperature could increase 0.9° - 1.3°C by 2050 and 1.6 -3.4°C by 2100.
Rainfall	Small decrease over land since 1970's Small increase over ocean since 1970's	Rainfall could either rise or fall. Most models predict an increase by 8-10 percent in 2050 and by about 20 percent in 2100, leading to more intense floods or droughts
Sea Level Rise	Relative sea level rise of 0.6 to 2.0 mm yr[-1] since 1950	Sea level could rise 0.2 meters (in the best-guess scenario) to 0.4 meters (in the worst-case scenario) by 2050. By 2100, the sea could rise by 0.5-1.0 meters relative to present levels. The impact would be critical for low-lying atolls in the Pacific, which rarely rise 5 meters above sea level. It could also have widespread implications for the estimated 90 percent of Pacific Islanders who live on or near the coast
El Nino		The balance of evidence indicates that El Niño conditions may occur more frequently, leading to higher average rainfall in the central Pacific and northern Polynesia. The impact of El Niño Southern Oscillation (ENSO) on rainfall in Melanesia, Micronesia, and South Polynesia is less well understood
Cyclones	Noticeable increase in frequency of category 4 and 5 cyclones since 1970	Cyclones may become more intense in the future (with wind speeds rising by as much as 20 percent); it is unknown, however, whether they will become more frequent. A rise in sea surface temperature and a shift to El Niño conditions could expand the cyclone path poleward, and expand cyclone occurrence east of the dateline. The combination of more intense cyclones and a higher sea level may also lead to higher storm surges

Source: (Bindoff et al., 2007; Cruz et al., 2007; Folland et al., 2003; Hay et al., 2003)

Table 2. Observed and predicted temperature, rainfall and Sea level rise.

Historical records on occurrences of extreme events like cyclones, storm surges, flooding and drought show that they are increasing in intensity or severity. Cyclones are expected to increase in intensity by about 5–20 percent. Storm frequency is likely to increase in the equatorial and northern Pacific. In general, the future climate is expected to become more El-Nino like, resulting in more droughts in the southern Pacific and more rain and consequent floods in the equatorial Pacific. Hurricane-strength cyclones; those with winds stronger than 63 knots or 117 km/hr have increased systematically in the southwest Pacific, a trend that has also been observed at the global level over the past 30 years (Emanuel, 2005; Webster et al., 2005). The region now experiences on average four hurricane-strength cyclones a year.

5. Impacts of climate change and climate variability on agriculture and crop production

Climatic change is already influencing agriculture and crop production in the PICs but Allen and Bourke (2009) caution that it is too early to draw conclusions since there is not enough information about probable changes in temperature and patterns of rainfall and rainfall extremes and furthermore agricultural responses to climate change will be complex because other factors affecting crop production will also at play. This may be true for effects of changing temperature and rainfall but other effects are obvious. The direct or immediate impacts of climate change on agriculture and crop production occur during or immediately after a natural hazard or extreme event, such as damage to crops, farmlands and agriculture infrastructure from cyclones and flooding. The World Bank (2006) reported that during the period 1950 to 2004, about 207 extreme events were recorded in the pacific region and the cost of climate-related disasters on the agricultural crops is estimated to range from US$13.8 million to US$14.2 million. Ten of the 15 most extreme events reported over the past half a century occurred from 1990 onwards as shown in table 3. There has been a substantial increase in the hurricane-strength cyclones since the 1950s with an average of four events in a year. Similarly, the number of reported disasters in the Pacific Islands region has also increased significantly since the 1950s and disasters are becoming more frequent with increasing intensity of extreme events. This period also registered 96 (50 percent) of the 192 minor disasters.

Cyclones are the most common extreme event and caused more disaster in the region. Cyclones accounted for 76 percent of the reported disasters from 1950–2004, followed by earthquakes, droughts and floods. The average cyclone damage to PICTs economies during this period was US$75.7 million in real 2004 value (World Bank, 2006). In New Caledonia, the estimated cost of damage to agriculture by cyclone Erica in March 2003 was US$13 million (Terry et al., 2008) while Cyclone Ami that hit Vanua Levu in Fiji in 2003, caused US$33 million loss (McKenzie et al., 2005), mainly due to flood damage of agricultural crops and infrastructure. In 2004 cyclone Ivy affected over 80 per cent of food crops in Vanuatu and cyclone Val in 1991, hit Samoa with maximum wind speeds of 140 knots causing massive damage; equivalent to 230 per cent of the country's real 2004 GDP (World Bank, 2006).

Drought, an extreme form of rainfall variability can affect the highest number of people per event. El Niño event in the past has resulted in water shortages and drought in some parts of the Pacific (e.g. PNG, Marshall Islands, Samoa, Fiji, Tonga and Kiribati), and increased

Event	Year	Country	Estimated losses
Cyclone Ofa	1990	Samoa	140
Cyclone Val	1991	Samoa	300
Typhoon Omar	1992	Guam	300
Cyclone Nina	1993	Solomon Islands	–
Cyclone Prema	1993	Vanuatu	–
Cyclone Kina	1993	Fiji	140
Cyclone Martin	1997	Cook Island	7.5
Cyclone Hina	1997	Tonga	14.5
Drought	1997	Regional	> 175[a]
Cyclone Cora	1998	Tonga	56
Cyclone Alan	1998	French Polynesia	–
Cyclone Dani	1999	Fiji	3.5

– Not available.
a. Includes losses of US$160 million in Fiji (Stratus 2000).

Table 3. Number of cyclones and cost of damage from 1990 to 1999(Source: World Bank, 2000)

precipitation, and flooding in others (e.g. Solomon Islands, and some areas in Fiji) (World Bank, 2000). In Fiji, the 1997/98 drought events resulted in 50 percent loss in sugarcane production and total losses in the industry were around US$50 million while other agriculture losses including livestock death amounted to around US$7 million (McKenzie et al., 2005). An extension of the dry season by 45 days has been estimated to decrease maize yields by 30 to 50 percent, and sugar cane and taro by 10 to 35 percent and 35 to 75 percent respectively (Hay et al., 2003). In Kiribati, breadfruit and banana crops suffer from drought stress resulting in lower yields (GoK, 2007). A drought associated with a severe El Nino-Southern Oscillation event in 1997, caused significant disruptions to village food and water supplies in PNG. There were severe shortages of food and water, with garden produce declining by 80 percent. Up to 40 percent of the rural population (1.2 million people) were without locally available food by the end of 1997 (Allen & Bourke, 2001). Crop production in many PICTs is effected by extreme drought events.

Increases in minimum and maximum temperature are already having a small influence on agricultural production and will have a greater influence in the future. For example in PNG, it was observed that tuber formation in sweet potato was significantly reduced at temperatures above 34 °C. Maximum temperatures in the lowlands of PNG are now around 32 °C, so an increase of 2.0–4.5 °C within a hundred years could reduce sweet production in lowland areas (National Agriculture Research Institute, 2011). In Solomon Islands, taro production has been reduced (less tubers and lower yields) in coastal areas over the years because of wave overtopping and warmer temperatures (GoSI, 2008). It was also observed that increase in temperature in PNG highlands has a severe impact on coffee production from *coffee rust* attack. Coffee rust is present in the main highland valleys at 1600–1800 m above sea level and a rise in temperature is likely to increase the altitude at which coffee rust has a severe impact on coffee production. Taro blight, a disease caused by the fungus *Phytophthora colocasiae*, also reduced taro yield at higher altitudes. The fungus is sensitive to temperature and a small rise in temperature could increase incidence of taro blight disease than occurs now. Some tree crops are bearing at higher altitude in the highlands but the lower altitudinal limit of some crops, such as Irish potato, Arabica coffee, and *karuka* (*Pandanus julianettii* and *P. brosimos*), will increase because of increasing temperatures(Allen & Bourke, 2009). In Fiji, the major concern of sugar production is the sporadic sucrose content in the yield which could be affected with increase in temperature, groundwater salinization and fluctuating soil moisture content. This is a concern because sugar is a major foreign exchange earner, accounting for about 40 per cent of the country's merchandise exports and 12 per cent of Fiji's Gross Domestic Product (Gawander, 2007). The scenario for sugar cane production in Fiji over the next 50 years will be in the following manner:

- 47 percent of the years will have the expected production of 4 million tonnes,
- 33 percent of the years will have half of the expected production,
- 20 percent of the years will have three-quarters of the expected production.

This was determine when using the period from 1992 to 1999, when Fiji was subjected to two El Niño events and an unusually high number of tropical cyclones as an analogue for future conditions under climate change. The outcome under this scenario would be an overall shortfall in excess of one quarter of expected production (GoF, 2005).

Hay et al. (2003) pointed out that for the Pacific region the smaller temperature increase relative to higher latitude locations is unlikely to place a severe limitation on crop production but the physiology of crops may be influenced in ways not yet identified. Using PLANTGRO a plant growth simulation model the following patterns were projected for Taro (*Colocassia esculenta*) and yams (*Dioscorea sp.*) in Fiji (GoF, 2005):

- Projected changes in mean conditions would have little effect on taro production, with the exception of the extreme low-rainfall scenario. It is likely that yam production will also remain unaffected, although if rainfall increases significantly, yam yields may fall slightly.
- When El Niño conditions are factored in, reductions in, production of 30-40% might be recorded in one out of three years, with a further one in five years affected by the residual effects of the ENSO events.

Agricultural productivity in PICTs is heavily dependent on the seasonal rainfall. About 70 percent of the gross cropped area in the Pacific Islands is geographically located so as to benefit from rains in the summer season (November – April). While the rainfall requirements and tolerance of extremes vary from crop to crop, a working figure for the south west Pacific is that a mean annual rainfall of 1800-2500mm is optimal for agricultural production and a mean annual rainfall of over 4000mm is excessive (Bourke et al., 2006). A significant (>50 percent) increases in rainfall on the windward side of high islands during the wet season may increase taro yields by 5 to 15 percent, but would reduce rice and maize yields by around 10 to 20 percent and 30 to 100 percent, respectively (Hay et al., 2003). In PNG, most of the rural population live and cultivate crops in areas where annual rainfall is in the range 1800-3500 mm. In mountainous locations where clouds form early in the day and reduce sunlight, human settlement and agriculture is generally absent. Localities where the annual rainfall is more than 4000 mm tend to be too wet and have too much cloud cover for good agricultural production. Yields of sweet potato and other crops tend to be lower on the southern sides of the main mountain ranges, for example, in Southern Highlands Province and mountainous parts of Gulf Province in PNG. This is because of both excessively high rainfall and high levels of cloudiness (Allen & Bourke, 2009)

Climate change predictions for the region suggest prolonged variations from the normal rainfall which can be devastating to agriculture. Shift of rainfall patterns affect planting time, growing stages, harvest periods, post harvesting storage and drastically reduce total yield. Agriculture and crop production is under stress from these climatic factors but it remains difficult to predict the likely outcomes with certainty because of limited empirical data for the Pacific region. Disruptions to food production and the economy may intensify in future, given the projections for more intense tropical cyclones and precipitation variations of up to 14 percent on both sides of normal rainfall (IPCC 4AR, 2007) by the end of the century. More so, in between climate extremes, altered precipitation and increased evapotranspiration (including its intensity as well as temporal and spatial shifts) will also be of concern as these changes take root. The increase in atmospheric carbon dioxide may benefit agriculture but these positive effects are likely to be negated by thermal and water stress associated with climate change (Lal, 2004) and changes in pests' voracity and weeds' growth; loss of soil fertility and erosion resulting from climatic variability being another problem. Increasing coastal inundation, salinization and erosion as a consequence of sea level rise and human activities may contaminate and reduce the size of productive agricultural lands and, thereby, threaten food security at the household and local levels.

The most destructive impact of excessive rainfall on agriculture infrastructure and crops are flooding and waterlogging. For example, flooding during 2004/2005 and 2006/2007 caused around US$76 million and US$11 million in damages, respectively in Fiji . The cane growers' direct and indirect costs from the 2009 flood are estimated to be US$13.4 million. The costs include losses in cane output, non-cane and other farm losses, and direct and indirect household (Lal et al., 2009).

Sea-level rise is affecting agriculture in three different ways: Coastal erosion resulting in loss of land and some areas are permanently inundated, making it unsuitable for agriculture production. Some areas are also subjected to periodic inundation from extreme events, including high tides and storm waves, contaminating the fresh water lens, with devastating effects especially on atolls; and seepage of saline water through rivers during dry seasons, resulting in increasing the salt level in soils. Storm surges and increased salt water intrusion limits the range of crops that can be grown. The small atolls in particular face serious problems for example, pit and swamp cultivation of taro is particularly susceptible to changes in water quality. In Tuvalu, groundwater salinization as a result of sea-level rise is destroying the traditionally important swamp taro pit gardens (Webb, 2006) and raises concern on the safety of drinking water (Tekiene, 2000). In Kiribati, coastal erosion reduces crop productivity such as of pandanus varieties and coconut. The pandanus fruit is used by people as long term preserved food but most trees are lost through coastal erosion (GoK, 2007).

		Observed climate change and extreme events impact on crops and agriculture production
Group 1 Relatively larger countries of Melanesia	Papua New Guinea	Tuber formation in sweet potato was significantly reduced at temperatures above 34 °C (Allan & Bourke, 2009, NARI, 2011).
	Solomon Islands	Taro production has been reduced (less tubers and lower yields) in coastal areas over the years because of wave overtopping and warmer temperatures. Cyclone Namu wiped out rice industry in 1986 (GoSI, 2009)
	Fiji Islands	Drought and cyclones in 1997 led to a decline of production to 2.2 million tons of cane and 275 000 tons of sugar from a peak of 4.1 million tons of cane and 501 800 tons of sugar in 1986 (Gawander, 2007) .The cane growers' direct and indirect costs from the 2009 flood are estimated to be US$13.4 million. The costs include losses in cane output, non-cane and other farm losses, and direct and indirect household (Lal et al., 2009)
	Vanuatu	Increased temperatures and variability of rainfall resulted in increased pest activities with yams being the crop most affected and in livestock there was increased incidence of intestinal problems in cattle often associated with pasture. Some plants flowering earlier than usual while others are fruiting much later than normal during the past 3–4 years (FAO, 2008).
	New Caledonia	The estimated cost of damage to agriculture by Cyclone Erica in March 2003 was US$13 million (Terry et al., 2008)
Group 2 Middle – sized countries of Polynesia	Samoa	The increasing threats from new diseases and pests for both livestock and crops are linked to cyclones, flooding and drought and other variations in climate. The increasing incidence of forest fires has led to the destruction of crops as evident in the past forest fires in rural communities (GoS, 2005).
	Tonga	Squash crop which had been producing 50% of the country's exports by value was more than halved
Group 3 Resource poor micro, predominantly atoll, states	Federated States of Micronesia	Taro pits on some islands and atolls have been contaminated by salt water associated with a depletion of fresh-water lenses, extended droughts and saltwater inundation/intrusion (FAO,2008)
	Kiribati	The pandanus fruit is used by people as long term preserved food but most trees are lost through coastal erosion due to sea level rise and breadfruit and banana crops suffer from drought stress resulting in lower yields (GoK, 2007).
	Marshall Islands	During the El Niño season of 1997–1998, there was significant reductions in most crop yields (FAO, 2008)
	Palau	Taro pits on some islands and atolls have been contaminated by salt water associated with a depletion of fresh-water lenses, extended droughts and saltwater inundation/intrusion (Burns, W., 2000)
	Tuvalu	Groundwater salinization as a result of sea-level rise is destroying the traditionally important swamp taro pit gardens (Webb, 2007)

Table 4. Observed climate change impact on agriculture and crop production

6. Current climate change adaptation activities in agriculture

A number of regional climate change adaptation initiatives to address agriculture and crop production are currently implemented through regional organizations like the Secretariat of the Pacific Community (SPC) and the Secretariat for Pacific Regional Environment Programme (SPREP) and supported by various donors. They include:

1. The Regional Programme on Adaptation to Climate Change in the Pacific Island Region (ACCPIR) which initially has three pilot sites in Fiji, Tonga and Vanuatu but now is being extended to other PICs. Activities in Vanuatu include introducing climate-resistant crops, breeding extreme weather- adapted livestock, developing community land-use plans, trialing new agroforestry and soil stabilisation methods, and undertaking innovative climate adaptation education programmes whilst in Tonga, the focus has been on land and forest management on more vulnerable islands including Eua and Vava'u. The projected increase in temperature and rainfall show 'Eua having the highest soil productivity and soil erosion risk level compared to the other islands.

2. The Land Resources Division (LRD) of SPC is conducting atoll agriculture research and development at the Centre of Excellence for Atoll Agricultural Research and Development in Tarawa, Kiribati. Areas of work include atoll soil management, water management, cultivar evaluation, and improving the resilience of food production systems to climate change. The centre is also documenting sustainable food production systems, and food preservation and utilisation methods for atolls.

3. The Centre for Pacific Crops and Trees (CePaCT) and Forestry and Agriculture Diversification under the Land Resources Division (LRD) of the Secretariat of the Pacific Community (SPC) is the Pacific regional gene bank based in Fiji. It plays an important role in climate change adaptation efforts, improving food security and supporting domestic and export trade in agriculture and forestry products. Since 2009, it has distributed 4,038 plants to 12 PICs. The crops/species distributed include taro, sweet potato, yam, banana, breadfruit, *Alocasia, Xanthosoma,* , cassava, potato and vanilla. The LRD, with the support of the AusAID International Climate Change Adaptation Initiative (ICCAI) and the US government, has established a 'climateready' collection of crops and varieties known to have suitable traits at CePaCT. The collection is now being evaluated in individual PICs for climate tolerant traits such as resistance to drought, salinity and water-logging. The collection is a dynamic one and will be modified according to the evaluation information received. The ICCAI is also supporting a number of other activities, such as salinity tolerance screening research; agrobiodiversity studies in Fiji and Palau; and collaboration with CSIRO in crop modeling (SPC, 2010).

4. The Pacific Adaptation to Climate Change Project (PACC) is implemented by SPREP and is focusing on climate change adaptation. Its objective is to enhance the resilience of a number of key development sectors (food production and food security, water resources management, coastal zone, infrastructure etc.) in the Pacific islands to the adverse effects of climate change. This objective will be achieved by focusing on long-term planned adaptation response measures, strategies and policies.

5. The FAO with financial support from Italy is supporting fourteen island countries in the region in food security through the regional programme: Food Security and Sustainable Livelihood programme in the Pacific Island Countries (FSSLP). The over-

arching goal of the regional programme is to help island people grow healthier by eating more nutritious local foods, while reducing the amount of processed imported food they eat. Some of the crops promoted under the programme include drought and salt tolerance, pest and disease resistant varieties that are adaptable to changing climate (FAO, 2009).

6. The five least developed countries in the region including, Kiribati, Samoa, Solomon Islands, Tuvalu and Vanuatu have placed food security as an important issue to address in their National Adaptation Programme of Action (NAPA) to climate change and are now in the process of implementing national projects addressing various aspects of food security including crop production. Other PICTs are also developing their national adaptation plans and food security is high on their agenda.

7. Other factors contributing to the vulnerability of agriculture and crop production

Simatupang and Fleming (2001) identified three major factors that affect food security in the PICTs have negatively impacted agriculture and crop production. First, a change of diet amongst Pacific Islanders from locally produced food to imported food. Most imported food items such as rice, wheat, sugar, meat, eggs, milk, canned meat and fish, coffee, tea, alcohol and soft drinks are superior to roots and tubers in some aspects of cooking and serving practicality, shelf life, and also with respect to social prestige, which tempts the indigenous people to substitute these foods in their traditional diet. In Marshall Islands, production of taro and sweet potato has fallen dramatically because of increased access to imported staples which are more convenient for preparation and storage (FAO, 2008). Many of the imported foods are, however, nutritionally inferior to the more traditional ones. Second, expansion of large plantations and smallholder commercial farms pulls a large area of land, and potentially much labour, out of traditional food crop production, reducing its output. Third, increasing land pressure has pushed food gardens to less fertile and marginal steep lands and further away from homes. A number of households, especially in urban areas, do not have sufficient access to sufficient land for food gardening. In Tonga for example, there is now insufficient land for all commoner males to obtain their own plots and indeed some 30 percent of Tongans now do not own land. Other factors include: increased incidences of weeds, pests and diseases, thus necessitating increased application of pesticide and herbicides, which may lead to other unintended environmental impacts occurring both on the sprayed site, and offsite, especially water systems; loss of traditional farming techniques traditional knowledge; loss of plant genetic diversity and inbreeding of livestock and other domestic animals; invasive species; lack of sustainable land management; and lack of capacity to manage farm animals (FAO, 2008).

These are other clearly identifiable, non-climate change factors contributing to low crop production and reduced crop yields. These often interact with each other so that climate change exacerbates existing problems and subsequently affects food security in the region. In recognition of the impact of climate change and other factors discussed in this section, the PICT's have formulated a "Framework for Action on Food Security in the Pacific". The plan guides countries in determining relevant, specific country-level activity addressing food security. The framework for action was prepared in response to a call for action on food security from Pacific leaders at the 39th Pacific Islands Forum, held in Niue in 2008 (Food Secure Pacific, 2010). Some PICTs are now implementing the framework at national level to address the factors affecting agriculture and crop production mention here but these efforts are challenged by threat pose to crop production by climate change.

8. Tools and methods for assessment of climate change impact on crop production

As Allen and Bourke (2009) pointed out, climatic change is already influencing agriculture and food production in the PICs, but it remains difficult to predict the likely outcomes with certainty. This is because there is not enough information about changes in temperature and patterns of rainfall and rainfall extremes and furthermore agricultural responses to climate change will be complex. This together with limited capacity in the region to comprehensively assess the impacts of climate change and variability on the production of major pacific crops like cassava, taro, sweet potato, banana, rice and sugar cane add to this complexity.

Most estimates of the impacts of climate change on agricultural production in other regions of the world are done using crop simulation models. They are important tools for predicting the likely crop production scenario in the future. Although climate change is a growing concern for decision-makers in the region, information on the impacts of climate change is often lacking or incomplete. In agriculture, crop production will be affected by a combination of factors such as effects of changes in temperature and precipitation regime on plant physiology, changes on growing season onset and length, CO_2 fertilization effect, technological improvements, water and availability of which interactions are rather complicated. Researchers in the past two decades have been focused on predictions of climate change and its possible impact on agriculture and food supply in the next couple of decades. This is a major gap in the Pacific region.

A combination of integrated modeling from different disciplines and appropriate research is needed to advance the understanding and prioritization of the challenges climate change pose on agriculture and food production. Use of the crop simulation models like Decision Support System for Agro-Technology (DSSAT) and Agricultural Production Systems Simulator (APSIM) tools are starting to be used in research that will enable researchers to understand the agro-management practices most conducive to cope with the impacts of climate variability and change on important food and economic crops in the Pacific. The APSIM model is being used to determine climate change impact on cassava yield in Fiji and there is also plan to use DSATT on cassava, taro and sugar cane to predict impact of climate change on crop growth and adjust management practices to mitigate the potential impacts in some Pacific Island countries.

The Pacific Food Security Toolkit "Building Resilience to Climate Change – Root Crop and Fisheries Production" which was produced by the University of the South Pacific, Secretariat of the Pacific Community, Secretariat of the Pacific Regional Environment Program and Food and Agriculture Organization is an important document. It contains six modules that cover climate change, overview of key Pacific food systems, ecosystems and food securities, Pacific root crops, Pacific fisheries and additional tools to support Researchers, Academics, Farmers and all stakeholders.

9. Challenges and opportunities for food production

Some of the major challenges to agriculture and crop production in PICTs include: (1) increasing population against weak economic growth and rapid depletion of natural resources base with unsustainable development; (2) low level of awareness on climate change perceptions and competing government priorities; (3) limited capacity at all levels (regional, national and community) to develop and effectively implement adaptation measures; (4) inadequate resources and weak socio-economic conditions; and (5) unavailability and problem accessing reliable data on climate change impact on the sector in the region.

The agriculture sector has not been given priority in terms of resource allocation and development planning by regional countries in the past although it supports the majority of the people's livelihood. The vulnerability of the sector to climate change is beginning to be recognized amongst Pacific islands leaders and governments because of their concern for the regions food security (Barnet, 2008). The development of regional and national strategies on climate change and sustainable development frameworks now focus more on natural resources sectors including agriculture and food security. There is some level of commitment from the Pacific regional governments and their international partners to address climate change impact on agriculture and crop production. Through implementation of regional and national adaptation strategies to climate change there are opportunities to address issues and challenges in agriculture and crop production.

10. Conclusions

Observed climate data in the Pacific region is showing evidence that the climate is changing. The immediate or direct threat to agriculture and crop production comes from extreme events like cyclones, storms, flooding and drought. These extreme events are increasing in severity or intensity and frequency and already causing substantial damage to food crops and associated infrastructure or often result in total collapse of the PICTs economies, especially the small island countries. Recovery efforts are often negatively impacted by continuous occurrences of these extreme events, thus most PICTs are "locked" into a vulnerable situation.

The effects of climate change on crop production through increase in temperature, changing rainfall patterns, salt water intrusion are less immediate but also complex. Agriculture and crop production is under stress from these climatic factors but it remains difficult to predict the likely outcomes with certainty because of limited empirical data for the Pacific region. There is a need to continue to monitor the impact and conduct studies PICTs to equip for the future. Use of crop simulation models is one option to generate relevant information for planning adaptive measures to address food production and food security in the region. Because of PICTs continues exposure to extreme events and limited adaptive capacity, they also need assistance to develop and implement relevant adaptation strategies to climate change, especially in the agriculture sector where most people derive their livelihoods.

11. References

Allen B.; & Bourke R.M. 2009. People, Land and Environment, In: *Food and Agriculture in Papua New Guinea*. R.M. Bourke & T. Harwood (Eds.), pp28-121, ANU E-press, Retrieved from http://epress.anu.edu.au/food_agriculture/pdf/

Allen, B. J.; & Bourke, R.M. 2001. The 1997 Drought and Frost in Papua New Guinea: Overview and Policy Implications, In: *Proceedings of Food Security for Papua New Guinea*, ISBN 1 86320 308 7, Lae, Papua New Guinea, June 2000.

Asian Development Bank (ADB). 2009. *Building climate resilience in the agriculture sector in Asia and the Pacific*. ISBN 978-971-561-827-4 Asian Development Bank, Retrieved from http://www.adb.org/Documents/Books/Building-Climate-Resilience-Agriculture-Sector/Building-Climate-Resilience-Agriculture-Sector.pdf

Barnet, J. 2008. Food Security and Climate Change in the South Pacific. *Pacific Ecologist* 14 32-36.

Bellwood, P. 1989. The Colonization of the Pacific, In: *The Colonization of the Pacific: A Genetic Trail*, 12/5/2011, Available from. http.//www.jrank.org/history/pages/6398/Pacific-Islands-Origins-Food-Production-In.html#ixzz1OYujxANG

Bindoff, N. L.; Willebrand, J.; Artale, V.; Cazenave, A.; Gregory, J.; Gulev, S.; Hanawa, K.; Le quere, C.; Levitus, S. ; Nojiri, Y.; Shum, C. K.; Talley, L. D.; & Unnikirshnan, A. 2007. Observations: Oceanic climate change and sea level, In: *Climate Change 2007: The Physical Science Basis. Contribution of Working Group I to the Fourth Assessment Report of the Intergovernmental Panel on Climate Change*. S. Solomon, D. Qin, M. Manning, Z. Chen, M. Marquis, K. B. Averyt, M. Tignor, and H. L. Miller (Eds.), pp 390-431, Cambridge University Press, ISBN 978 0521 70596-7, Cambridge, UK

Bourke, R.M.; McGregor, A.; Allen, M.G.; Evans B.R.; Mullen, B.F.; Pollard, A.A.; Wairiu, M.; & Zotalis, S. 2006. *Main Findings - Solomon Islands Smallholder Agriculture Study*, Volume 1. AusAID, ISBN 1 920861 68 8, Canberra, Australia

Burns, W. 2003. The Impact of Climate Change on Pacific Island Developing Countries in the 21st Century, In: *Climate Change in the South Pacific: Impacts and Responses in Australia, New Zealand, and Small Island States*, A. Gillespie, & W. Burns (Eds.), pp 233-250, Advances in Global Change Research Volume 2, Kluwer Academic Publishers, Netherlands.

Cruz, R. V. O.; Harasawa, H.; Lal,M.; Wu,S.; Anokhin, Y.; Punsalmaa,B.; Honda, Y,; Jafari, M.; Li, C.; & Huu Ninh, N. 2007. Asia In: *Climate Change 2007: Impacts, Adaptation, and Vulnerability. Contribution of Working Group II to the Fourth Assessment Report of the Intergovernmental Panel on Climate Change*. M. Parry, O. F. Canziani, J. Palutikof, P. J. van der Linden, and C. E. Hanson, (Eds.) pp 469-506, Cambridge University Press, ISBN 978 0521 70597-4, Cambridge, UK.

Emanuel, K. 2005. Increasing destructiveness of tropical cyclones over the past 30 years. *Nature*, 436: 686-688.

Food and Agriculture Organisation (FAO). 2009. Stories of Hope from the Pacific -How Better Food Security is Making Island Life Healthier. 12/6/2011, Available from: http://www.fao.org/fileadmin/templates/tc/spfs/pdf/FAO_SAP_brochure_final _2_.pdf

Food and Agriculture Organisation (FAO). 2008. Climate Change and Food Security in Pacific Island Countries. 10/4/2011, Available from: http://www.fao.org/climatechange/1700302529d2a5afee62cce0e70d2d38e1e273.pdf

Feresi, J.; Kenny, G.; Dewet, N.; Limalevu, L.; Bhusan, J.; and Ratukalou, L. 2000. Climate Change and Vulnerability and Adaptation Assessment for Fiji. 15/5/2011, Available from: http://hdl.handle.net/10289/1568

Folland, C.K.; Renwick, J.A. ; Salinger, M.J.; & Rayner, N.A. 2003: Trends and Variations in South Pacific Islands and Ocean Surface Temperatures. *Journal of Climate.*, 16, 2859-2874.

Food Secure Pacific. 2010. Towards a Food Secure Pacific Framework for Action on Food Security in the Pacific. 21/6/2011, Available from: http://www.foodsecurepacific.org/documents/FINAL%20TOWARDS%20A%20F OOD%20SECURE%20PACIFIC_June1.pdf

Gawander, J. 2007. Impact of Climate Change on Sugar Cane Production in Fiji. WMO Bulletin 56 (1) p34-39

Government of Fiji (GoF). 2011. Draft National Climate Change Adaptation Strategy for Land-Based Resources (2012 – 2022), Suva, Fiji.

Government of Fiji (GoF). 2005. First national Communication to UNFCCC, Suva, Fiji.

Government of Kiribati (GoK). 2007. National Adaptation Programme of Action. Tarawa, Kitibati

Government of Samoa (GoS). 2005. National Adaptation Programme of Action. Apia, Samoa

Government of Solomon Islands (GoSI). 2008. National Adaptation Programme of Action. Honiara, Solomon Islands

Griffiths, G.M.; Salinger, M.J. & Leleu, I. 2003. Trends in Extreme Daily Rainfall Across the South Pacific and Relationship to the South Pacific Convergence Zone. *J. Climatol.*, 23, 847-869

Huey-Lin, L. 2009. The impact of climate change on global food supply and demand, food prices, and land use. *Paddy Water Environ* 7:321–331

Jansen, T.; Mullen B.F.; Pollard, A.A.; Maemouri, R.K.; Watoto, C.; & Iramu, E. 2006. Subsistence Production, Livestock and Social Analysis, *Solomon Islands Smallholder Agriculture Study*, Volume 2. AusAID, ISBN1 920861 47 5, Canberra, Australia

Hay, J.E.; Mimura, N.; Campbell, J.; Fifita, S.; Koshy, K.; McLean, R.F.; Nakalevu, T. Nunn, P.; & Neil de Wet. 2003. Climate Variability and Change and Sea-level Rise in the Pacific Islands: *Region A Resource Book for Policy and Decision Makers, Educators and other Stakeholders.* SPREP, Apia, Samoa

Intergovernmental Panel on Climate Change (IPCC). 2007. Climate Impacts, Adaptation and Vulnerability, Working Group II to the Fourth Assessment Report of the IPCC. Climate Change 2007. Cambridge University Press, ISBN 978 0521 70596-7, Cambridge, UK.

Lal, M. 2004. Climate change and small island developing countries of the south Pacific. Special issue on sustainable development. *J. of Contemporay Fiji*. 2, 15 -31

Lal, P.N.; Rita, R.; & Khatri, N. 2009. Economic Costs of the 2009 Floods in the Fiji Sugar Belt and Policy Implications. IUCN, Gland, Switzerland.Manton, M.J.; Dellaa-Marta, P.M.; Haylock, M.R; Hennessy, K.J.; Nicholls, N.; Chambers,

L.E.; Collins, D.A.; Daw, G.; Finet, A.; Gunawan, D.; Inape, K.; Isobe, H.; Kestin, T.S.; Lefale, P. ; Leyu, C.H.; Lwin, T.; Maitrepierre, L.; Oprasitwong, N.; Page, C.M.;

Pahalad, J.; Plummer, N.; Salinger, M.J.; Suppiah, R.; Tran, V.L.; Trewin, B.; Tibig, I.; & Yee, D. 2001: Trends in Extreme Daily Rainfall and Temperature in Southeast Asia and the South Pacific: 1961-1998. *J. Climatol.*, 21, 269-284.

Mataki, M.; Lal, M.; & Koshy, K. 2006. Baseline Climatology of Viti Levu (Fiji) and Current Climatic Trends. *Pacific Sci.* 60:49–68.

McGregor, A. 2006. *Pacific 2020 Background Paper: Agriculture.* Common Wealth of Australia, Retrieved from
    http://www.ausaid.gov.au/publications/pdf/solomon_study_vol2.pdf

McKenzie, E.; Kaloumaira, A.; & Chand, B. 2005. The Economic Impacts of Natural Disasters in the Pacific. *Technical Report,* University of the South Pacific (USP) and the South Pacific Applied Geoscience Commission (SOPAC), Suva, Fiji.

National Agriculture Research Institute (NARI). 2011. *Production Issues in Sweet Potato Needs Addressing.* 3/6/2011, Available from:
    http://nariweb.nari.org.pg/2010/production-issues-of-sweet-potato-needs-addressing/

Preston, B. L., R. Suppiah, I. Macadam, and J. Bathols. 2006. Climate Change in the Asia/Pacific region: A consultancy Report Prepared for the Climate Change and Development roundtable. Commonwealth Scientific and Industrial Research Organisation (CSIRO), Clayton,Victoria, Australia.

Simatupang P, Fleming E. 2001 Food Security Strategies for Selected South Pacific Island Countries. Working Paper 59, CGPRT Centre, JapanSecretariat of the Pacific Community (SPC). 2011. *Pacific Regional Information System (PRISM)* 15/6/2011, Available from: http://www.spc.int/prism/

Secretariat of the Pacific Community (SPC). 2010. *LRD Annual Report 2010.* Secretariat of the Pacific Community -Land Resources Division, ISBN: 978-982-00-467-2, Suva, Fiji

Secretariat of the Pacific Community (SPC).2008. Fish and food security. *Policy Brief,* Secretariat of the Pacific Community, :LRD, Suva Fiji.

Secretariat of the Pacific Regional Environment Programme (SPREP). 2009. *Climate Change, Variability and Sea- level Change.* 15/4/2011, Available from:
    http://www.sprep.org/topic/climate.htm

Tekinene, M and Paelate, A. 2000. Tuvalu's vulnerability and adaptation assessment. Government of Tuvalu, Funafuti, Tuvalu

Terry, J. P.; Kostaschuk, R. A.; & Wolting, G. 2008. Features of Tropical Cyclone-induced Flood Peaks on Grande Terre, New Caledonia. *Water and Environment Journal* 22: 177-183.

Webb, A. 2007. Assessment of Salinity of Groundwater in Swamp Taro (*Cyrtosperma chamissonis*) "pulaka" pits in Tuvalu. *EU EDF8-SOPAC Project Report 75 Reducing Vulnerability of Pacific ACP States,* Pacific Islands Applied Geoscience Commission, Suva Fiji.

Webster, P.J., Holland, G.J., Curry, J.A., & Change, H.R. 2005. Changes in Tropical Cyclone Number, Duration and Intensity in a Warming Environment'. *Science,* 309: 1844-1846.

World Bank. 2006. Not if but when - Adapting to natural hazards in the Pacific Islands Region. *A policy note.* The World Bank, East Asia and Pacific Region.

World Bank. 2000. Cities, Seas and Storms: Managing Pacific Island economies: Vol. IV, *Adapting to climate change.* East Asia and Pacific region, Papua New Guinea and Pacific Island Countries Management Unit, Washington DC, USA.

5

Permanent Internal Migration as Response to Food Shortage: Implication to Ecosystem Services in Southern Burkina Faso

Issa Ouedraogo[1], Korodjouma Ouattara[1],
Séraphine Kaboré/Sawadogo[1], Souleymane Paré[1] and Jennie Barron[2]
[1]Institut de l'Environnement et de Recherches Agricoles
(INERA), Ouagadougou,
[2]Stockholm Environmental Institute (SEI), Stockholm,
[1]Burkina Faso
[2]Sweden

1. Introduction

Mankind's exploitation of ecosystems for services such as food, shelter, fuel and fresh water has had profound effects on the natural environment for millennia (Achard et al., 2002; Bottomley, 1998, Ouedraogo, 2010). Further, since the 1800s, humans have had increasingly dramatic effects on the global environment following massive increases in the global population coupled with intense agrarian and industrial development. Indeed, man has become the most powerful, universal instrument of environmental change in the biosphere today (Meyer & Turner, 1994; Miller, 1994; Ojima et al., 1994). This has resulted in global climate change, forest and soil degradation, and loss of biodiversity, among other changes, to the extent that the sustainability of our planet's ecosystems is threatened (Lambin et al., 2003; Sala et al., 2000; Trimble & Crosson, 2000; Vitousek et al., 1997).

Among all zones in the world, the tropical environment is by far the most affected by the deforestation and forest degradation. In recent decades increasingly large areas of grasslands, woodlands and forests in the tropic have been converted into croplands and pastures (Mayaux et al., 2005; Lambin et al., 2003; Reid et al., 2000; Houghton, 1994). While in the tropical humid forest alone, the annual loss of forest was estimated at 5.8 million hectares (Achard et al., 2002), the tropical dry forest however, representing 42% of the forest in the tropics, has been severely fragmented, disturbed, and in many areas it has been severely depleted (Hartter et al., 2008).

In Burkina Faso (West Africa) for instance, the tropical dry forest, mainly located in the southern, eastern and western zones of the country, has experienced rapid deforestation process (Ouedraogo, 2010). Due to the repetitive severe droughts started in the 1980s in the Sahel, which caused important loss of crops and domestic animals, many farmers and breeders from the drought affected areas have been continuously migrating towards the southern, eastern and western zones of the country where food production and grazing facilities are still available. Such migration might have contributed to the rapid loss of forest ecosystem.

2. Statement of the research problem

Southern Burkina Faso has experienced rapid population increase since the 1980s resulting from a positive natural population growth and more importantly from a large immigration of farmers from drought-affected areas of the northern and central regions of the country (Ouedraogo *et al.*, 2009, 2010, 2011a, 2011b; Henry *et al.*, 2003). Prior to immigration, the southern Burkina Faso was less populated and were naturally endowed with a significant stock of dry forest (Howorth and O'Keefe, 1999). Furthermore, there was a peaceful co-existence between ethnic groups who were practicing sound agricultural activities with less impact to environment (Howorth and O'Keefe, 1999). However, with the growing migration, different farming techniques are taking place in southern Burkina Faso, since farmers move with their secular culture along with them. Implementation of new farming techniques together with the increasing demand for food to feed the growing population may have contributed to important ecosystem degradation in southern Burkina Faso.

Many works have been carried out in southern Burkina Faso at different spatial scales, highlighting the roles of population growth on deforestation (Ouedraogo *et al.*, 2009, 2010), the trajectories of forest cover change (Ouedraogo *et al.*, 2011a), forest cover transition processes (Ouedraogo *et al.*, 2011), land use dynamics (Paré et al, 2008; Ouedraogo, 2006), access to forest products (Coulibaly-Lingani *et al.*, 2009), agro-silvo-pastoral activities (Ouedraogo, 2003), etc. However, study that assesses the implication of the food production systems to forest ecosystem sustainability in southern Burkina Faso is lacking. Therefore, there is a need to carry out such study which is essential for sustainable regional and national land resource managements and for sound and environmentally harmless food production in Burkina Faso.

3. Objectives

The main objective of the study is to generate knowledge to support sound and informed decision making for sustainable food production systems in southern Burkina Faso. The study explores specifically:

1. The dynamics of cultivated land in southern Burkina Faso from 1986 to 2006
2. The dynamics of the population of southern Burkina Faso from 1986 to 2006
3. The relationships between cropland area and population densities
4. The dynamics of farming practices in southern Burkina Faso from 1976 to 2006

4. Study area

The study was carried out in southern Burkina Faso (Figure 1). The study area lies between latitudes 10° 58′N to 11° 52′N and longitudes 2° 40′W to 1° 12′E. It is characterized by a low relief with an average altitude of 300 m a.s.l. Phytogeographically, the area is situated in the Sudanian regional centre of endemism in the south Sudanian zone (Fontes & Guinko, 1995). The natural vegetation comprises mostly dry forest and tree savanna community types. The climate is tropical with a unimodal rainy season, lasting for six months (May to October). Based on data collected from the nearest *in situ* mini-weather station at Léo, the provincial city of Sissili, the mean (±Standard Error) annual rainfall from 1976 to 2007 was 883±147 mm. Mean daily minimum and maximum temperatures ranged from 16 to 32 °C in January

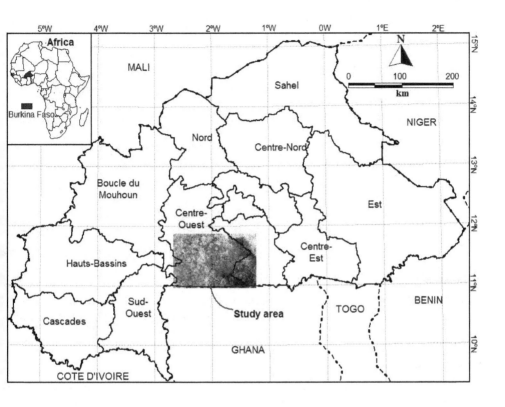

Fig. 1. Study area

(the coldest month) and from 26 to 40 °C in April (the hottest month). According to the FAO soil classification system (Driessen *et al.*, 2001), the most frequently encountered soil type is Lixisol (tropical ferruginous soils), which is poorly to fully leached, overlying sandy, clayey-sandy and sandy-clayey material.

The population is comprised of four main ethnic groups: Nuni, Wala, Mossi and Fulani. The Nuni and Wala groups have been living in the area for centuries and are considered indigenous, while the Mossi, who originate from the Central Plateau in Burkina Faso, and the Fulani, herders from the northern region of the country, are considered migrants. The latter two groups were attracted to southern region during the 1980s in search of arable land and green pasture, respectively (Howorth & O'Keefe, 1999). The dominant agricultural production methods in the study area are traditional subsistence farming systems with cereals (such as sorghum, millet and maize), tubers (yam and sweet potatoes) and animal husbandry. However, over the last ten years, there has been intense competition for land between the traditional farming systems and more lucrative production systems.

5. Materials and methods

The methodology used for the study combines land cover change detection, analysis of population dynamics and, inventory and analysis of farming techniques.

5.1 Land cover detection

Land cover change detection was based on time-series satellite images processing (Landsat: 1986, 1992, 2002; ASTER: 2006). All images were geometrically and radiometrically corrected and then classified using the maximum likelihood classifier based on training samples, available topographic maps and in-situ observations. Three appropriate classification schemes of the study area were used to assign pixels to land use classes: cropland, open woodland and dense forest.

5.2 Population data

We extracted the population data of the study area from the national population census reports (1975, 1985, 1996 and 2006). To estimate the inter-census period population data (1992 and 2002), we used the population projection methods (Weeks, 1999).

5.3 Inventory of farming practices

A household survey was performed in selected villages using semi-structured questionnaire to record information related to farming techniques, production acreage, crop yields and reasons for migrating if respondent was migrant.

5.4 Data analysis

To relate land cover change with population dynamics, Pearson correlation analysis was performed for each land cover class. Correlation test was also performed between area of cultivated land and population density. Data from the household survey was analyzed using descriptive statistics. All statistical analyses were performed with SPSS 18.

6. Results

6.1 Land use dynamics

Results from the image processing revealed a significant change within land use classes. In general, there was an increase in area of cropland at the expense of shrinking open woodland and dense forest covers in Southern Burkina Faso (Table 1, Figure 2). Comparing the area of cropland for the four series of assessment, it appears that there was an increase in cropland over time. In 1986, at the onset of the study period, cropland occupied 7.5% of the study area. It was more than doubled in 1992 and increased by three-fold in 2002. In 2006, it increased to nearly 27%. Over the study period, the annual rate of increase in area of cropland was 0.96% (Table 1).

Dense forest land covered ca. 70% of the study site in 1986 and decreased to 40% at the end of the study period in 2006. The rate of decrease in dense forest cover during the study period was estimated at 1.45% per annum (Table 1). The area of open woodland decreased slightly between 1986 and 1992, but the rate of increase in this land use type was nearly

doubled in 2002 (Table 1). Over the study period the overall annual increase rate was 0.5%
(Table 1).

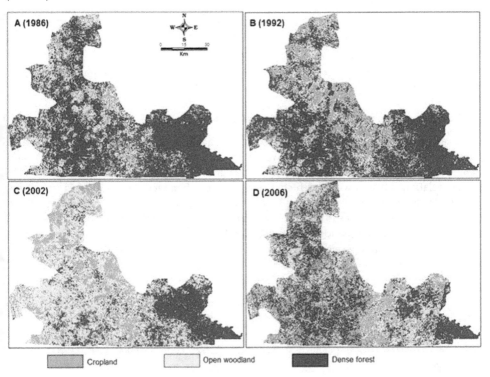

Fig. 2. Pictorial representation of land cover dynamics in Southern Burkina Faso during the
study period.

Years	Cropland (%)	Dense Forest (%)	Open woodland (%)
1986	7.50	69.68	22.82
1992	16.48	63.09	20.44
2002	24.73	31.39	43.90
2006	26.62	40.61	32.08
Annual change rates	0.96	-1.45	0.46

Table 1. Land cover change from 1986 to 2006 in Southern Burkina Faso

6.2 Population dynamics

The population of Southern Burkina Faso has increased during the study period. Estimated
at 119352 inhabitants in 1986, the population has nearly doubled in 20 years (Table 2) with a
means annual increase of 4664 people. The density of the population has also doubled with
an annual increase rate of 0.6 inhabitants/km² (Table 2). The proportion of migrant people
in the selected villages were 3%, 27%, 34%, 42% and 56% in 1976, 1986, 1992, 2000 and 2006,
respectively.

Years	population	Density
1986	119352	16.78
1992	134821	18.96
2002	178540	25.11
2006	212628	29.90
Annual increase	4664	0.55

Table 2. Dynamics of the population size and population density during the study period

6.3 Relationship between land use conversion and population density

The Pearson correlation analysis indicated that change in land cover could be linked to population growth. In general, there was a strong relationship between population and change in areas of cropland ($r^2 = 0.90$; $p < 0.001$) and dense forest land ($r^2 = 0.56$; $p = 0.03$); but the relationship was weak with change in area of open woodland ($r^2 = 0.11$; $p = 0.42$). Significant relationships between population and areas of cropland and open woodland were observed throughout the four time series (Table 3). The higher correlations occurred in 2002 and 2006 with cropland, and in 1986 and 1992 with open woodland ($p < 0.001$). However, the correlations were generally low between population and dense forest land and, especially it was not significant in 2002 ($p = 0.145$).

	Cropland				Dense forest				Open woodland			
	1986	1992	2002	2006	1986	1992	2002	2006	1986	1992	2002	2006
r^2 (%)	85.2	76.8	91.7	96.1	63.2	56.6	20.5	70.3	91.1	92.1	87.5	66.5
F	41.3	24.2	78.3	174.6	13.0	10.1	2.8	17.6	72.4	82.6	50.1	14.8
P-value	0.001	0.003	0.000	0.000	0.011	0.019	0.145	0.006	0.000	0.000	0.000	0.000

Table 3. Coefficient of determination (r^2), together with F-statistic and p-values for significant Pearson correlation to examine relationship between areas of each cover type and population density.

6.4 Farming practices over time and environmental implication

Results from the farmers interviews revealed the presence of important migrant people in Southern Burkina Faso. They were from different origins and mostly from 14 different provinces of the central and the northern regions of the country (Figure 2). Home provinces from the central regions included Kadiogo, Boulkiemde, Sanguie, Ganzourgou, Kouritenga, Oubritenga, Kourweogo, and the northern provinces included Soum, Bam, Yatenga, Passore, Zondoma, Sanmatenga and Namentenga. From the central regions, migrants were predominantly coming from Boulkiemde and Oubritenga while from the northern regions they were mostly coming from Yatenga, Sanmatenga and Bam. Among reasons provided for migrating to Southern Burkina Faso, the respondents mentioned the declining soil fertility in the pushing village (92% of the respondents), scarcity of arable land (76%), erratic rainfall (73%), need to make income (55%), and the politico-economic unrest in Côte d'Ivoire (12%) started in 2000, which caused the return of Burkinabé from the coffee, cocoa and banana plantation areas.

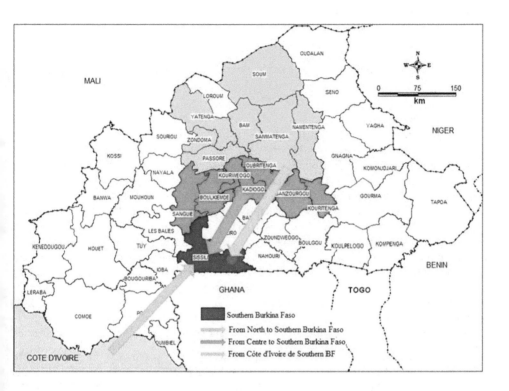

Fig. 3. Migration flow to Southern Burkina Faso

Four main sources of income generation were available in the study area, namely crop
production (63%), livestock husbandry (27%), non timber forest products (6%), and
wood/charcoal production (4%). The mean farm size of the migrants changed from 3.0 ha to
3.7 ha during the period from 1986 to 2007 for a mean household size of 6 ± 2 persons.
During the same period, the farm size of the native population, with the same household
size, changed from 2.0 ha to 3.1 ha. The main agricultural tools (Figure 3) in use were "Daba"
(local traditional rudimentary tool based on human force) and plough (based on animal
force). In the 1980s, about 95% of the respondents were using "Daba", but they shifted
progressively to the use of plough. The change was more pronounced among migrants; in
2007, more than 83% of the migrants were using the plough while this figure was only 59%
among natives.

Food production has importantly increased from 1986 to 2007 as reported by respondents
and has met the food security in the region. Half of crop produced by household is send to
markets for income generation. At the same time, crop yield per hectare has reduced. To
produce more food, one needs to clear more lands.

7. Discussion

Results showed a net increase in area of cropland at an annualized rate of 0.96% which is much higher than estimation at the national level (0.2%) by FAO (2001). This indicates that Southern Burkina Faso is facing serious deforestation. The present cropland increase rate is rather in line with some previous study undertaken in the same region. For instance, Ouedraogo (2006b) and Paré *et al.* (2008), estimated the annual increase of cropland at 1.03% and 0.7% in Bieha district (in Sissili province) and in Sissili and Ziro provinces, respectively. Furthermore, a study made in the Volta Region of Ghana, geographically close and ecologically similar to our study area, found a comparable result regarding cropland change i.e., 1.1% per annum (Braimoh, 2004). The substantial increase in the area of cropland during the study period could be explained by the growing interest in maize cultivation and cotton production due to their high economic values in the country. Maize is one of the cereals commonly used in the country as a main source of food and is principally grown in the southern, southwestern and eastern parts of the country where fertile soil and rainfall are abundant (Ouattara *et al.*, 2008). In the 2000s, the government introduced an agricultural policy to increase cotton production. This was made by providing agricultural incentives such as ploughs and fertilizers to some farmers. In the context of technology and market improvements in the agricultural system, cultivated land is likely to increase (Bilsborrow and Carr, 2001; Lambin *et al.*, 2003; Gray, 2005). More importantly there has been a growing expansion of agribusiness in Southern Burkina Faso since 2000. The actors in the agribusiness involve individual investors (Ouedraogo, 2003; Ouedraogo, 2006b; Paré *et al.*, 2008) who use machinery (mostly tractors) and casual labor to farm large areas (40-100 hectares each). They grow maize during the first few years and thereafter plantations of cashew trees.

The average annual degradation of the dense forest (1.45%) is closely comparable to the estimations made by Ouedraogo (2006b) for Bieha district from 1986 to 2002 (1.13 per annum) and also by Braimoh (2004) for the Volta Region of Ghana from 1984 to 1999 (2.24% per annum). This fairly high deforestation of dense forest could be linked to the observed exploitation of firewood and charcoal in southern Burkina Faso, most likely to meet the energy demand in biggest cities such as Ouagadougou and Koudougou (Krämer, 2002; Ouedraogo, 2006a, Ouedraogo *et al.*, 2009, 20010, 2011a, 2011b).

The large population growth observed in Sissili province was amplified by farmers' migration. Migration of farmers towards the eastern (Reenberg and Lund, 1998), south-western (Gray, 1999, 2005) and southern (Howorth and O'Keefe, 1999; Henry *et al.*, 2003; Ouedraogo, 2003; Ouedraogo, 2006b; Paré *et al.*, 2008; Ouedraogo *et al.*, 2009, 2010) regions of Burkina Faso originated from the 1980s when severe drought hit Sahelian countries. During that period, farmers and herders in the arid zones lost a substantial quantity of crops and domestic animals. For survival reasons, most of the affected people moved to more humid areas in the south. This massive mobility of farmers could have negatively affected forest sustainability. According to Geist and Lambin (2001) and Lambin *et al.* (2003), migration in its various forms is the most important demographic factor causing land use change both spatially and temporally. Migration operates as a significant driver with other non-demographic factors, such as government policies, change in consumption patterns, economic integration and globalization (Fearnside, 1997). Some tenure policies initiated in the 1980s have provoked the migration or were intrinsically linked with increased migration. The land reform, adopted in

1984 in Burkina Faso and revised several times, aimed at promoting wide scale migration of farmers from the drought-affected regions to the sparsely populated regions of the south and southwest (Faure, 1996; Reenberg and Lund, 1998; Gray, 1999).

The high variability in time and space of the population could indicate that multiple factors contributed to the population growth. In the 2000s, for instance, the political crisis in Côte d'Ivoire amplified the migration flow in the province with the return Burkinabé who were working in the plantations (Ouedraogo *et al.*, 2009). Most of them were settled by the government in Sissili province between 2000 and 2005.

The strong correlation between population and change in cropland in the whole province indicates the prevalence of shifting cultivation and the weak technological improvement for a large number of farmers, as new areas were cleared to increase crop production rather than improving current farming techniques. This feature is very common in the tropical regions (Reenberg and Lund, 1998; Lambin *et al.*, 2003; Ningal *et al.*, 2008).

Migrant people in Southern Burkina Faso came mainly from the central Plateau and northern region of the country. Explanation to this could be that these two regions have specific demographic and ecological characteristics which push people to migrate as pointed by migrant respondents. The central regions accounted for more than 46% of the total population of the country (INSD, 2007) from which more than 90% were farmers (Breusers, 1998). This region is nowadays crowded and the capacity of the lands to sustain agriculture and grazing under extensive subsistence practices is almost exceeded (Gray, 1999, 2005; Reij *et al.*, 2005). In such conditions, the easiest way is to migrate towards new frontiers where land is still available (Boserup, 1972; Bilsborrow and Carr, 2001). In the northern region, the rains are insufficient and unreliable resulting in an increasing aridity. The mean annual rainfall ranges from 400 to 600 mm within a rainy season which does not exceed four months (June to September). To face these conditions, farmers developed secular techniques (Figure 4) known as *Zaï* and *Demi-lunes* (plant-pit systems) (Slingerland and Stork, 2000;

Daba   **Locally used plough**

Fig. 4. Agricultural tools used in Southern Burkina Faso

Sorgho *et al.*, 2005) and their success is a function of the spatial and temporal distribution of the rains. These techniques were seen as more and more laborious and hazardous by some local farmers. Therefore, they see out-migration of one or more family members as a mean of earning cash income and diversifying risk (Raebild *et al.*, 2007; Youl *et al.*, 2008). Unfortunately, during their first years of settlement in the attracted area, the first activity they practice to make rapid income for survival is to cut wood (Figure 5) for charcoal production (Ouedraogo, 2006a). This drastically impacted the sustainability of forest in the south of Burkina Faso.

A. Typical landscape in northern Burkina Faso   B. *Demi-lunes* techniques

C. Stone-lines techniques   D. Zaï techniques

Fig. 5. Water harvesting techniques for soil water holding capacity improvement in Northern Burkina Faso

The results indicated that migrants had larger farmlands and used environmentally harmful techniques (shifting cultivation, slash and burning techniques) in their land use systems while native population tended to take more care of land and environment by intensifying the production within the same croplands instead of cutting forest to make space for new croplands. The justification for this could be that the native people have a strong and secular relationship with their ever-changing environment developed over several years (Howorth and O'Keefe, 1999, Ouedraogo *et al.*, 2009, 2010). Therefore, despite the recent introduction of cash crop productions (cotton mainly), this community has been inventive and adaptive in their resource use patterns and survival strategies. Inversely, the migrant people who came to work in a new environment have two main objectives to meet as expressed by Ouedraogo *et al.* (2009): in the one hand, they had to secure their income and domestic food, in the other hand; they had to produce more to meet also the food shortages and chronic food insecurity that their parents face in the home village. To do so, with the few labour available, migrants had to use animal traction (ploughs), thus, cutting large forest areas to make space for agriculture as compare to the natives.

Fig. 6. The way migrant people clear forest to make quick cash and space for agriculture

Results revealed that crop production has increased during the study period. This is fundamental for the food security in Burkina Faso. Cereals (mainly used for food) produced in Southern Burkina Faso are dispatched in the central and northern regions of the country to secure food access to all population in Burkina through markets. This is the reason why the government has named Southern Burkina Faso, the *grenier du Burkina Faso* meaning the "food storehouse of Burkina Faso".

8. Conclusion

Results from the present study disclosed a rapid cropland increase at the detriment of a shrinking forest covers in Southern Burkina Faso. Total population also exhibited a rapid increase in size as a result of important migration of farmers. Change in land cover types correlated with population growth which implies that more people is synonymous to more land clearance for agriculture and more deforestation and forest degradation to meet primary needs in Southern Burkina Faso. Food production has importantly increased as a result of large space exploitation for agriculture. While increased food production is a good sign for food security in the entire Burkina Faso, the induced deforestation and forest degradation are per se an indicator of unsustainable forest ecosystem management and unsecured mobility policy which may threaten the environmental balance and bring in the future conflicts due to completion for space between native and migrant population. Therefore, there is urgent need for agricultural intensification-related policy initiatives to discourage expansion of cultivated lands and its associated fragmentation of forested areas.

9. Acknowledgements

Funding for this study was provided by the Challenge Programme for Water and Food (CPWF), V1 (Targeting and Scaling out).

10. References

Achard, F., Eva, H.D., Stibig, H.J., Mayaux, P., Gallego, J., Richards, T. & Malingreau, J.P. (2002). Determination of deforestation rates of the world's humid tropical forests. *Science* 297(5583), 999-1002.

Bilsborrow RE, Carr DL. 2001. *Population, agricultural land use and the environment in Developing Countries*. In Lee D., Barrett C., eds. Tradeoffs or synergies? Agricultural intensification, economic development and the environment. CAB International Wallingford, UK, 35-55p.

Boserup E. 1972. Conditions of Agricultural Growth - Reply to Sheffer. *American Antiquity* 37: 447-447.

Bottomley, B.R. (1998). Mapping rural land use & land cover change In Carroll County, Arkansas utilizing multi-temporal Landsat Thematic Mapper satellite imagery. Diss. Arkansas:University of Arkansas.

Braimoh AK. 2004. Seasonal migration and land-use change in Ghana. *Land Degradation & Development* 15: 37-47.

Breusers M. 1998. *On the Move: Mobility, land use and livelihood practices on the Central Plateau in Burkina Faso*, Thesis. Wageningen Agricultural University, Wageningen; Netherlands, 349p.

Coulibaly-Lingani, P., Tigabu, M., Savadogo, P., Oden, P.C. & Ouadba, J.M. (2009). Determinants of access to forest products in southern Burkina Faso. Forest Policy and Economics 11(7), 516-524.

Driessen, P., Deckers, J. & Spaargaren, O. (2001). Lectures notes on the major soils of the world. FAO World Soil Resources. Report- 94. Food and Agriculture Organization of the United Nations. Rome.

FAO. 2001. *Global Forest Resource Assessment. Main report*. FAO Forest Paper 140. Food and Agriculture Organization of the United Nations, Rome, Italy.

Faure A. 1996. *Private Land Ownership in Rural Burkina Faso*. IIED, Dryland Network Working Paper No. 59.

Fearnside PM. 1997. Carbon emissions and sequestration by forests: Case studies of developing countries - Guest editorial. *Climatic Change* 35: 263-263.

Fontes, J. & Guinko, S. (1995). Carte de végétation et de l'occupation du sol du Burkina Faso: Projet Campus. UPS, ICIV Toulouse, France.

Geist HJ, Lambin EF. 2001. *What Drives Tropical Deforestation? A Meta-analysis of Proximate and Underlying Causes of Deforestation Based on Subnational Case Study Evidence*. LUCC International Project Office, University of Louvain; Louvain-la-Neuve; Belgium

Gray LC. 1999. Is land being degraded? A multi-scale investigation of landscape change in southwestern Burkina Faso. *Land Degradation & Development* 10: 329-343.

Gray LC. 2005. What kind of intensification? Agricultural practice, soil fertility and socioeconomic differentiation in rural Burkina Faso. *Geographical Journal* 171: 70-82.

Hartter, J., Lucas, C., Gaughan, A.E. & Aranda, L.L. (2008). Detecting tropical dry forest succession in a shifting cultivation mosaic of the Yucatan Peninsula, Mexico. Applied Geography 28(2), 134-149.

Henry, S., Boyle, P. & Lambin, E.F. (2003). Modelling inter-provincial migration in Burkina Faso, West Africa: the role of sociodemographic and environmental factors. Applied Geography 23(2-3), 115-136.

Houghton, R.A. (1994). The Worldwide Extent of Land-Use Change. Bioscience 44(5), 305-313.

Howorth, C. & O'Keefe, P. (1999). Farmers do it better: Local management of change in southern Burkina Faso. Land Degradation & Development 10(2), 93-109.

INSD 2007. *Résultats préliminaires du recensement général de la population et de l'habitat de 2006*. Institut National des Statistiques et de la Demographie (INSD). Direction de la Demographie, Ouagadougou. Burkina Faso.

Krämer P. 2002. *The fuel wood crisis in Burkina Faso, solar cooker as an alternative*, Solar cooker archive. Ouagadougou, Burkina Faso.

Lambin, E.F., Geist, H.J. & Lepers, E. (2003). Dynamics of land use and land cover change in tropical regions. Annual Review of Environment and resources 28, 205-241.

Mayaux, P., Holmgren, P., Achard, F., Eva, H., Stibig, H. & Branthomme, A. (2005). Tropical forest cover change in the 1990s and options for future monitoring. Philosophical Transactions of the Royal Society BBiological Sciences 360(1454), 373-384.

Meyer, W. & Turner, B.L. (1994). Changes in land use and land cover: a global perspective: University Press, Cambridge, UK.

Miller, G.T. (1994). Sustaining the earth: an integrated approach. Belmont, California: Wadsworth Publisher Company.

Ningal T, Hartemink AE, Bregt AK. 2008. Land use change and population growth in the Morobe province of Papua New Guinea between 1975 and 2000. *Journal of Environmental Management* 87: 117-124.

Ojima, D.S., Galvin, K.A., Turner, B.L., II, Houghton, R.A., Skole, D.L., Chomentowski, W.H., Salas, W.A., Nobre, A.D., Kummer, D.M., Moran, E.F., Mausel, P., Wu, Y., Ellis, J., Riebsame, W.E., Parton, W.J., Burke, I.C., Bohren, L., Young, R., Knop, E. & Brondizio, E. (1994). Global impact of land-cover change. Bioscience 44(5), 300-356.

Ouattara K, Ouattara B, Nyberg G, Sedogo MP, Malmer A. 2008. Effects of ploughing frequency and compost on soil aggregate stability in a cotton-maize (*Gossypium hirsutum-Zea mays*) rotation in Burkina Faso. *Soil Use and Management* 24: 19-28.

Ouedraogo B. 2006a. Household energy preferences for cooking in urban Ouagadougou, Burkina Faso. *Energy Policy* 34: 3787-3795.

Ouedraogo I. 2006b. Land use dynamics in Bieha district, Sissili province; southern Burkina Faso, West Africa. *Umoja: Bulletin of the African and African American Studies* 1: 18-34.

Ouedraogo, I., Savadogo, P., Tigabu, M., Cole, R., Odén, P.C. & Ouadba, J.M. (2009). Is rural migration a threat to environmental sustainability in Southern Burkina Faso? Land Degradation & Development 20(2), 217-230.

Ouedraogo, I., Tigabu, M., Savadogo, P., Compaoré, H., Oden, P.C. & Ouadba, J.M. (2010). Land cover change and its relation with population dynamics in Burkina Faso, West Africa. Land Degradation & Development DOI: 10.1002/Idr.981, p n/a.

Ouedraogo I., Savadogo P., Tigabu M., Cole R., Odén PC., Ouadba JM., 2011a. Trajectory Analysis of Forest Cover Change in the Tropical Dry Forest of Burkina Faso, West Africa. Landscape Research, Vol. 36, No. 3, 303–320.

Ouedraogo I., Savadogo P., Tigabu M., Dayamba SD., Odén PC., 2011b. Systematic and Random Transitions of land cover types in Burkina Faso, West Africa. International Journal of Remote Sensing, DOI: 10.1080/01431161.2010.495095.

Ouedraogo, M. (2003). New stakeholders and the promotion of agro-silvo-pastoral activities in Southern Burkina Faso: false start or inexperience.: IIED, Dryland Programme, 118p; Issue Paper).

Paré, S., Söderberg, U., Sandewall, M. & Ouadba, J.M. (2008). Land use analysis from spatial and field data capture in southern Burkina Faso, West Africa. Agriculture, Ecosystems & Environment 127(3-4), 277- 285.

Raebild A, Hansen HH, Dartell J, Ky JMK, Sanou L. 2007. Ethnicity, land use and woody vegetation: a case study from south-western Burkina Faso. *Agroforestry Systems* 70: 157-167.

Reenberg A, Lund C. 1998. Land use and land right dynamics - Determinants for resource management options in Eastern Burkina Faso. *Human Ecology* 26: 599-620.

Reid, R.S., Kruska, R.L., Muthui, N., Taye, A., Wotton, S., Wilson, C.J. & Mulatu, W. (2000). Land-use and land-cover dynamics in response to changes in climatic, biological and socio-political forces: the case of southwestern Ethiopia. Landscape Ecology 15(4), 339-355.

Reij C, Tappan G, Belemvire A. 2005. Changing land management practices and vegetation on the Central Plateau of Burkina Faso (1968-2002). *Journal of Arid Environments* 63: 642-659.

Sala, O.E., Chapin, F.S., Armesto, J.J., Berlow, E., Bloomfield, J., Dirzo, R., Huber-Sanwald, E., Huenneke, L.F., Jackson, R.B., Kinzig, A., Leemans, R., Lodge, D.M., Mooney, H.A., Oesterheld, M., Poff, N.L., Sykes, M.T., Walker, B.H., Walker, M. & Wall, D.H. (2000). Biodiversity - Global biodiversity scenarios for the year 2100. Science 287(5459), 1770-1774.

Slingerland MA, Stork VE. 2000. Determinants of the practice of Zai and mulching in North Burkina Faso. *Journal of Sustainable Agriculture* 16: 53-76.

Sorgho MM, Sylvain K, Karim T. 2005. Burkina Faso: the Zai technique and enhanced agricultural productivity. *IK Notes, No.80.*

Trimble, S.W. & Crosson, P. (2000). Land use - US soil erosion rates - Myth and reality. Science 289(5477), 248-250.

Vitousek, P.M., Mooney, H.A., Lubchenco, J. & Melillo, J.M. (1997). Human domination of Earth's ecosystems. Science 277(5325), 494 - 499.

Weeks, J.R. (1999). Population: an introduction to concepts and issues. New York: Wadsworth publishing Company.

Youl S, Barbier B, Moulin CH, Manlay RJ, Botoni E, Masse D, Hien V, Feller C. 2008. Modélisation empirique des principaux déterminent socio-économiques de la gestion des exploitations agricoles au sud-ouest du Burkina Faso. *Biotechnologies Agronomies Societé et Environnement* 12: 9-21.

6

Strengthening Endogenous Regional Development in Western Mexico

Peter R.W. Gerritsen
Departamento de Ecología y Recursos Naturales - Imecbio,
Centro Universitario de la Costa Sur,
Universidad de Guadalajara,
Mexico

1. Introduction

Much has been written on the negative effects of globalization in the Mexican countryside, showing the multi-dimensionality of the problems generated, noting not only the economic effects, but also the socio-cultural and ecological repercussions, which have to do with various issues related to rural life and production, such as the quality of rural producers' life, identity and traditional practices, or sustainable natural resource management (Cortez et al., 1994; Schwentesius *et al.*, 2003; Esteva & Marelle, 2003). Additionally, reference is also made to the trans-national nature of the problem, emphasizing the involvement of processes that go beyond the regional and national territory (Halweil, 2000; Schwentesius et al., 2003). Table 1 presents an overview of (some of) these negative effects, as identified from bibliography.

• Fomenting export agriculture and contract farming
• Lack of attention to peasant and indigenous farming
• Un-fair trade
• Poverty and natural resource degradation
• Disarticulation of peasant economies from larger national and international economies
• Migration to urban centers and to the United Status
• Displacement of traditional by hybrid and genetically modified varieties
• Loss of traditional and popular identity and culture
• Trans-nationalization of the agro-industry

Table 1. An overview of some of the negative effects of economic globalization as identified from bibliography (Cortez et al., 1994; Schwentesius et al., 2003; Esteva & Marelle, 2003; Jansen and Vellema, 2004; Halweil, 2000)

These negative effects contest the process of economic globalization as related to rural actor's livelihood strategies, and as related to the natural resources on which they depend (Toledo, 2000; Carabias & Provencio, 1993). It also questions the capacities of governmental institutions to offer pertinent solutions. Moreover, it gives rise to the idea that overcoming poverty, increased production, appropriate technology development, and farmer participation requires profound social and institutional adjustments. More specifically, this requires a development model that is able to respond to the specific necessities of the rural sector, as well as one that is able to develop strategies that strengthen governmental intervention in favor of sustainability (Muñoz & Guevara 1997; Gerritsen et al., 2003).

Although it goes beyond discussion to deny the negative effects of economic globalization, an increasing number of social actors can be identified at the local levels, which look for alternative strategies to restrain from the negative effects (Toledo 2000; Gerritsen et al., 2004). These actors have developed a number of responses and strategies, which can be conceptualized as efforts to construct viable and sustainable alternatives for a different social order in the local territory (Cortez et al.,1994; Morales, 2004; Waters, 1995).

This chapter's main interest and focus lies on describing and understanding those local experiences that hold the possibility to develop and strengthen development models that may serve as a platform for the designing new agricultural policies that better respond to the social, economic and environmental challenges present in different territories (cf. van der Ploeg & Long, 1994). It presents an overview of an action-research program that has been implemented by personnel of the Rural Development Group of the Department for Ecology and Natural Resources (DERN-IMECBIO, according to its Spanish acronym) at the South Coast University Campus (CUCSUR, according to its Spanish acronym) of the University of Guadalajara. The work of our group has consisted of supporting local initiatives to construct viable alternatives in the South Coast of Jalisco, located in western Mexico.[1]

In the following pages I will presen some theoretical notions, followed by four different thematic working fields of local development that have been supported and studied by us. The chapter finishes with a discussion and a conclusion, aiming at locating the different thematic experiences within current theoretical debates on the local effects of economic globalization.

2. Endogenous rural development and the farm enterprise

In general terms, farm production makes up the mobilization of resources, the enrollment of farmers in the agro-food chain in order to sell their produce, and contributes to the creation and maintenance of rural identity and rural social organization (van der Ploeg et al., 2002) (Figure 1).

Originally, farming production was based on diversified, rather autonomous strategies (Toledo, 1990). In other words, endogenous forces were at the heart of rural development (van der Ploeg, 1994). Furthermore, one can also speak of endogenous rural development. Endogenous rural development is understood here as a development model that departs from local natural resource use and management, local actors´ capacities and abilities to develop strategies for the appropriation of material and non-material resources, and where there exists local actors´ control on (the value of) the farming production (van der Ploeg & Long, 1994).

---

[1] This chapter compiles part of fifteen year's research and extension work experience of the Community Development Group. The author of this paper took up the task to compile the experience gained over the years in this second English written version (see Gerritsen, 2006 for the first version). This paper was published before in Spanish in different versions and forms, see Villalvazo et al. (2003); Gerritsen et al. (2004, 2005a. 2005b); and Figueroa et al. (2003, 2004). The theoretical foundations of the work are described in Gerritsen (2010).

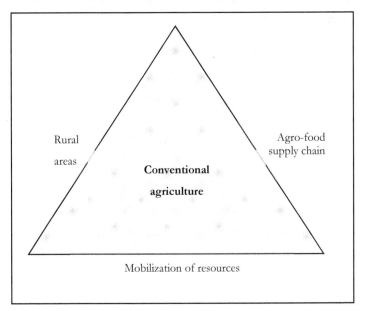

Fig. 1. The three sides of the agricultural enterprise (van der Ploeg et al., 2002: 12)

Generally speaking, there are four dimensions of rural development (either endogenous or not) and they have to deal with the available natural resources in the locality (that is, the domain of farm production and the domain of farm reproduction), the interaction of farmers with other (external) actors and institutions, and the incorporation in different markets (that is, the domain of economic and institutional relations), the maintenance of traditions and certain identity, and finally, an idiosyncratic and specific vision of the world in general and agricultural work in particular (that is, the domain of family and community relations) (van der Ploeg, 1990) (Figure 2).

In the local territory, endogenous potential can be understood as a specific configuration of the different farming domains (and the great many different farming activities that each one of these domains includes), created and coordinated purposefully by local actors. The outcome of this domain coordination is the conformation of specific farming styles, which are to be understood as purposeful socio-productive strategies. A central characteristic of farming styles is the close bond between manual and mental labor, i.e. the close relationship between farming discourse and practice (van der Ploeg, 1994).[2]

Aside from its close bond with farmer strategies, endogenous potential for rural development can also be located in different links of the productive chain[3]. The endogenous potential of productive chains refers to those products that arise from the local territory, and

---

[2] Due to this close relationship, endogenous rural development and farming styles are studied by us by applying an actor-oriented perspective (Long & Long, 1992; Long, 2001).
[3] A productive chain is understood here as: *"a system constituted by interrelated actors […] and by a succession of operations of production, transformation and commercialization of a product, or a group of products, [developed] in a given surrounding"* (Heyden et al., 2004: 11).

where production, transformation and commercialization activities are carried on with an actor's own resources and by departing from trans-generational knowledge, and, moreover, where the consumers recognize these products as typical or authentic for the region.[4]

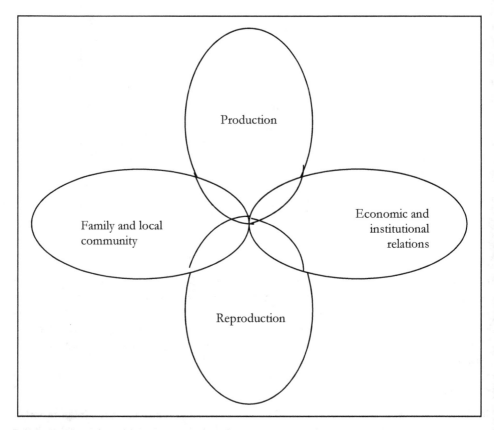

Fig. 2. The domains of farming (van der Ploeg, 1990: 29)

Development models implemented as part of economic globalization have impacted in different grades and in different moments on rural and indigenous production[5] and have aimed at substituting these with agro-industrial methods[6], causing the ecological, social,

---

[4] Opposed to productive chains that are disconnected from the local territory, where inputs are external and knowledge is based mainly on science and technology. It is these kinds of productive chains where the great majority of the high socio-environmental costs of production, transformation and transportation can be located, with the goal to reach consumers all over the world (Halweil, 2000).

[5] The peasant (or indigenous) production mode is understood here as those production systems that are based on diversification, local resource use, family labor, and subsistence-oriented production (Toledo, 1995).

[6] The agro-industrial (or modern) production mode is understood here as those production systems that are based on specialization, external resource use, the use of hired farm laborers, and where a commercial production orientation prevails (Toledo, 1995).

economic and cultural consequences mentioned before in Table 1 (Calva, 1993; Saxe-Fernández, 1998; Morales, 2002). From the perspective of sustainability, the agro-industrial development model for consolidating the *"modernization of the Mexican countryside"* can be considered non-viable and extremely harmful in all its dimensions (Toledo, 2000).

Economic globalization processes affect the endogenous potential of a specific territory by changing the locally-specific ecological, social and socio-institutional, cultural, and ethical conditions. In others words, it induces a reconfiguration of the social relations of production, as well as the social and material bases of production. As a consequence, new – intraregional - farming styles can arise. At the same time, it can change the conditions of the different links of the productive chain, affecting the patrimonial values of the region. Figure 3 illustrates this by presenting normatively defined development patterns for farming styles in relation to diversity in natural resources.

In order to strengthen regional sustainable development initiatives that depart from the endogenous potential in local territories, it is necessary to design and implement productive alternatives that built upon existing social, economic, cultural and environmental processes, which implies strengthening traditional knowledge, improving traditional technology for family agriculture, generating viable peasant economies, reactivating regional economic dynamics, strengthening a culture of self-sufficiency and one that builds upon and strengthens a harmonious interrelation and respect for living nature (Morales, 2001; Toledo, 2003). In our work, efforts have been directed at four thematic fields of action that relate to these different fields of attention: agroecology and fair trade (aimed at strengthening farm reproduction and production), gender and natural resource management (aimed at incorporating the gender issue in our work), appropriate technology development and, finally, regional production and territorial value recuperation. Within these different thematic fields, various theoretical-methodological tools have been applied.[7] In the following sections, I will describe our different experiences in these fields.

3. Agroecology: Strengthening the reproductive and productive base of farm enterprises[8]

In the reproduction (and production) link of the productive chain, agro-ecology has been the starting point of our work, and, in lesser degree, the related field of fair trade. Following Guzman *et al.* (2000), the restoration of essential ecological processes of agro-ecosystems is the first step in the revalorization of traditional farming systems, where endogenous elements predominate over the exogenous ones.

This thematic field originates in the beginning of 2001, as a response to the social demands of an indigenous community in the region where we live and work. Until now, activities to

---

[7] This is explained by Rist (2004), who mentions that in the literature on endogenous development, attention goes mostly to *"what to obtain"*, rather than describing *"how to do it"*. Therefore, a great variety of methodologies is potentially useful for identifying, strengthening and evaluating endogenous development initiatives, whose choice depends on the specific characteristics of the local territory (Broekhuizen & van der Ploeg, 1995).

[8] Based on Figueroa et al. (2004).

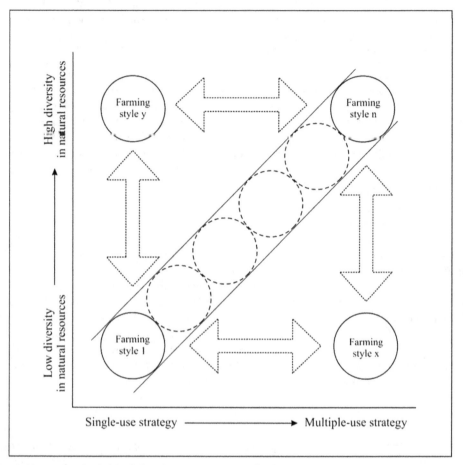

Fig. 3. Normatively defined development patterns for farming styles in relation to resource diversity (Gerritsen, 2002: 27)

improve the soil and the quality of local foods have been carried out, and have spread to other communities. Farmer training workshops are an important part of the work and are realized by (professional) advisers and by farmer extensionists.

The recuperation of traditional ecological knowledge and practice is oriented towards the revalorization of local knowledge, in addition to the advisory work, allowing a permanent dialogue between the different parties. Likewise, following Restrepo (1998), it is the farmers themselves, who discover the main problems that affect them. Moreover, it is these farmers who set the priorities and look for solutions, or demand a solution to external institutions, when not locally available.

A central activity has been the development of organic fertilizers elaborated with local materials and knowledge. The diverse ways in which farmers have are prepared organic fertilizers reflects the non-existence of fixed recipes. Additionally, it reflects farmer experience, their problem-solving creativity and the traditional knowledge put to use in daily practice, demonstrating great skill and capacity for self-management.

In addition to the preparation of organic fertilizers, the importance of traditional knowledge on nature observation, and the importance of the moon in agriculture, that is, each one of its phases and consequent influence on plant and the animal growth, has been strengthened by thematic workshops. Furthermore, the importance placed on preventing the burning of crop residues has favored crop association recovery and allowed reflection on the necessity to diversify not only crops, but also animal use within the household. Likewise, level curve ploughing was realized by promoting the "A-apparatus" to avoid erosion in steep areas.

The different agro-ecological proposals for sustainable development have been promoted by us as suggestions, recommendations, or as a guide, and can be used and developed according to the specific conditions of participating farmers, without creating an artificial separation of agricultural work from the daily of rural communities.

Several years after having initiated the activities described, a number of lessons can be learned, of which we want to mention the following:

- *The importance of the recovery of crop diversity*: we found a greater diversification in maize cultivation fields of participant farmers, amongst others, due to the abolishment of chemical herbicides. Associations of crops that have been established include: maize, bean, Jamaica (*Hibiscus* sp.), cucumber, tomato, pumpkin, banana and blackberry.
- *The abolishment of fire as a practice to eliminate crop residues and weeds*: this practice has been reduced by those farmers that have participated in the different workshops, and has favored the incorporation of organic matter into the soil.
- *The trace of level curves*, with the use of A-apparatus. This practice diminishes soil erosion, together with the use of dead and live fences.
- *The incorporation of organic fertilizers*, elaborated by participants.

It is important to mention that these practices have obtained relevance in the sense that they have reinforced farmer autonomy and independence and have contributed to a revalorization of manual (family) labor. For example, all the cleaning in the parcel was performed manually, a fact that furthermore reinforces the incorporation of organic matter to the soil and replaces the use of (chemical) fertilizer and pesticides. The adoption of the organic fertilizers by the producers of several communities has many perspectives to continue advancing, due to the flexibility in the use of supplementary local resources to elaborate the organic fertilizers, as we already mentioned before.

4. Gender relations in small-scale projects[9]

Rural development policies do not only affect male, but also female farmers, although the majority of the policies is directed mainly at the former (Moser 1993; Kabeer 1998). It is for this reasons that our work has included a gender dimension, in order to strengthen the position of women in different societal spaces, such as: citizen participation and decision making, community leadership, and access, use and control of natural resources, among others (Enríquez, 2000).

---

[9] Based on Reyes et al. (2004).

Since 1995, several actions have been established with a group of women farmers in one of the indigenous communities in the region, directed at reinforcing production strategies based on the reality and the necessities suggested by the women. These actions have consisted in supporting women solely, but also include mixed - male and female - farmer groups, in developing small-scale projects, such as: organic coffee, embroidery, natural conserves, organic honey, and traditional medicine.

Within the projects, women are not perceived as passive objects of development, but as active actors who have capacities and abilities to modify their socio-material surrounding, based on their knowledge of the specific characteristics of their natural surroundings, and their definition of strategies in order to interact with different local and governmental actors (cf. Long & Long, 1992; Long, 2001).

The actions within this thematic field are oriented to reinforce coherent strategic necessities of the participant women farmers. This includes the following activities:

- Supporting the organization of the group of women;
- Training the different groups, both women solely and mixed in order to strengthen production, administration and commercialization capacities;
- Generation of mechanisms for active and constructive participation of women, allowing a change of the women's social position in their community.

As mentioned before, these strategies were translated into a number of small-scale development projects, which for the sake of feasibility were established initially within the frame of the family production. Not only economic benefits have been derived from the productive projects, but there has also been the systematization of a series of social indicators that allow us to establish a retrospective reflection of the process of the on-going work with the women.

Furthermore, the inclusion of a gender approach in our actions has allowed a valorization of the contribution of women farmers to be strengthened further. Although the small-scale projects have turned around family production, throughout time it has become clear that the strategic needs of the women farmers have to be addressed also in the training workshops implemented.

5. Endogenous rural development and technology development[10]

Endogenous rural development not only requires the strengthening of the reproductive and productive base of the farm enterprise, or giving attention to the role of male and female farmers. It also has to deal with the development of appropriate technology (Pretty 1995). Generally speaking, technology is understood here as a combination of a number of techniques and the knowledge to apply these techniques.

Firewood is one of the most frequently used renewable resources in rural areas and one of the aspects to be considered here is the type of technology used for its burning. Traditional stoves are inefficient in their energy consumption and have a negative impact on human health. For this reason we started to promote improved firewood stoves, called "Estufas

---

[10] Based on Figueroa et al. (2003).

Lorena". The Lorena stoves use less firewood and have decreased the negative health effects associated with fire smoke. They can be considered as a technology that is appropriate to local conditions.

Our work in promoting the Lorena stoves in the Southern Coast of Jalisco has increased the understanding of the problem with firewood collection, especially related to women farmers, as well as the inconveniences of traditional stoves. Collection and the nuisance of smoke in the kitchen were mentioned as two major problems regarding firewood and its burning.

Until now, a total number of approximately 250 Lorena stoves have been built in the region, illustrating the possibilities of fomenting appropriate technology amongst peasant and indigenous farmers.

According to results, Lorena stoves reduce firewood use up to 70%, in comparison to traditional stoves. Besides, the improved firewood stoves have substantially improved the health of farmers, especially women. While the introduction of the Lorena stove has been a success, farmers do not directly relate its functioning with natural resource conservation (through there was an observed decrease in the firewood necessities). This stresses the need for incorporating a environmental education component in our work.

6. Endogenous development and regional agricultural and artisanal production[11]

There are a great diversity of products, which are produced in the region. These products, called here regional products, can be local foods and beverages, with typical recipes that have a limited geographic distribution with respect to his elaboration. We also include ceramics, embroidery, furniture and other handmade products (cf. van der Meulen, 1999).

Regional production can be divided into three categories. First, many of the regional products are the product of agricultural work, elaborated by individuals, families or by groups of farmers who live in rural zones. Secondly, in urban zones we found the second category, i.e. to be the typical occupations, for example, baker or butcher. In these cases, importance is attributed to preparation or elaboration of certain product, rather than the origin of the resources. Finally, the third category consists of those regional products in urban zones that are commercialized by small entrepreneurs. These can consist of products that are traditional for a region or products that carry the name of the region, showing their origin in a given region (*ibid.*). It is worth noting that the different categories are not mutually exclusive, and it is not always totally clear in which of the three categories a specific product belongs.

So far, we have described 32 regional products, which represent a first approximation in the Southern Coast of Jalisco (and part of the neighboring state of Colima) in regards to understanding of how actors respond to economic globalization that affects them in one way or another in the local territory.

With respect to the characteristics of regional production, a statistical analysis suggests the existence of two large groups. On the one hand, we found a group of producers and

---

[11] Taken from Gerritsen et al. (2004, 2005a).

businessmen that use an advanced process of industrialization, and who are located mainly in urban zones. Products of this group are characterized by a high exchange value. Additionally, specialist dominates this group. On the other hand, we found a second group comprised of those actors whose products are characterized by a high use value. These actors do not use a very advanced industrialization process; rather they take advantage of locally available resources. In other words, the logic of this group is geared toward home consumption, and they only commercialize surplus produce. Additionally, the activities of this group are more related to farming, and, as such, this group is mainly found in rural zones.

Our analysis further suggest that the activities of the first group (the specialists in urban areas) requires a considerable economic investment and constant acquisition of inputs for production and commercialization, meanwhile most products in the rural zones are developed more or less in a context of economic limitations and with a predominant use of their own resources.

Many of the regional products can be considered as "novelties" (Swagemakers, 2002). With this term, reference is made here to specific changes made by an actor in the production process with the purpose of reaching a desired situation. In order for a novelty to develop, a "niche" is required, that is specific local conditions where the idea can emerge and the fabrication of a regional product can take place. It does not only require a favorable context, but also time so that the development of a regional product can mature. Third, the development of a regional product is closely related with insecurity in or vulnerability of livelihood strategies, in both production and commercialization (Wiskerke & van der Ploeg, 2004). As such, many of regional products are a result from constant experimentation. These three elements, *novelty - niche - insecurity* are highly related with the life histories of the social actors who produce and commercialize these regional products.

7. Understanding endogenous rural development initiatives

In the foregoing, we described the four thematic work fields we have been working with, which aim to support and strengthen endogenous rural development initiatives in the Southern Coast of Jalisco. These different experiences can be considered as different strategies of valuation, and use and management of natural resources in local territories. Furthermore, these strategies depart from the existing endogenous potential in the local space, which can be characterized by a productive autonomy, maintenance of identity and local culture, and strong roots to the local area. In relation to the latter, a fundamental difference is observed with those strategies, where the mobilization of resources is based on organization outside of the local area. In these latter cases, resources are not only mobilized outside the territory, but also the production (and reproduction), transformation and commercialization, including economic and institutional relations and the family or community relations that accompany it. The latter implies that any effort to strengthen endogenous rural development initiatives allows revaluing the social, economic and ecological characteristics of local territories, i.e. it involves a revaluation of multi-functional nature of these areas.

The endogenous rural development initiatives as developed by the social actors supported show a farming domain coordination (Figure 2) that is different than that of conventional producers (Figure 1). Compared to the latter, analytically one can state that endogenous development initiatives are characterized by a "broadening", a "deepening" and a "re-grounding" of the different farming activities, as illustrated by Figure 4.

Following van der Ploeg et al. (2002), the broadening of farming activities refers to the incorporation of new farming activities into the farming domains, such as agri-tourism, care activities, diversification or nature and landscape management activities. Deepening refers to the aggregation of new values to the farm, as a consequence of the shift to organic farming, the elaboration of high-quality and regional products, and the shortening of the supply chains. Finally, the re-grounding of farming practice refers to the substitution of certain values for others, such as, for example, new forms of cost-reduction, off-farm income sources and new forms of co-operative management.

In our case, a deepening and re-grounding of farming practice characterize the endogenous development initiatives in the Southern Coast of Jalisco. Deepening refers to the elaboration of organic and regional products, as well as the search for fair trade commercialization initiatives. Furthermore, re-grounding takes place by the substitution of chemical by organic fertilizers, as well as the use of Lorena stoves. It also takes place by through the association of women farmers around common productive issues.

Retaking Figure 2, a (qualitative) assessment of the sustainability of endogenous rural development initiatives can also be made, as compared to conventional agricultural practices (Gonzalez, 2006). Figure 5 presents a qualitative comparison of four crops that are important in the region, including the underlying livelihood strategies, as well as their environmental impact.

Comparing organically cultivated peanuts and maize with agro-industrially produced melon and agave in relation to single-use vs. multiple-use strategies (i.e. specialization vs. diversification) and the presence of a low vs. high diversity in natural resources (i.e. enrichment vs. degradation of the natural environment) shows that the endogenous rural development initiatives are characterized by diversification and environmental enrichment, while melon and agave are specialized and degrading farming systems.

8. Towards a new emergent rurality

Until now, we presented a general description of the endogenous rural development program, as well as the different initiatives supported by us, through training, organization and applied research. However, the strengthening of endogenous rural development initiatives goes beyond mere training and organization of different farmer groups with whom we work. It has relates to the topic of "rurality", that is to say, our understanding of and actions associated with "the rural". Generally speaking, we understand "rurality" referring to the social representations of rural zones, which relate to existent dynamics and socio-materials processes (van der Ploeg, 1997). In the case of Mexico, reflecting about existing rurality is important, due to the deep transformations to which the Mexican countryside is submitted, and which are due to neo-liberal development and economic globalization processes (Schwentesius et al., 2003; Esteva & Marielle, 2003).

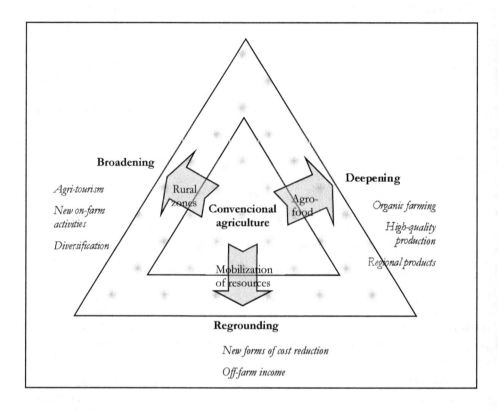

Fig. 4. The structure of rural development at farm enterprise level (van der Ploeg et al., 2002: 12)

Our work in the South Coast of Jalisco suggests the presence of a new emergent rurality where the endogenous properties form the essence. As such, we understand rurality as being the result and expression of a co-production process, i.e. the on-going interactions and mutual transformations between man and nature in the local area, and where farmers play a strategic role. It is through the process of co-production that the typical (cultural) landscape of a region comes to life, or where an unique (agro-) biodiversity is being created (van der Ploeg, 1997; Gerritsen, 2002). Moreover, it has multiple benefits, such as the generation of local employment; the generation of income articulated to the regional economy, the maintenance of culture and local identity, and the conservation of natural resources, amongst others.

We assume that this new rurality is emergent as it appears there is an increasing number of producers who look for developing alternative models of production, transformation, and industrialization. The strategies of these actors include an appropriation of local territory's patrimonial values and the (re)valuation of its natural resources (cf. Casablanca & Linck, 2004). In this context, the absence of a favorable institutional context to strengthen these experiences is noticeable. Those who have played a role of facilitator in this process have been all non-governmental organizations.

Finally, the consolidation of this new rurality not only requires an appropriation of the patrimonial values by the regional society, but also the protection of these values.[12] In addition, it requires a deep change in the thinking of civil society towards farming practices in general and food products in particularly.

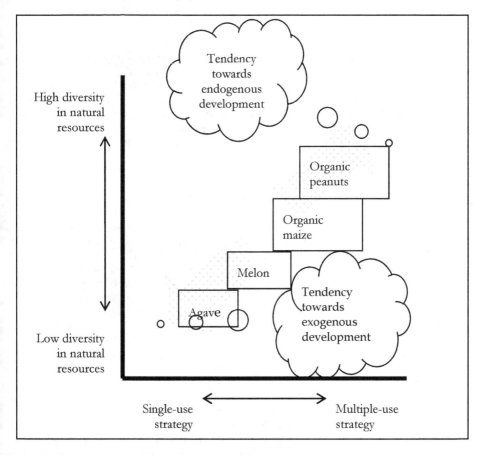

Fig. 5. Qualitative sustainability assessment of four crops

9. Acknowledgements

The author acknowledges the University of Guadalajara and the international program on research collaboration NCCR North South for their support.

---

[12] Some of the mechanisms to protect the different initiatives of endogenous rural development might be the origin denominations in the case of the productive process, or the certification of the final products (Casablanca and Linck 2004).

10. References

Broekhuizen, R. & J.D. van der Ploeg. (1995). *Design methods for endogenous rural development* Wageningen. Agricultural University, Internal report, Wageningen

Calva, J.L. (1993). ´El Modelo de desarrollo agropecuario impulsado mediante la ley agraria y el TLC´. In: *Alternativa para el campo mexicano,* Calva J.L. (coord.), 15-43, Edición Fontamara, ISBN 9684761600,México, D.F.

Carabias, J & E. Provencio (1993). ´Hacia un modelo de desarrollo agrícola sustentable´. Pp. 45-59 in: Calva J.L. (Coord.). *Alternativas para el campo mexicano. Tomo II.* Fontamara-FES. México.

Casablanca, F. A. & T. Linck. (2004). *Tipificación de alimentos y apropiación de recursos patrimoniales.* Paper presented at the Congreso Internacional Agro-empresas rurales y Territorio (ARTE), Toluca, Estado de México, 1-4 de diciembre de 2004.

Cortez, R. C., Concheiro, B. L. & León L. A. (1994). *Los pueblos indios frente a la globalidad: efectos y respuestas.* Paper presented at the Coloquio, Impacto de la Modernización en el Ámbito Laboral a Finales del Siglo XX. Xalapa, Veracruz. Universidad Veracruzana.

Enríquez, M. (2000). *Los proyectos productivos para mujeres: Del discurso del desarrollo a las experiencias vividas.* El Colegio de la Frontera Sur. Tesis de maestría.México: San Cristóbal de las Casas:

Esteva, G. & C. Marielle (Coord.). (2003). *Sin maíz no hay país.* Comité sin Maíz no hay País/CONACULTO. ISBN: 970-35-0434-5, México, D.F.

Figueroa B., P., G. Cruz S., V.M. Villalvazo L., & P.R.W. Gerritsen. (2003). *Tecnología apropiada para el desarrollo endógeno: la estufa Lorena y el ahorro de leña en dos comunidades rurales del Occidente de México.* Presentation delivered at the X Simposio sobe Conservación, Manejo de Recursos Naturales y Desarrollo, organizado durante los días Noviembre, 24-27, en Autlán, Jal., México.

Figueroa B., P. R. Moreno, P.R.W. Gerritsen & V. Villalvazo. (2004). *Desarrollo endógeno y seguridad alimenticia: experiencia del ejido de Ayotitlánen el Occidente de México.* Paper presented at the Congreso Internacional Agro-empresas Rurales y Territorio (ARTE), Toluca, Estado de México, 1-4 de diciembre de 2004.

Gerritsen, P.R.W. (2010). *Perspectivas campesinas sobre el manejo de los recursos naturales.* Mundiprensa/Universidad de Guadalajara. ISBN: 9786077699095, Mexico city.

Gerritsen, P.RW. (2006). *On endogenous rural development and new images of rurality in western Mexico.* Paper presented at the XI International Congress of the Latin American Studies Association, San Juan, Puerto Rico, March 15-18, 2006.

Gerritsen, P.R.W. (2002). *Diversity at Stake. A farmer's perspective on biodiversity and conservation in Western Mexico.* Wageningen Studies on Heterogeneity and Relocalization 4. ISBN: 9067546836, Wageningen.

Gerritsen, P.R.W., M. Montero C. & P. Figueroa B. (2003). ´Percepciones campesinas del cambio ambiental en el Occidente de México´. *Economía, Sociedad y Territorio* Vol. II, Núm. 14, Julio-Diciembre de 2003. ISSN: 1405-8421

Gerritsen, P.R.W., G. Cruz, V. Villalvazo, & P. Figueroa. (2004). *"Productos regionales" en el Occidente de México: ¿respuestas locales frente a la globalización económica?* Paper presented at the Congreso Internacional Agro-empresas rurales y Territorio (ARTE), Toluca, Estado de México, 1-4 de diciembre de 2004.

Gerritsen, P.R.W., V. Villalvazo L., P. Figueroa B., G. Cruz S. & J. Morales H. (2005a). *Productos Regionales y Sustentabilidad: Experiencias de la Costa Sur de Jalisco.* Paper presented at the Vto Congreso de la Asociación Mexicana de Estudios Rurales (AMER), Ciudad de Oaxaca, Oaxaca, 25 al 28 de mayo de 2005.

Gerritsen, P.R.W, V. M. Villalvazo, P. Figueroa B. & G. Cruz S. (2005b). *Fortaleciendo procesos endógenos en la Costa Sur de Jalisco: Imágenes de una nueva ruralidad emergente.* Paper presented at the I Foro Académico Interinstitucional: "Diálogos sobre el Sur de Jalisco: Actualidad y Futuro del Desarrollo", Zapotlán el Grande, Jalisco, 22 de febrero de 2005.

Gónzalez, F., R. (2006). *Analisis de sustentabilidad de cuatro cultivos en el ejido La Cienéga, Municipio de Limón.* Autlán: Centro Universitario de la Costa Sur, Universidad de Guadalajara. B.Sc.-thesis.

Gúzman C., G., M. González de Molina & E. Sevilla G. (2000). *Introducción a la agroecología como desarrollo rural sostenible.* Ediciones Mundi-Prensa. ISBN: 8471148706, Madrid/Barcelona/México.

Halweil, B. (2002). *Home grown. The case for local food in a global market.* Washington, D.C.: Worldwatch Paper 163.

Heijden, D. van der, P. Camacho, C. Marlin & M. Salazar G. (2004). *Guía metodológica para el análisis de cadenas productivas.* Quito: Ruralter/SNV/CICDA/Intercooperation.

Jansen, K. & S. Vellema. (2004). *Agribusiness and society. Corporate responses to environmentalism, market opportunities and public regulation.* London/New York: Zed Books.

Kabeer, N. (1998). "Conectar, extender, trastocar: El desarrollo desde una perspectiva de género", en: Naila Kabeer, (ed.). *Realidades trastocadas. Las jerarquías de género en el pensamiento del desarrollo.* México. Universidad Autónoma de México.

Long, N. (2001). *Development sociology. Actor perspectives.* London and New York: Routledge.

Long, N. & A. Long (Eds). (1992). *Battlefields of knowledge. The interlocking of theory and practice in social research and development.* London y New York: Routledge Publishers.

Malischke, T.K., R. González F. & P.R.W. Gerritsen. (2005). *Percepciones campesinas sobre la degradación ambiental Una comparación de agricultores orgánicos y convencionales en el ejido de La Ciénega, Municipio de El Limón, Jalisco, México.* Paper presented at the Vto Congreso de la Asociación Mexicana de Estudios Rurales (AMER), Ciudad de Oaxaca, Oaxaca, 25 al 28 de mayo de 2005.

Meulen, H. van der. (1999). *Streekproducten in Nederland. Inventarisatie, criteria, certificering en case studies.* Wageningen: Leerstoelgroep Rurale Sociologie, Wageningen Universiteit.

Morales H., J. (2001). *Construyendo la sustentabilidad desde lo local: la experiencia de la Red de Alternativas Sustentables Agropecuarias de Jalisco.* Guadalajara: ITESO. Technical report.

Morales H., J. (2003). *Desarrollo rural alternativo en el sur de Jalisco: experiencias hacia la sustentabilidad rural.* ITESO: Informe final de investigación.

Morales H., J. (2004). *Sociedades rurales y naturaleza, En busca de alternativas hacia de la sustentabilidad.* Guadalajara: ITESO/Universidad Iberamericano.

Moser, C.O.N. (1993). *Gender planning and development. Theory, practice and training.* London and New York: Routledge Press.

Muñoz, C. & A. Guevara. (1997). *Pobreza y Medio Ambiente.* Pp. 165-149 in Martínez, G. (Comp.) *Pobreza y Política Social en México.* México, D.F.: Fondo de Cultura Económica /ITAM.

Ploeg, J.D. van der. (1990). *Labor, markets and agricultural production.* Boulder, San Francisco and Oxford: Westview Press.

Ploeg, J.D. van der. (1992). 'The reconstitution of locality: technology and labour in modern agriculture'. Pp. 19-43 in Marsden, T., R. Lowe and S. Whatmore (Eds) *Labour and locality: uneven development and the rural labour process.* London: David Fulton Publishers. Critical perspectives on rural change series, IV.

Ploeg, J.D. van der. (1994). 'Styles of Farming: an introductory note on concepts and methodology'. Pp. 7-30 in Long, A. and J.D. van der Ploeg (Eds) *Born from within. Practice and perspective of endogenous rural development.* Assen: Van Gorcum Publisher.

Ploeg, J.D. van der, A. Long & J. Banks. (2002). *Living countrysides. Rural development processes in Europe: the state of the art.* Doetinchem: Elsevier.

Ploeg, J.D. van der. (1997). 'On rurality, rural development and rural sociology.' Pp. 39-73 in Haan, H. de and N. Long (Eds) *Images and realities of rural life. Wageningen perspectives on rural transformations.* Assen: Van Gorcum Publishers.

Ploeg, J.D. van der & A. Long. (Eds.). (1994). *Born From within. Practice and perspectives of endogenous rural development.* Assen, Paises Bajos: Van Gorcum.

Pretty, J.N. (1995). *Regenerating agriculture: policies and practice for sustainability and self-reliance.* London: Earthscan Publications Ltd.

Restrepo, J (1998). *La mejora campesina, una opción frente al fracaso de las granjas integrales didácticas,* Servicio de Información Mesoamericano sobre Agricultura Sostenible, Colección Agricultura Ecológica para Principiantes 4.

Reyes G., C., V.M. Villalvazo L, M. Enríquez M. & P. Figueroa B. (2004). *Mujeres en la búsqueda de alternativas productivas en la comunidad indígena de Cuzalapa, reserva de la biosfera Sierra de Manantlán.* Paper presented at the II Coloquio Nacional de la Red de Estudios de Genero del Pacifico Mexicano, durante los días 26 y 27 de marzo de 2004 en la ciudad de La Paz en Baja California Sur.

Rist, S. (2004). ´Endogenous development as a social learning process.´ *COMPAS Magazine* Sept. 2004: 26-29.

Saxe-Fernández, J. (1998). ´Neoliberalismo y TLC: ¿Hacia ciclos de guerra civil?´. Pp. 87-124 in: de Pina, G. J.P y J. Alba, G. (Eds). *Globalización, Crisis y Desarrollo Rural en América Latina* México, D.F.: Colegio de Postgraduados Universidad Autónoma de Chapingo. Memorias de sesiones plenarias del V Congreso Latinoamericano de Sociología Rural.

Schwentesius, R., M.A. Gómez C., J.L. Calva T. & L. Hernández N. (Coord.). (2003). *¿El campo aguanta más?* Texcoco: Universidad Autónoma de Chapingo / La Jornada.

Swagemakers, P. (2002). *Verschil maken. Novelty-productie en de contouren van een streekcooperatie.* Wageningen, Países Bajos: Circle for European Studies/Leerstoelgroep Rurale Sociologie.

Toledo, V.M. (1990). 'The ecological rationality of peasant production.' Pp. 53-60 in Altieri, M.A. and S.B. Hecht (Eds) *Agroecology and small farm development.* Boca Raton/Ann Arbor/Boston: CRC Press.

Toledo, V.M. (1995). *Campesinidad, agroindustrilidad y sostenibilidad. Los fundamentos ecológicos e históricos del desarrollo rural.* México City: Interamerican Group for Sustainable Development of Agriculture and Natural Resources. Report No. 3.

Toledo, V.M. (2000). *La Paz en Chiapas. Ecología. Luchas indígenas y modernidad alternativa.* Mexico City: Ediciones Quinto Sol/UNAM.

Toledo, V.M. (2003). *Ecología, espiritualidad y conocimiento. De la sociedad del riesgo a la sociedad sustentable.* México, D.F.: Universidad Iberamericana/PNUMA.

Villalvazo L., V.M., P.R.W. Gerritsen, P. Figueroa B. & G. Cruz S. (2003). ´Desarrollo rural endógeno en la Reserva de la Biosfera Sierra de Manantlán, México.´ *Sociedades Rurales. Producción y Medio Ambiente* 4 (1): 41-50.

Villalvazo L., V.M., P.R.W. Gerritsen, P. Figueroa B., R. Ramirez P. & L. Córdoba R. (2005). *Alternativas productivas y desarrollo endógeno en el Occidente de México.* Paper presented at the I Congreso Internacional "Casos Exitosos de Desarrollo Sostenible del Trópico, Xalapa, Veracruz, México del 2 a 4 de mayo de 2005.

Waters, M. (1995). *Globalization.* London and New York: Routledge.

Wiskerke, J.S.C. & J.D. van der Ploeg. (2004). *Seeds of transition. Essays on novelty production, niches and regimes in agriculture.* Assen: Royal van Gorcum.

7

Achieving Household Food Security: How Much Land is Required?

P. Ralevic[1,3,*], S.G. Patil[2] and G.W. vanLoon[1,*]

[1]School of Environmental Studies,
Queen's University, Kingston, Ontario,
[2]University of Agricultural Sciences (Raichur),
[3]Faculty of Forestry, University of Toronto, Ontario,
[1,3]Canada
[2]India

1. Introduction

Global hunger is on the rise (FAO, 2009). The number of those that are undernourished has increased steadily over the past decade reaching 1.02 billion people in 2009 (FAO, 2009), and remaining near this total over the past two years. The Millennium Development Goal to reduce world hunger by half by 2015 will likely not be met with current observed trends according to a recent UN report on food insecurity and world hunger (FAO, 2009). Limited land and growing population (Gerbens-Leenes and Nonhebel, 2005), global yield decline (Alston *et al.*, 2009), financial resources and the economic downturn in 2008 are a few reasons that have led to the recent trend in world hunger.

As of 2010, some 237 million or 21% of the total population in India remain undernourished (FAO, 2011). While the percentage decline is an improvement over the 25% in 1990; a large portion of the population still does not consume even the minimum recommended daily intake of energy and protein. Further, a 2001 survey of the diet and nutritional status of India's rural populations indicated that across all age and physiological groups, the consumption of most foods was below the recommended daily intake as set out by the Government of India (NNMB, 2002). The problem was underlined in a recent report of the International Food Policy Research Institute, which recognizes that some 40% of India's children are malnourished, with mortality rates of 2.5 million per year attributed to inadequate food consumption (von Braun *et al.*, 2008). This represents one out of every five such deaths in the world; the rate of malnutrition is double that of Sub-Saharan Africa and five times that of China (von Braun *et al.*, 2008). Several reports indicate that there is an urgent need for a comprehensive nutrition strategy including incentive-oriented policies that involve community and household participation in order to ensure adequate production and retention of food within communities (McIntyre *et al.*, 2001; von Braun *et al.*, 2008). For rural regions of India, this is particularly true as villages are often far removed from other sources of income and from the larger agricultural markets in urban centres.

---

* Corresponding Authors

As India's population continues to increase, the pressure on land resources to supply the food and fodder demands of households will be ever greater, calling for a need to utilize on-farm resources more efficiently. While yield is a key aspect to improved nutrition (FAO, 2009), optimal allocation of crops and livestock within a given land area is also an important factor towards food security.

The present study seeks to develop methodology that can be employed to determine how much land is required to satisfy the basic human nutritional requirements of a household, and in other instances livestock requirements in terms of fodder demand. Primary data on land and population inventories are used to compare the minimum land area requirements among the various different landholding and crop yield categories. The aim of the present study is to inform policy on the essential land requirements necessary to meet the nutritional demands of households.

2. Methodology

The micro-level household survey of mixed cropping / livestock systems and associated land use was carried out in the semi-arid region of northern Karnataka state, in southern India, during October and November in 2007 (Figure 1). In this part of the state, two crops

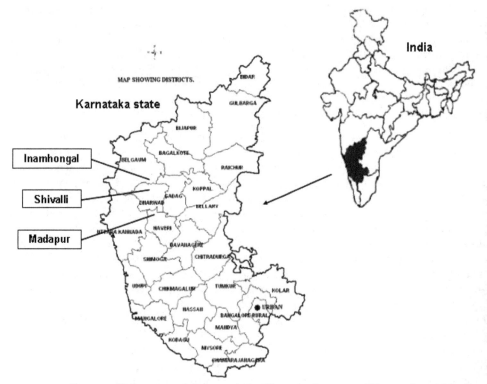

Fig. 1. Map showing the location of the sampled villages in the state of Karnataka within the respective districts (GOK, 2005 from Ralevic *et al.*, 2010).

can be grown, depending on availability of water: the *kharif* crop cultivated during and after the monsoon season and the *rabi* crop during the drier winter months. Three villages were sampled: Inamhongal (N 15° 37.623', E 75° 04.551', Zone-III) Belgaum district, Madapur (N 15° 03.250', E 75° 18.095', Zone-VIII) Haveri district and Shivalli (N 15° 28.135', E 075'08.435', Zone-VIII) Dharwad district (Figure 1). Data were gathered in order to determine the base year crop distribution, production and needs for food and fodder of individual households within various landholding categories: landless, marginal (0-1 hectares), small (1.01-2 ha), medium (2.01-4 ha) and large (>4.01 ha). These base year data describe the situation for the *kharif* crop harvested in late 2006 and the *rabi* crop harvested in 2007. A detailed account of the region, survey methodology, primary data collection procedures, and social indicators can be found in Ralevic *et al.* (2010).

In the present study, the definition of household size or population is based on consumptive units (CU) as set out in Table 1, and from this the nutritional demands were calculated (Ralevic *et al.*, 2010). All livestock-related calculations are similarly done on a livestock unit (LSU) basis, where 1 livestock unit represents an equivalent of 1 bullock, buffalo or lactating dairy cow. The LSU equivalents are used to determine the fodder demands which for 1 LSU (lactating dairy cattle) is assumed to be 2.17 t dry / y .

Group	Consumptive unit
Adult male (moderate work)	1.2
Adult female (mod. w.)	0.9
Adolescents (12-21 yrs)	1
Children (9 to 12 yrs)	0.8
Children (7 to 9 yrs)	0.7
Children (5 to 7 yrs)	0.6
Children (3 to 5 yrs)	0.5
Children (1 to 3 yrs)	0.4

Table 1. Human nutritional consumptive units (adopted from Gopalan *et al.*, 1996).

Using the collected demographic data, the average household nutritional needs within the landholding categories were determined. A linear optimization model was then used to run various land area requirement scenarios using a variety of constraints and objectives as in Ralevic *et al.* (2010). The linear model was constructed in excel using the What'sBest LINDO software ad-ins for optimization. The objective function in all scenarios modeled was to minimize land area. In the present paper, the following scenarios are evaluated:

1. The minimum land area required to satisfy basic human food energy and protein needs of individual households under presently observed yields and three cropping intensities; 200% (upper) , 164% (present average situation) and 150% (lower).

Objective function:

Minimum land area = Land area in the *kharif*[1], *and* where

---

[1] Field data has shown that nearly all land is cultivated in the *kharif* season. Therefore, the minimum land area is based primarily on the cultivation within the *kharif* season, with anything over a 100% intensity being cultivated in the *rabi* season.

Land area in *rabi* = Land area in the *kharif* x (Cropping Intensity ratio -1)

$$MinLA = \sum_{cv} L_{cv,s1} \text{ , where } \sum_{cv} L_{cv,s2} = \sum_{cv} L_{cv,s1} x(CI_R - 1) \tag{1}$$

where LA = Land Area, cv = 1,2,3...14, $s1$ = *kharif*, $s2$ = *rabi*, R = 1, 2,3.

Cropping intensity ratio= (land area used in the *kharif* season + land area used in the *rabi* season) / total available land area (also expressed in percent).

?   The minimum land area required to satisfy the fodder demands in addition to food requirements under current yields and under three cropping intensities of 200% (upper), 164% (present situation) and 150% (lower).

Fodder demands were determined based on the number of livestock units per household.

The food energy and protein demand constraints, as well as the fodder demand constraints and the input data are outlined in Ralevic *et al.* (2010).

3. The findings

3.1 Households and land in the study area

For the three villages, the base year data on average household size and average land area are presented in Table 2. The average household population in the villages ranged from 5 to 6.5 CUs with a weighted average of 5.8 consumptive units. The CU value differed depending on the size of landholding, with larger landholders tending to support larger families compared with those who had little or no land. Average holding of land by households within the four categories ranged from 0 ha/hh (landless households) to 8.2 ha/hh (large households). More detailed demographic information can be obtained from Ralevic *et al.*, (2010).

Land category	Land area (ha/hh)	Population (CU/hh)	Livestock units (LSU/hh)	Sample-wide landholdings (% of total)
Landless	0	5	0.09	17
Marginal	0.63	5.3	0.53	21
Small	1.55	5.7	1.10	25
Medium	2.85	6.5	2.35	20
Large	8.16	6.5	4.08	16
Weighted total	2.38	5.80	1.63	~100

CU = consumptive unit, LSU = livestock unit, hh = household

Table 2. Population and land area data for sampled households presented as a weighted average among landholding categories.

3.2 Mix of crops and cropping intensity in the base year

The study area is in a region of the country that supports a considerable diversity of crops grown during the two agricultural seasons. Variations both in precipitation and in soil types among the villages influence cropping patterns and intensity of cropping (UAS, 1985). Table 3 shows the cropping intensity and the crop types that were cultivated in the three villages during the base year. Depending on the village, about three quarters of the land was planted to a variety of pulses, oilseeds and commercial crops, while the remaining one quarter was devoted to cereals in all three cases.

	Village		
	Inamhongal	Madapur	Shivalli
Total cultivable land area (ha)	1805	704	1117
Cropping intensity (%)	182	118	192
Cereals (%)	25.8	23.4	23.4
Pulses (%)	53.8	1.1	26.6
Oilseeds (%)	1.4	23.3	24.6
Commercial crops (%)	18.9	40.4	25.2
Other[1]		11.8	

[1] Includes non-principal crops reported by a small number of households (i.e. coconut, banana etc.).

Table 3. Percent of total cultivable land devoted to crops during the base year among the sampled villages for the *kharif* and *rabi* seasons.

The intensity of cropping was highest in Shivalli, at 192%, followed by Inamhongal at 182% and Madapur at 118%.

In Inamhongal, the principal crops in the *kharif* season were maize, horsegram, greengram and onion, and in the *rabi* season, sorghum, wheat, maize and bengalgram. Cotton was also cultivated throughout the *kharif* and *rabi* seasons. Unlike in the other two villages, chilli, a high value commercial crop commonly grown in this area, was not cultivated here, given that chilli is grown in red soils (UAS, 1985) and is therefore unsuitable for Inamhongal's black soils. Similar to chilli, nuts and oilseeds are predominantly absent due to nuts such as groundnut favouring growth in light red soils. Of the three villages, Inamhongal had the most diverse cropping pattern.

The principal crops cultivated in Madapur were mainly commercial in nature, including chilli, cotton, and groundnut. There was only limited production of foodgrains in Madapur and there was much less village-wide crop diversity as was observed in Inamhongal. The high value and assured market for the commercial crops heavily contributed to the observed pattern of cultivation. With regard to the low cropping intensity, the under-utilization of land was due to a combination of factors, including the inability of farmers to purchase the material inputs, such as high quality seed and fertilizer, necessary to cultivate the additional land (Singh *et al.*, 2007), as well as the limited availability of labour (Singh and Marsh, 1994; Suryanarayana, 1997). Discussions with farmers in Madapur pointed out that there was a shortage of labour and lack of irrigation that could have been accessed during periods of limited rainfall particularly in the *rabi* season.

In Shivalli, a diverse cropping pattern was observed that included crops from all four of the major crop groups: cereals, pulses and legumes, nuts and oilseeds and commercial crops. The principal crops cultivated during the *kharif* season were horsegram, greengram, groundnut, onion and chilli, and in the *rabi* season, sorghum, wheat, bengalgram and sunflower were grown.

In each of the villages, compared with other categories, grain crops were generally more productive in terms of amount of food energy generated per hectare. Crop yields (Table 4) of specific crops were similar within all three villages and are comparable to yields in other dryland parts of India during the same year.

Crop variety	Yield (t/ha)	
	Kharif	*Rabi*
Sorghum	1.28	0.81
Wheat		0.66
Maize	2.24	
Bajra	0.76	
Greengram	0.57	
Redgram	0.42	
Horsegram	0.77	
Bengalgram		0.76
Groundnut	1.14	
Sunflower	0.30	0.30
Safflower	0.25	0.40
Chilli	0.67	0.15
Onion	5.08	2.95
Cotton	0.73	0.73

[1] Primary data.

Table 4. Average crop yields for selected seasons of cultivation in 2006-2007 within the sampled villages (see Ralevic *et al.*, 2010). A blank cell indicates that the particular crop was not cultivated during the indicated season.

Given the central importance of providing for household food security, it is interesting to separately examine the types of crops grown by farmers who had only limited amounts of land, i.e. those having marginal and small landholdings (Table 5).

Crop category	Marginal (%)		Small (%)	
	kharif	*rabi*	*kharif*	*rabi*
Cereals	2.7	15.9	9.3	36.8
Pulses	35	18.5	36.2	22.6
Oilseeds	19.3	8.7	16.2	8.5
Commercial crops	28.6	20.5	34.0	9.7
Other crops	8.1	0	2.5	1.6
Fallow land	6	36	1	20
Total (%)	~100	~100	~100	~100
Average land area (ha)	0.63		1.55	

Table 5. Crop categories cultivated by farmers having marginal and small landholdings for the base year as a weighted average for all sampled villages.

For marginal households, commercial and other cash crops (pulses and oilseeds) made up about 88% and 75% of total cultivated land during the *kharif* and *rabi* seasons respectively while for small households, the same figures were 86% and 53% during the base year. This meant that, on average, only a small portion of the limited land area for these households was used to produce food grains, the basic component of their dietary requirements. This portion was considerably less than that of the medium and large landholders.

3.3 Current nutritional status of the population

Using the information regarding average household size and production data for farmers in the various landholding categories, it is possible to determine the extent to which fundamental energy and protein nutrient requirements were provided on farm for the average case. Such base year calculations show that a large portion of the population currently living within the villages was unable to achieve food security under cropping conditions during the year under investigation (Figure 2).

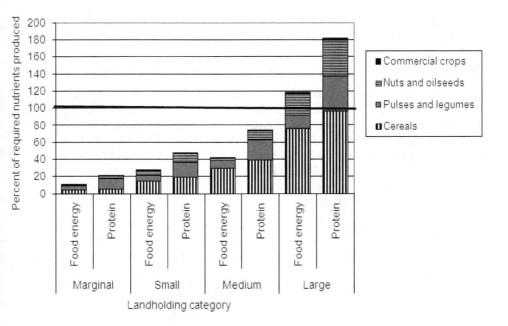

Fig. 2. Total calculated household food energy and protein produced within the landholding categories during the base year, in relation to the required nutrient level for a basic diet, as a percentage of total requirements (retrieved from Ralevic *et al.*, 2010). For each landholding type, a breakdown of nutrient production by crop category is presented.

According to these results and the population distribution within landholdings (Table 2), total household production made up 40 % of the villages' caloric requirement and 66% of the protein requirement excluding milk production. When milk production is included in the calculation, total household production made up 48% of the villages' calorific requirement and 79% of the protein requirement. The average values indicate that only the 14% of households having land holdings greater than 4 ha could produce sufficient food to satisfy nutritional demands. The remaining 86% could not achieve household food security through their individual agricultural activities, even under an assumption that all of the food that is produced is consumed within the village. These assertions are based on the assumption that the village average crop distribution was practiced in each case. Individual personal consumption of foodstuffs was not measured during the field survey, but the data indicate that the farmers with large landholdings sold surplus production of cereal-based crops to households within the village and to the wider market. This is also supported by work of Babu *et al.* (1993).

Therefore, on average, large landholding households would be able to consume an adequate amount of on-farm food energy and protein, while landless, marginal, small and even many medium landholdings must depend in large part on purchase of food to meet these requirements (Babu *et al.*, 1993). Field observations suggested that even when it was necessary to purchase food, persons in these lower income households generally appeared malnourished. Therefore, present land area constraints, low yields and high population demands all contribute to some level of undernutrition.

3.4 Minimum land area needed to provide for household food security

It is clear then, that in the base year using the average number of household members within the villages, most farmers would not be able to satisfy their household food requirements under the current cropping conditions. Improvements in yield and/or increased cropping intensity would be two obvious ways in which the food needs of the local populations could be more closely met. Our interest in this research, however, was to examine whether rational design of crop selection could also contribute to a larger and more appropriate supply of food. A question that we wished to address is "Can optimizing crop selection contribute to improved nutrition within this area?"

With this goal and using the objective of finding the minimum land area required to satisfy nutritional requirements as defined above, a new optimal cropping pattern was calculated for each village. Examples of the recommended crop distributions for a household having 5.8 consumptive units and 0 to 3 livestock units are given in Table 6. In the optimized model, cereals are given prominence especially in the highly productive *rabi* season, because of their high energy yield and nutritional status. Given their high fodder yield, they become especially important in cases where the household keeps cattle.

The table shows that, with a cropping intensity of 1.5, the minimum land area needed to provide for food security when crop distribution is optimized is just over 2 ha. This places the area requirement at the border between the small and medium landholding categories. In the study area, only 36% of the households had an area of land of 2 ha or more.

When the household has no livestock, the protein requirements are met by producing and consuming a good supply of pulses, mainly greengram and horsegram, during the *kharif* season. When the household had one milk-producing animal, there was less need to produce pulses as a protein source .The reduction in pulses could then be substituted by increased production of commercial crops. Therefore, with the same land area, having one cattle would be able to increase the net value of products. As the number of cattle is increased to two or three, the same minimum land area could provide for food security, but more of the land would need to be planted to cereal crops to cover for the fodder needs, thus reducing the commercial crops.

| Crop category | Livestock units | | | | | | | |
| | 0 | | 1 | | 2 | | 3 | |
	kharif	*rabi*	*kharif*	*rabi*	*kharif*	*rabi*	*kharif*	*rabi*
Cereals	5.4%	96.7%	5.4%	96.7%	26.6%	96.7%	60.6%	96.7%
Pulses	55.2	3.3	49.6	3.3	43.6	3.3	37.6	3.3
Nuts and oilseeds	1.8	0	1.8	0	1.8	0	1.8	0
Commercial crops	37.5	0	43.4	0	28.3	0	0	0
Total (%)	~100	~100	~100	~100	~100	~100	~100	~100
Optimal land area (ha)	2.03	1.02	2.03	1.02	2.03	1.02	2.03	1.02
Net value of products (Rs)[2]	27,650		38,710		44,320		54,180	

[1] Consists of 2 adult males, 2 adult females, 1 adolescent, and 1 child who is 5-7 years of age.

[2] Includes the value of all crops, milk, manure and fuel that may have been produced from the cropping pattern and livestock combinations.

Table 6. Calculated optimal cropping pattern by crop category under a minimum land area scenario for achieving food security for the average household having a CU[1] of 5.80 and cropping intensity of 150%.

Using a similar optimization strategy, the minimum land requirements can be recalculated for situations where the cropping intensity is varied. Table 7 provides such values when the cropping intensity is 150%, 164% (the current average intensity in the three villages) and 200%. Once again, it is important to note that the estimates are based on employing current yield values in the calculation.

As shown in the table, when the need to provide fodder for livestock is taken into consideration together with the need to satisfy basic human requirements, the minimum land area generally may increase depending among other things, on cropping intensity and the number of livestock. Note that for 0 livestock the land area required in most cases is the same as if there were 1 or 2 livestock.

	Cropping intensity (%)	Livestock units (all cattle)			
		0	1	2	3
Minimum land area (ha)	200	1.02	1.02	1.18	1.51
	164	1.59	1.59	1.59	1.83
	150	2.03	2.03	2.03	2.03

Table 7. Minimum land area needed to provide for the nutritional needs of humans and cattle. Throughout the study area, the current average land use intensity is 164 %. CI = cropping intensity, CU=consumptive units/household.

Figure 3 shows and extends the results under the three cropping intensity scenarios for situations where greater numbers of cattle are kept within the household.

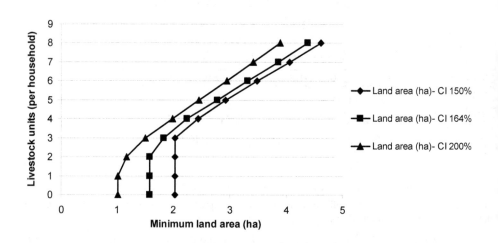

Fig. 3. Minimum land area based on livestock units as well as on average consumption units on a household basis for marginal, small, medium and large households.

The results shown in Figure 3 further illustrate the relation between human and animal nutritional requirements.Under all cropping intensities the minimum land area does not increase when there are a small number of livestock, since sufficient fodder is available in the form of the secondary biomass from the principal crops such as sorghum and maize. In the case of the 164% intensity, up to 2 livestock units can be supported on the same amount of land that is required to support only the household members. Above 2 LSUs, the demand for fodder drives the minimum land area requirement, similar to what Gerbens-Leenes and Nonhebel (2005, 2002) have found.

In the present situation, it was found that small and marginal households kept on average 1.1 and 0.53 livestock units respectively each, equivalent to each or every second household having one cattle or buffalo. From the field observations presented here, such households are unable to independently feed themselves and also are unable to produce the required fodder needed by the animal. Such households will be dependent on other larger farms for food as well as additional land for grazing .

In comparison, using optimal crop selection and the current 164% cropping intensity, small households having the average land area of 1.55 ha could potentially satisfy both the basic nutritional requirements within their households as well as the nutritional needs of up to 2 livestock units equivalents.

4. Minimum land area requirements: Comparisons with other studies

Other studies (Gerbens-Leenes et al., 2002; Gerbens-Leenes and Nonhebel, 2005, 2002) have quantified the minimum land area that is required to satisfy food requirements of a given population. These studies calculate the amount of land needed to provide nutrients through 'singular foods', or single crops, as well as through several crops in line with the consumption patterns of the population. Although the studies deal with the Dutch population and, as recognized by the authors, cannot be applied as a direct comparison to India's population, they nonetheless highlight important relationships with respect to food and land area that are also evident in the present study. First, the studies found that a shift in diet to one containing animal foods or products requires more land. Gerbens-Leenes et al., 2002 and Gerbens-Leenes and Nonhebel, 2005, 2002 also found that increases in land area are required to satisfy the fodder demands of the livestock, similar to the results of this study.

Second, when consumption patterns are included such that a variety of crops are grown, land area requirements increase. For instance, Gerbens-Leenes et al. (2002), and Gerbens-Leenes and Nonhebel (2005, 2002) found that when the consumption pattern was not considered, a singular high-yielding crop would be 'optimized' for production. In other words, when constraints[2] for the various crops to be consumed are removed, the required land area to maintain a minimum level of food energy and protein decreases. However, including a more diverse consumption pattern and requirements for agrobiodiversity leads to more realistic results and should also be considered in the optimization model. The diversity constraint was set in this study by fixing the consumption of certain crops under each food category such that: 45 % from the required cereal consumption would have to be made up by sorghum, 25 % by maize, 30 % by wheat. Under the required pulse consumption, 30 % would have to come from greengram, 30 % from bengalgram, and 40% from horsegram (see also Ralevic et al., 2010).

The cropping pattern also illustrates that mixed cropping or more diverse cropping provides the necessary balanced diet for a household (Gerbens-Leenes et al., 2002; Gerbens-Leenes and Nonhebel, 2005, 2002). In this study, the results should be taken to mean that a certain proportion of different food groups should be cultivated, but not necessarily the

---

[2] Constraints requiring certain proportion of cereals and certain proportion of pulses to be consumed by the household.

exact mix of species in land area as suggested. For instance, using data from Table 6, consider a household having two hectares of land and one cow that wishes to satisfy nutritional requirements and be self-sufficient, given the current yields. This household could cultivate mostly cereals in the *rabi* season, and a mixture made up substantially of pulses and commercial crops on the reduced amount of land cultivated during the *kharif* season. The specific choice of grains and legumes could depend on the farmers' preference.

To our knowledge no studies could be found that calculate the minimum land area for the Indian household production system in the dryland region applying optimization methodology. Previous optimization studies (Parikh and Ramanathan, 1999; Kanniappan and Ramachandran, 1998; Raja *et al.*, 1997; Painuly *et al.*, 1995; Singh and Marsh, 1994; Parikh, 1985; Parikh and Kromer, 1985) did not optimize for minimum land area.

Present day pattern of land distribution within the sampled villages has very important implications in achieving food security. Seventeen percent of the population is landless and therefore are highly vulnerable in terms of undernutrition. And there is great disparity in terms of landholdings amongst the peasant farmers. The average *per capita* land area for large landholdings at 1.41 ha is nearly 12 times larger than the 0.12 ha/capita available to marginal household occupants. In fact, 14% of the village population representing larger landholdings on average owns 65 % of the land area within the villages. If a comprehensive land redistribution were undertaken, the average available land per household (including the presently landless community) is 2.4 ha. As can be seen in the data provided here, this would be sufficient to provide for basic food security for the entire village population.

5. Other strategies for increasing food security

5.1 Increasing cropping intensity

As is clear from the data reported in this paper, one potential avenue for enhancing food security is to increase the intensity of cultivation. In Madapur, for example, the cropping intensity was only 118%, owing to the lack of adequate rainfall during certain periods of the year and the limited access to supplemental irrigation as described above. A substantial increase in year-round cropping as is practiced in Inamhongal would make possible the production of greater quantities of grains and pulses. The environmental consequences resulting from the enhanced intensity of cropping (Kuniyal, 2003) must, however, be critically examined and evaluated further.

If the cropping intensity were to be 200% in all villages and under an optimal cropping pattern, this could reduce the land area required to achieve food security within the villages by approximately 1710 ha (of 3625 ha), or 47%. For example, in the case of Madapur, increasing the cropping intensity from 118% to 200% could reduce the required land area under an optimal cropping pattern by nearly half, 47%, of total present cultivable land area.

5.2 Increasing yield

The low yields in the dryland semi-arid regions of southern India can be attributed to what Singh *et al.* (2009) refer to as inadequate traditional management practices coupled with the erratic and highly variable inter-annual precipitation. According to research from the

International Crops Research Institute for the Semi-Arid Tropics (ICRISAT), field experiments show that when improved watershed management and natural resource management is integrated in with agriculture, such that land is properly graded and the seedbed prepared so that rainfall has adequate time to infiltrate the soil, the yield of crops can be more than doubled (Singh *et al.*, 2009). If yield were to be doubled, this would reduce the minimum land area that is required to satisfy the human nutritional demands and up to 3 LSUs for a marginal household from 1.8 ha (present) to 0.9 ha (CI = 164%). While the reduction is still not adequate to ensure food security for marginal households, there is a clear nutritional gain from increasing yield.

6. Implications for biofuel production

Given the present difficulty of meeting household food and fodder demand, especially for marginal and small households, it is important to discuss the potential implications of biofuel crops.

There is growing concern that diversion of agricultural land for biomass plantations or the direct conversion of food to fuel could lead to decreased availability of land for food production, particularly among low-income countries (Boddiger, 2007; Hellengers *et al.*, 2008; Ignaciuk *et al.*, 2006; Peters and Thielmann, 2008). Additionally, diversion of land for biofuels crops, when done without proper assessment, can lead to food shortages and increased costs of staple crops such as maize and rice (Koh and Ghazoul, 2008). Nonhebel (2005) also showed that in developing countries there is insufficient land to meet the needs for both food and energy when biomass plantations are substituted for arable land. While this study does not directly assess the impact of land for biofuels conversion, it is apparent from the lack of land to meet basic human and livestock demands that land diversion for biofuel crops could lead to further food insecurity, especially among marginal and small landholdings. While large landholdings could potentially support biofuel production, the reduction in by-products such as fodder would directly impact lower income households who depend on the surplus fodder from larger landholdings.

7. Acknowledgments

The authors are grateful to the Centre for Energy within the Ontario Centres of Excellence, and the School of Environmental Studies at Queen's University for funding the field work. The authors are also thankful to Mr. S. Benoit, Dr. U.K. Shanwad and Dr. Debali for their assistance in the planning and logistics regarding survey implementation, and the field assistants and graduate students from the University of Agricultural Sciences, Kiran T.R., Ramesh N., Hemanth, Shwetha B.N., K.M. Raghu, Sidram, V. Yadahalli and M. Kulkarni for their careful and at times difficult data collection. Many thanks are also extended to village accountants and government officials for their assistance in data collection. We are also very grateful to the hundreds of farmers who participated in this study, and the village communities as a whole for their exceptional hospitality.

8. Glossary

Variables (capital letters)

$L_{cv,s1}$ = calculated land area of crop variety cv in the *kharif* season (hectares)

$L_{cv,s2}$ = calculated land area of crop variety cv in the *rabi* season (hectares)

CI_R = cropping intensity ratio

Running index (subscripts)

cv = crop varieties cultivated in the study area: 1. sorghum, 2. wheat, 3. Maize, 4. pearl millet, 5. greengram, 6. redgram, 7. horsegram, 8. bengalgram, 9. groundnut, 10. sunflower, 11. safflower, 12. chilli, 13. onion, 14. cotton.

$s1$ = season in which crops are cultivated: s1. *kharif*

$s2$ = season in which crops are cultivated: s2. *rabi*

R = cropping intensity: 1. Ratio of 2 (or 200%), 2. Ratio of 1.64 (or 164%), 3. Ratio of 1.50 (or 150%).

9. References

Alston, J.M., Beddow, J.M., & Pardey, P.G. (2009). Agricultural research, productivity and food prices in the long term. *Science,* Vol. 325, pp. 1209-1210.

Babu, S.C., Thirumaran, S., & Mohanam, T.C. (1993). Agricultural productivity, seasonality and gender bias in rural nutrition: Empirical evidence from South India. *Social Science and Medicine,* Vol. 37, pp. 1313-1319.

Boddiger, D. (2007). Boosting biofuels crops could threaten food security. *The Lancet,* Vol. 370, pp. 923-924.

FAO (Food and Agriculture Organization of the UN). (2011). Country Profile - India. Available at http://www.fao.org/countryprofiles/index.asp?lang=en&ISO3=IND.

FAO (Food and Agriculture Organization of the UN). (2009). The state of food insecurity in the world 2009: Economic crisis- impacts and lessons learned. Rome, Italy. Available at ftp://ftp.fao.org/docrep/fao/012/i0876e/i0876e.pdf.

Gerben-Leenes, W., & Nonhebel, S. (2005). Food and land use. The influence of consumption patterns on the use of agricultural resources. *Appetite,* Vol. 45, pp. 24-31.

Gerbens-Leenes, P.W., & Nonhebel, S. (2002). Analysis: Consumption patterns and their effects on land required for food. *Ecological Economics,* Vol. 42, pp. 185-199.

Gerbens-Leenes, P.W., Nonhebel, S., & Ivens, W.P.M. (2002). A method to determine land requirements relating to food consumption patterns. *Agriculture, Ecosystems and Environment,* Vol. 90, pp. 47-58.

GOK, 2005. Maps - Districts of Karnataka. Available at http://www.mysterytrails.com/DistrictsOfKarnataka_Map.htm. (accessed 10.05.09.).

Gopalan, C., Sastri, B.V.R., & Balasubramanian, S.C. (1996). Nutritive value of Indian foods (7th ed.). Hyderabad, India: National Institute of Nutrition.

Hellengers, P., Zilbermann, D., Stedto, P., & McCornick, P. (2008). Interactions between water, energy, food and environment: evolving perspectives and policy issues. *Water Policy,* Vol. 10, pp. 1-10.

Ignaciuk, A., Vohringer, F., Ruijs, A., & van Ierland, E.C. (2006). Competition between biomass and food production in the presence of energy policies: a partial equilibrium analysis. *Energy Policy,* Vol. 34, pp. 1127-1138.

Kanniappan, P., & Ramachandran, T. (1998). Optimization model for energy generation from agricultural residue. *International Journal of Energy Research,* Vol. 22, pp. 1121-1132.

Koh, L.P., & Ghazul, J. (2008). Biofuels, biodiversity and people: Understanding the conflicts and finding opportunities. *Biological Conservation,* Vol. 141, pp. 2450-2460.

Kuniyal, J.C. (2003). Regional imbalances and sustainable crop farming in the Uttaranchal Himalaya, India. *Ecological Economics, Vol.* 46, pp. 419-435.

McIntyre, B.D., Bouldin, D.R., Urey, G.H., & Kizito, F. (2001). Modeling cropping strategies to improve human nutrition in Uganda. *Agricultural Systems,* Vol. 67, pp. 105-120.

Nonhebel, S. (2005). Renewable energy and food supply: will there be enough land? *Renewable and Sustainable Energy Reviews* Vol. 9, pp. 191-201.

NNMB. (2002). Diet and nutritional status of rural population. National Institute of Nutrition 2002. NNMB Technical Report No.21.

Painuly, J.P., Hemlata, R., & Parikh, J. (1995). A rural energy-agriculture interaction model applied to Karnataka state. *Energy* Vol. 20, pp. 219-233.

Parikh, J.K. (1985). Modeling energy and agriculture interactions – I: A rural energy systems model. *Energy,* Vol. 10, pp. 793-804.

Parikh, J.K., & Kromer, G. (1985). Modeling energy and agriculture interactions- II: Food-fodder-fuel-fertilizer relationships for biomass in Bangladesh. *Energy,* Vol. 10, pp. 805-817.

Parikh, J.K., & Ramanathan, R. (1999). Linkages among energy, agriculture and environment in rural India. *Energy Economics,* Vol. 21, pp. 561-585.

Peters, J., & Thielmann, S. (2008). Promoting biofuels: Implications for developing countries. *Energy Policy,* Vol. 36, pp. 1538-1544.

Ralevic, P., Patil, S.G. & vanLoon, G .(2010). Integrated agriculture production systems for meeting household food, fodder and fuel security. *Journal of Sustainable Agriculture,* Vol. 34, pp. 878-906.

Raja, R., Sooriamoorthi, C.E., Kanniappan, P., & Ramachandran, T. (1997). Energy planning and optimization model for rural development- A case of sustainable agriculture. *International Journal of Energy Research,* Vol. 21, pp. 527-547.

Singh, S., & Marsh, L.S. (1994). Optimization of biomass energy production in a village. *Biomass and Bioenergy,* Vol. 6, pp. 287-295.

Singh, S.P., Gangwar, B., & Singh, M.P. (2007). Farming systems diversification: a study on marginal holders in western Uttar Pradesh. *Agricultural Economics Research Review,* Vol. 20, ISSN 0971-3441

Singh, P., Pathak, P., Wani, S.P., & Sahrawat, K.L. (2009). Integrated watershed management for increasing productivity and water-use efficiency in semi-arid tropical India. *Journal of Crop Improvement,* Vol. 23, pp. 402-429.

Suryanarayana, M.H. (1997). Food security in India: Measures, norms and issues. *Development and Change,* Vol. 28, pp. 771-789.

UAS (University of Agricultural Sciences). (1985). *Package of Practices for high yields.* Dharwad, India: University of Agricultural Sciences.

von Braun, J., Ruel, M., & Gulati, A. (2008). Accelerating progress toward reducing child malnutrition in India: A concept for action. (Research Brief 12) Washington, D.C.: International Food Policy Research Institute (IFPRI).

8

Enhanced Food Production by Applying a Human Rights Approach – Does Brazil Serve as a Model of Best Practice?

Hans Morten Haugen
Diakonhjemmet University College, Oslo
Norway

1. Introduction

The ignorance of agriculture in development policy circles that prevailed in most of the 1980s and 1990 has finally come to an end. Agriculture, including the right use of the land, is considered central for the following objectives:

a. Overcoming poverty, as GDP growth from agriculture is found to be overall twice as effective in reducing poverty as GDP growth derived from other sectors; and GDP growth originating in agriculture benefits the poorest half of the population substantially more (World Bank, 2007, 6)
b. Mitigating climate change, by promoting 'low greenhouse gas agriculture' (Food and Agricultural Organization (FAO), 2009, Nellemann et al. 2009, IPCC 2007, chapter 8, IAASTD, 2009, 538).
c. Reducing imports of food, in order to avoid becoming vulnerable to volatile prices on the world market; note in this context that Brazil, which relies on a global market for food, recognizes the concept of food sovereignty, most notably in Article 5 of the Law 11.346 of 2006 on a National Food and Nutrition Security System (SISAN), which reads (extract): "countries must be guaranteed the supremacy of their own decisions regarding food production and consumption."
d. Enhancing energy security, as in the case of Brazil, where ethanol from sugarcane has bypassed petrol as the dominant fuel, while the sugarcane currently occupies less than one per cent of (the best) agricultural lands in Brazil.

It might be difficult to achieve all these objectives simultaneously, and in particular the latter seems difficult to reconcile with the three former. FAO, however, promotes integrated food-energy systems (Bogdanski et al. 2011), which makes frequent references to Brazil, including Brazil's processing flexibilities in producing either food (sugar) or fuel (ethanol), depending on the overall supply and market conditions.

This chapter analyzes Brazil, primarily as its efforts towards enhanced food production has been widely endorsed simultaneously as the human right to food has been strengthened, both through legal amendments and specific programmes. Brazil's policies are referred to when other states' agricultural and social policies are reviewed (Economist, 2011a; see also

Economist, 2010a). At the same time, the expansion of agriculture has taken place in a way that severely threaten vulnerable ecosystems and indigenous peoples, also outside of the Amazon. The overall question for the chapter can be formulated as: *"To which extent are the Brazilian policies for enhanced food production based on a human rights approach and to which extent is the overall Brazilian agricultural strategy recommendable to other developing countries?"*

On this background, Brazil's policies are analyzed in light of requirements of international human rights treaties and national legislation. Relevant human rights provisions are interpreted in light of "any subsequent practice in the application of the treaty which establishes the agreement of the parties regarding its interpretation", as specified by Article 31.3(b) of the Vienna Convention on the Law of Treaties (United Nations, 1969).

The chapter will proceed as follows. First, the specifics of the three Brazilian agricultural models will be outlined, and the specific measures to promote them will be analyzed. Then, there will be an outline of the relevant legislation in Brazil, both concerning food production, fuel production and land issues. These laws will be seen in relation to relevant provisions from the International Covenant on Economic, Social and Cultural Rights ('ICESCR', United Nations, 1966a), which is the international treaty that recognizes both everyone's human right to food and peoples' rights to natural resources, and prescribes specific measures relating to food production, in Articles 1 and 11, respectively. Relevant clarifications have been done in General Comment 12 (CESCR, 1999) and in Voluntary Guidelines to Support the Progressive Realization of the Right to Adequate Food in the Context of National Food Security ('Voluntary Guidelines') (FAO, 2004). Finally, the negative consequences of the rapid growth of the Brazilian agricultural area will be reviewed, based on an assessment of the availability of arable land, forests and water, and the livelihood of indigenous peoples living outside of the Amazon.

2. Brazil

This section will map the central characteristics of Brazil, starting with an outline of the most pertinent concerns in the Brazilian agriculture, emphasizing the tensions between agricultural expansion and forest preservation, before moving on to a description of the three agricultural models that are found in Brazil. The most relevant legislations will be presented, first applying to agriculture and forests and then to the three agricultural models.

2.1 Agricultural expansion in Brazil

The Brazilian dilemmas are related to several facts. First, the inherited agricultural model is one in which the inequalities in land ownership is one of the most unequal in the world. This reflects a historical pattern of unequal distribution that no Brazilian government has been able to change; the Gini coefficient for land distribution was 0.857 in 1996 and 0.872 in 2006 (IPSNews 2009). It must, however, be recognized that the Gini coefficient regarding income distribution has been reduced from 0.607 in 1998 (CIA Fact Book) to 0.554 in 2008 (Sua pesquisa (no date)). Second, the most profitable and productive crops are plantation crops, most notably soy, sugarcane and palm oil. Third, increased mechanization of plantations will make much manual labor abundant, hence creating social insecurity. Fourth, while Brazil has available degraded and underutilized land, the pressure on original forests in the Amazon and other vulnerable biomes continues and policies for restricting agricultural use of these areas are not efficient, leading the authors to observe: "Public and private sector players in Brazil and neighboring countries now recognize that agricultural investment and expansion pose serious environmental challenges" (Deininger and Byerlee, 2010, xxix).

Moreover, deforestation has also been a concern for the Committee on Economic, Social and Cultural Rights (CESCR), recommending Brazil to "take the necessary measures to combat continued deforestation in order to ensure the effective enjoyment of economic, social and cultural rights, especially by indigenous and vulnerable groups of people" (CESCR, 2009, para. 26). This was actually the only paragraph among the CESCR's concerns which addressed food, which might be interpreted to imply that the CESCR did not view other aspects of Brazil's policies in the realm of food as equally critical for realizing the right to food in Brazil. Among the 'positive aspects' of Brazil's policies, the National School Food Programme is explicitly referred to (ibid, paragraph 3(d)).

The United Nations' Special Rapporteur on the right to food has stated that he "was impressed by the level of commitment of Brazil..." (Special rapporteur on the right to adequate food, 2009a, paragraph 32). The same Special rapporteur has recently addressed issues relating to land acquisition (2009b, 2010b) and the potentials of agroecology and the productivity gains by investing in small-scale agriculture (2010a). These approaches might seem to be at odds with the prevailing perception of the form of agriculture practiced in Brazil. The perception is that the Brazilian agricultural model is characterized by large units, high mechanization and monoculture.

2.2 Three agricultural models

While Brazil's diversity makes it difficult to make adequate categorizations, it is nevertheless possible to identify three agricultural models: agroindustry, family farming, and traditional harvesting and low scale-agriculture.

Agroindustry dominates in terms of area, covering approximately ¾ of all agricultural area in Brazil (Brazilian Ministry of Agrarian Development, 2009, 4) and representing 62 per cent of (registered) gross production value (ibid, 5). The average size of agroindustrial units, as reported in the 2006 Agricultural and Livestock Census, was 309.18 ha (ibid, 5). Overall agricultural policies are implemented by the Ministry for Agriculture, Livestock and Food Supply (MAPA). Much of this production is exported. Agricultural exports represent 35 per cent of Brazil's export earnings, and Brazil is the world's largest exporter of ethanol, sugar, coffee, chicken and beef – and the second largest exporter of soy. This is also the result of agricultural scientific efforts, primarily Embrapa (Brazilian Agricultural Research Corporation), which has resulted in an increase of average yield per hectare from 1528 kg in 1990 to 2850 kg in 2006 (Embrapa Agrienergy (no date)).

Family farming is crucial for domestic food security. Moreover, family farms represent 84.4 per cent of all (registered) farms (Brazilian Ministry of Agrarian Development, 2009, 4). The number of units has increased by approximately 10 per cent from 1996 to 2006, to more than 4.5 million (ibid, 10). The distinction between a family and a non-family farm is four fiscal units. The size of one fiscal unit varies according to agricultural conditions (National Institute for Colonization and Agrarian Reform (INCRA), 1980) and four fiscal units can represent anything between 20 and 440 ha. Hence, in some parts of the Amazon, a farm can be more than 400 ha and still be recognized as a family farm. In order to promote family farming, a separate Ministry for Agrarian Development (MDA) was established during the Cardoso Government, and the National Family Farming Program (PRONAF) has been in operation since 1995. PRONAF has several components, like the More Food Program, which has improved access to credit for the purchase of agricultural technology and equipment.

Finally, traditional agriculture and harvesting has also been promoted. The Brazilian government acknowledges that traditional communities, being both indigenous peoples and quilombolas (descendents of slaves) live in an area covering almost one quarter of Brazil (Government of Brazil, 2010), The total number of persons belonging to indigenous peoples is estimated to be 700.000. The number of persons belonging to quilombola communities is approximately 2 million, and there are also other traditional communities (Santilli, 2009, 191), but only indigenous and quilombola communities are entitled to collective land rights. Presently, 1711 quilombola communities are recognized (Fundação Cultural Palmares, 2011). By 2011 there were 343 areas for indigenous communities and 87 qiulombola territories which were *registered* (International Labour Organization (ILO), 2011, 790), but this does not imply full regularization and titling. The process on land demarcation is a potentially very long and complex one, and consists of four phases: Identification or delimitation (by the National Indigenous Foundation (Funai); declaration (by Ministry of Justice); demarcation (boundaries are clearly marked in the terrain); homologation (by the President) (Brazil, 2008, 20, note 11). Specifically on food production, Embrapa has initiated more than 20 projects – in coordination with Funai – to seek to improve the productivity of indigenous peoples' agriculture. Moreover, the Ministry for Social Development and Combat of Hunger (MDS) – by social safety programmes – seek to enhance the food security among indigenous peoples.

In summary, while the agroindustry is prevailing and will continue to expand, it must also be acknowledged that the Brazilian authorities have sought to strengthen also the two other agricultural models. This is a challenge to Paul Collier, who sees large-scale industrial agricultural as the only solution to the world's hunger crisis (Collier 2008). It is therefore reasonable to state that Brazil has a comprehensive approach towards agricultural productivity in general, but also that there is almost a separation between the various agricultural models, illustrated by the fact that they sort under different ministries.

As the general obligation provision of the ICESCR specifies "the adoption of legislative measures" (ICESCR, Article 2.1), it is highly relevant to review relevant Brazilian legislation, applying to each of the three models.

2.3 Legislation applying to agricultural expansion vs forest preservation

As we saw above, the CESCR has expressed concerns for deforestation in Brazil. The basis for this concern is that those persons and communities which depend on the harvesting from the forest will be negatively affected by the conversion of forests to other uses. Hence, before analyzing legislation applying specifically to the three agricultural models, the most relevant laws regarding forests will be briefly analyzed.

Specifically on the Amazon, Act 11.952 of 2009 regularized properties which had been occupied by farmers before 2004, even if the occupations and subsequent utilization of the relevant property were illegal. If the property is smaller than 100 hectares, it will be given for free, and if it is between 100 and 1500 hectares it will be sold at subsidized rates. Larger properties than 1500 hectares would have 'only' 1500 hectares regularized. As stated in a World Bank study "this law could encourage speculative land occupation and deforestation in expectation of future regularization" (Deininger & Byerlee, 2010, 121).

Applying to the whole of Brazil, amendments to the Forest Act (Codigo Florestal) were approved by the Lower House in May 2011. The basic elements of these amendments are to provide for an amnesty for forest clearance beyond what is allowed for each property – but only for cutting before July 2008 – and to reduce the proportion of each property that must be preserved. Increased deforestation as a result of this legislative proposal has been reported by the Brazilian Environmental Agency (IBAMA) (The Rio Times 2011). The proportions under the current Act are 80 % in the biological Amazon; 35 % in the Cerrado within the legal Amazon (9 states); and 20 % in rest of Brazil. It is hard to get figures on the number of persons who have been accused of violating the Forest Act by clearing more forests than permitted, but the mere need for an amnesty for those involved in illegal clearance shows that the law enforcement is not effective.

In this context of facilitating deforestation, a reference can be made to Article 186 of the Brazilian Constitution, which specifies the social function of rural land, with four requirements that must be met simultaneously: "I. rational and adequate use; II. adequate use of available natural resources and preservation of the environment; III. compliance with the provisions which regulate labor relations; IV. exploitation which favors the well-being of the owners and workers." While the emphasis is on the use of the land, the social function cannot be met by removing forests, as this does not meet requirement II.

Cattle raising is the type of agriculture which is most threatening to the vulnerable biomes in the Amazon and the Cerrado (south and east of the Amazon) for the time being. This form of agriculture has low productivity per hectare, but is a most effective way of gaining control over land. Moreover, the Rural Real Estates Act No. 4,504 of 1964 introduced the phrase temporary possession ('posse temporários da terra'), enabling anyone to obtain a title of land simply by starting to use this land. Article 97.II specifies that anyone who has served for a year the claimed land can get a title to this land. The 1988 Constitution, on the other hand, establishes in Article 191 a higher requirement of the duration of use:

The individual who, not being the owner of rural or urban property, holds as his own, for five uninterrupted years, without opposition, an area of land in the rural zone, not exceeding fifty hectares, making it productive with his labour or that of his family, and having his dwelling thereon, shall acquire ownership of the land.

The difference between the two provisions is that the former requires purchase ('adquirir'), while the latter simply says that five years uninterrupted an unopposed use enables one to acquire the land. The whole process of obtaining documents is characterized by non-transparency, fraud and legal uncertainty. According to a study, private possession of 100 million hectares of land in the legal Amazon – which represents 509 million hectares – is based on fraudulent documents (Wilkinson et al., 2010, 15; quoted in Sauer and Pereira Leite, 2011, 4).

The vanguards in any land occupation activities are the so-called 'grileiros' or land grabbers, but these are not the only actors driving the Brazilian deforestation (Fearnside 2008). The name grileiros is derived from an insect that can do damage to paper documents so that they appear older than what they actually are, in order to facilitate the falsification of land title documents. It is fair to say that the grileiros are inadequately regulated, and that even violence and murder committed by them is underinvestigated and unpunished.

2.4 Legislation applying to the three agricultural models

Regarding the agro-industry model it is also relevant to notice that the social function of rural land provision of Brazilian Constitution has been a basis for a program of expropriation of unproductive lands, transferring some of it to landless persons and communities. There will be different opinions on how comprehensive this policy of land distribution policy is. As we have seen, the Gini coefficient for land distribution in Brazil illustrates the skewed distribution, and INCRA finds that 120 million hectares – or 27 per cent of all agricultural lands in Brazil – is unproductive (Sauer and Pereira Leite, 2011, 5). Finally, of relevance for understanding how Brazil's authorities facilitate large-holdings, Brazilian law has never defined any maximum size for properties.

Addressing family farms, several laws have recently been adopted. These laws provide for enhanced public purchases from family farms, including to the school feeding programme (PENAI; law 11,947 of 2009) and the national biodiesel programe (PNPB; law 11,097 of 2005). The latter establishes a 'social fuel seal', given to purchasers that buy a given share of their total input for biodiesel production, varying between 15 and 30 per cent, depending on the region. Processors without this seal will not be allowed to participate in the regular auctions for biodiesel, where 80 % of the national demand is purchased at slightly higher prices. Moreover, a Family Farming Act (11.326 of 2006) has been adopted, as specified by Decree 6.882 of 2009.

The policy for promoting purchase from family farms within the national biodiesel programme was a response to the situation in the bioethanol sector, which is dominated by sugarcane large-holdings. The national biodiesel programme sought to facilitate the gradual expansion of family farming and avoid that the introduction of biodiesel resulted in more plantations and monocropping. While the volume targets for the purchase have not fully been met on a national level, it is still possible to talk of a gradual integration of family farming into the national biodiesel programme. Section 4 will analyze in greater detail whether this shift represents a challenge for the production of food and food security.

Also for the traditional model of agriculture, several laws have been enacted recently. The seed law 10.711 of 2003, as specified by Decree No. 5.153 of 2004, exempts traditional cultivars from the requirements in Article 11, saying that "production, processing and marketing of seeds and seedlings are conditioned to prior registration" in the National Register for Cultivars (RCN), and Article 48 prohibits any restrictions on the distribution and exchange of such traditional cultivars.

In order to reduce the incentives for deforestation, the sociobiodiversity programme of the National Food Supply Company (Conab) includes the 'Policy of Minimum Guanteed Price', established by law 11,775 of 2008, applying to everyone who harvests traditional food from the forests. By ensuring such a minimum price it is believed that persons are more motivated to preserve and harvest from the forests than to destroy and convert the forests. This Act must be considered to be a good example of how Brazil seeks to promote social inclusion and reduce the incentives for deforestation by those living in or from the forests.

There are, however, serious problems facing many of the indigenous peoples living outside of the biological Amazon, as their traditional territory has been subject to agricultural expansion, and if territories have been demarcated at all, they are too small to maintain an adequate standard of living. This is despite efforts by Funai and the Federal Prosecution Office to strengthen the land rights of indigenous peoples in accordance with Articles 231 and 232 of the Brazilian Constitution, the latter with a mandate in Article 129.V of the Constitution to defend the rights and interest of the Indian populations.

While a full review of all relevant legislation in Brazil is obviously not possible, this section has highlighted the laws which are believed to be the most relevant in order to understand the Brazilian agricultural system, and also provided some insights into the context in which they operate. The section has not addressed the laws which regulate the agricultural cooperatives, which are important, in particular in the South, and neither has the section addressed the laws relating to the marketing of agricultural products. It is found that the laws have not been adequately effective in order to ensure that forest land is not converted, or that non-productive land is effectively used, but that they have been used to promote family farming and traditional harvesting and seed exchange.

3. Substantive human rights and human rights principles

Before introducing the most relevant human rights for the purpose of the analysis, a brief outline of the most central approaches for understanding human rights, particularly economic, social and cultural rights, will be provided. The most innovative aspect of this section is nevertheless believed to be the analysis of human rights principles, which are highly relevant for assessing the overall conduct of the relevant policies, and hence the quality of the day to day policies.

3.1 Approaches for understanding economic, social and cultural human rights

While most central characteristics apply to all human rights, and there should be no artificial distinction between the various types of human rights, certain approaches have been developed that apply particularly to economic, social and cultural rights. These approaches have been introduced by legal scholars in order to get a better understanding of primarily the state obligations derived from the economic, social and cultural rights. These approaches have subsequently been applied by UN human rights bodies, international organizations and states.

The typology of respect, protect and fulfill (facilitate and provide) is the most frequently applied (Eide 2007). Under this approach, the state shall avoid interfering in the enjoyment of any of the recognized rights, but also preventing others' interference. The state is, moreover, obliged to undertake measures in all relevant policy spheres for the effective realization of the human rights, and – if the situation requires – ensure that the necessary goods and services are provided to those who have no other opportunity to exercise the human rights by their own efforts. No provision of such goods and services should continue beyond what is necessary to ensure enjoyment of human rights (CESCR, 1999, paragraph 39).

As specified by Welling (2008, 950-953), one can also apply a core content approach, seeking to identify the core of the specific human rights provisions (CESCR, 1991, paragraph 10), or the analytical categories approach, focusing on structure, process and outcome. Hence, an adequate understanding of the societal context affecting the realization of the human rights, and the conduct that is applied for the realization of human rights are most relevant for the actual realization of human rights.

When analyzing the right to food as a social human rights, it is relevant to note how the CESCR understands the realization of – and the obligations deriving from – this and other rights recognized in the ICESCR, stating that:

"the obligation differs significantly from that contained in article 2 of the International Covenant on Civil and Political Rights ['ICCPR', United Nations, 1966b] which embodies an immediate obligation to respect and ensure all of the relevant rights. Nevertheless, the fact that realization over time, or in other words progressively, is foreseen under the Covenant should not be misinterpreted as depriving the obligation of all meaningful content (ibid, paragraph 9 (extract)."

The CESCR itself identifies two obligations which are of immediate effect, namely to ensure that the rights are exercised without discrimination and to take "deliberate, concrete and targeted" measures for the realization of the ICESCR (ibid, paragraphs 1 and 2). These two obligations of immediate effect can be read out of the ICESCR Article 2.2 and Article 2.1, respectively. We will come back to these provisions below.

3.2 Substantive human rights

While criticisms are raised against the human right to food (Economist, 2010c, 2011a), this chapter will not be used to discuss whether the right to food actually exists. The strong and unanimous endorsements by various bodies of the United Nations of the reports by the Special Rapporteur on the right to food (Special rapporteur 2009b, 2010a, 2010b), confirm that the states consider the right to food to be a human right. Both Article 11.1 and Article 11.2 of the ICESCR recognize the right to food, the latter specified as the right to be free from hunger. In brief terms, the right to be free from hunger can be understood as the core content of the right to food. As Article 11.2 is more comprehensive, the emphasis will be on this provision, which is also justified by the simple fact that hardly any states can claim that none of its inhabitants are facing hunger, hence claiming that the corresponding obligations are not applicable.

Article 11.1 embeds the right to food in the right to an adequate standard of living, including "the continuous improvement of living conditions." It is reasonable to state that while the latter element of the right applies generally, the obligations derived from this right must be particularly observed with due concern for the most marginalized persons (Haugen, 2007, 123).

Article 11.2 has two subparagraphs; the latter applying to international trade (Haugen, 2009, 271-271) and the former specifying that the states shall take measures which are needed:

To improve methods of production, conservation and distribution of food by making full use of technical and scientific knowledge, by disseminating knowledge of the principles of nutrition and by developing or reforming agrarian systems in such a way as to achieve the most efficient development and utilization of natural resources.

The qualification in the introductory part of Article 11.2 saying that the measures are 'needed' cannot be interpreted to reduce the emphasis of the mandatory term 'shall'. Both the fact that the listed measures fall within highly relevant policy spheres, the fact that only one state, Pakistan, abstained from the vote when Article 11.2(a) was adopted (Haugen, 2007, 128, note 64), and the fact that the drafters decided – after a discussion – not to include the term 'necessary measures' in the introductory paragraph of Article 11.2 (ibid, 128-129) demonstrates that the states cannot avoid considering these measures.

A previous chairperson of the CESCR has identified the structure of this provision in the following manner, after acknowledging that the paragraph "is a relatively confused and by all means no all-embracing mixture of means and ends" (Alston, 1984, 23). He finds that in Article 11.2(a) there are three objectives:

1. to improve methods of food production;
2. to improve methods of food conservation;
3. to improve methods of food distribution.
4. Moreover, the following measures are identified: making full use of technical and scientific progress;
5. disseminating knowledge of the principles of nutrition;
6. developing or reforming agrarian systems.

Furthermore, the final phrase 'in such a way as to achieve the most efficient development and utilization of natural resources' applies to the whole paragraph, and not only to the last part of it (ibid, 35).

Three elements of this paragraph will now be analyzed: the relationship between improved production and improved distribution; the reform of agrarian systems; and the development and utilization of natural resources.

3.2.1 The relationship between improved production and improved distribution

First, while realization of the right to food by necessity depends on the fact that enough food is actually being produced, the relationships between food production and consumption is not necessarily straightforward, even if one does not consider agricultural exports and imports. The CESCR addresses these complex relationships by introducing the terms availability and accessibility. Improved methods of food production – in other words increased food *availability* (CESCR, 1999, paragraphs 8 and 12) will be beneficial for the realization of the right to food provided that the most relevant conditions are in place in order to facilitate food *accessibility* (ibid, paragraphs 8 and 13). In addition to food prices, these conditions relate to ownership structures, or in the words of the CESCR "acquisition pattern or entitlement", identifying a particular vulnerability among "indigenous population groups whose access to their ancestral lands may be threatened" (ibid, paragraph 13). The Brazilian Constitution recognizes indigenous peoples' property rights in Articles 231 and 232.

Moreover, measures for enhanced ecological sustainability, better infrastructure, market systems and adequate public extension services in order to provide new and improved seeds, are also crucial for food availability and accessibility. Much more emphasis should be placed on the adequate distribution of food-producing resources (Haugen, 2007, 140), including by various forms of participatory breeding involving the farmers themselves. Stated differently, by improving food production in a socially inclusive manner, this will enhance the food self sufficiency of local communities. This will in turn reduce vulnerabilities and dependencies. It must be noted, however, that study by the International Fund for Agricultural Development (IFAD) also finds that "[w]here farmers are consulted about their priorities, they often select priorities other than yield" (IFAD, 2001, 136).

No specification of what is understood by the phrase "improve methods of production ... of food" is provided by General Comment 12, including which food producing technologies that are most appropriate, but rather states the obvious that production of food is crucial for the food system (CESCR 1999, paragraph 25). Hence, it does not specify how new agricultural products are to be developed and made available. Improved agricultural productivity is an obvious objective for all involved in food production, and the yield gap identifies the gap between the current and the potential yield. The yield gap differs considerably between various regions (Neumanna et al., 2010). When identifying policies that could close the yield gap, FAO's High Level Panel of Experts on Food Security and Nutrition identified four interventions: revitalized extension services, improved markets, strengthened property rights and enhanced infrastructure (FAO, 2011, 25). Obviously, all of these measures require high-quality public measures, relating to legislation, institutions and budget priorities. It is fair to say that Brazil is relatively advanced in these sectors, illustrated by the expansion of the MDA's National Rural Extension Policy, initiated in 2003 (Special Rapporteur on the right to food, 2010a, 15).

3.2.2 Reform of agrarian systems

Reforming agrarian systems is the second element of ICESCR Article 11.2(a) to be analyzed, and applies to land reform and strengthened property rights. No serious attempt of clarifying this phrase is done in General comment 12, but "right to inheritance and the ownership of land and other property" is specified as elements of women's equal access to economic resources (CESCR 1999, paragraph 26; see also Voluntary Guidelines 8B). While property rights function differently in diverse socio-economic contexts, the following observations must be considered to apply across these different contexts:

"Vulnerable groups suffer most from a lack of property rights. Indigenous peoples are frequently victims of property discrimination; collectively held indigenous lands have often been declared public or unoccupied lands (and collectivity can be retained in formalizing property rights). Women, who constitute half of the world's population, own very little of the world's property – as little as two percent in some countries (Commission on the Legal Empowerment of the Poor, 2008, 36; see also Eide, 2009)."

We see that collective ownership is emphasized. In addition to the property rights to land, there must also be a regulation of the natural resources found on – and under – the land. In most states, mineral and other sub-surface resources are regulated differently from other resources, as illustrated by the Brazilian Constitution Article 231.3. Moreover, as seen in Section 2.2 above, the whole process of obtaining land ownership for indigenous communities in Brazil can be long and cumbersome, with possibilities for legally challenging each of the decisions in the various phases of the process.

The transfer of property rights from a community to an actor seeking to establish a development or investment project is also bound to be contentious. The basic approach as formulated in ILO Convention 169 concerning Indigenous and Tribal Peoples in

Independent Countries ('ILO 169', ILO 1989) Article 6.1(a) is that any such transfer must only take place after a "through appropriate procedures, and in particular through their representative institutions..." (see also UN Declaration of the Rights of Indigenous Peoples ('UNDRIP'), United Nations, 2007, Article 18). What these 'appropriate procedures are, have been sought clarified. Two complaints against two of Brazil' neighbor states, Peru and Surinam, will be provided as examples. On the one hand, the Human Rights Committee, monitoring the ICCPR, "considers that participation in the decision-making process must be effective, which requires not mere consultation but the free, prior and informed consent of the members of the community" (HRC 2009, paragraph 7.6). Three terms are central: 'effective', 'free, prior and informed consent' (FPIC) and 'the members'. An analysis of FPIC will be undertaken in the following section; now the emphasis will be on the term 'members'. As applied by the HRC the term 'members' applies to everyone in the community, hence being a comprehensive approach.

While also emphasizing effective participation and FPIC, the Inter-American Court of Human Rights (IACHR) deviates from the HRC concerning the scope of participation. The IACHR has confirmed that the customary tradition of the relevant peoples must be decisive when determining what is effective participation (IACHR 2008, paragraphs 13, 19 and 26), implying that the state must consult with those representatives that the peoples themselves have chosen. This illustrates the distinction between effectiveness and representativeness. For indigenous peoples, it must be expected that the least influential in the community are also those who are most dependent upon natural resources, and have the highest overall vulnerability. This gives a basis for questioning whether an effective participation involving only those who are said to represent the respective peoples, is adequate and able to avoid future tensions with those claiming not to have been adequately represented.

3.2.3 Development and utilization of natural resources

The phrase "development and utilization of natural resources" must be understood in light of the emerging understanding of natural resources that has been witnessed the last decades, as also specified by the World Trade Organization's (WTO) Appellate Body in US - Shrimps when clarifying the term 'exhaustible natural resources' (WTO 1998, paragraph 131). Hence, the utilization of such resources must take place in the context of the principle of sustainable development, implying that all decisions must be taken with due regard to how they affect nature and current and subsequent generations.

The term natural resources are also found in two other provisions of the ICESCR and ICCPR; both having an identical wording. First, common Article 1.2 reads (extract): "All peoples may, for their own ends, freely dispose of their natural wealth and resources..." An even stronger wording is found in Article 25 and Article 47 respectively: "Nothing in the present Covenant shall be interpreted as impairing the inherent right of all peoples to enjoy and utilize fully and freely their natural wealth and resources." By applying the term 'inherent' this must be understood to imply that the right cannot be traded away. The fact that the term 'peoples' includes indigenous peoples is confirmed by UNDRIP, Article 3.

Natural resources preservation is not addressed in detail in the CESCR's General comment on the right to food (CESCR 1999, paragraphs 12, 25, 26 and 27). As an example of a natural resource issue that has emerged recently and which has created tensions both within and between countries is how the decision-making process shall be conducted when concerning access to genetic resources and the subsequent sharing of benefits resulting from the utilization of these resources. The patenting of such resources is allowed by the International Treaty on Plant Genetic Resources for Food and Agriculture ('ITPGRFA', FAO, 2001a), Article 12.3(b), provided that these resources have been subject to a changed 'form'; as emphasized by numerous declarations by representatives of industrialized countries delivered after the adoption of the ITPGRFA (FAO 2001a). While the ITPGRFA is not a human rights treaty, farmers' rights are recognized in Article 9. Protection of traditional knowledge and rights to participate in decision-making and benefit-sharing are components of farmers' rights, and these elements are also emphasized in the Voluntary Guidelines (FAO 2004, Guideline 8D).

3.2.4 Summary, emphasizing the international and social dimension

The above analysis shows that the realization of the right to food by necessity must take into account a wide range of issues. While some of them are obviously costly, such as the extension services and improvements in infrastructure, it is also highly relevant to note that the drafters foresaw that realization of the right to food depending upon international assistance. The general obligation provision of the ICESCR, Article 2.1, emphasizes international assistance and co-operation, implying that international cooperation and assistance is central for the realization of all the recognized rights of the ICESCR, but there are explicit references to international assistance in both Article 11.1 and Article 11.2. Hence, it is reasonable to state that international assistance is particularly important for the realization of the right to food.

Particularly for the purpose of developing and providing new varieties, the international research cooperation has been successful. The rate of return from investments in national agricultural research services (NARS) is 60 per cent, and from investments in international agricultural research centers (IARC) is more than 70 per cent (von Braun et al., 2008, 10; see also World Bank, 2007, 6). Any developing country should facilitate the cooperation with these centers and provide for interaction between farmer associations and breeders and these centers.

Finally, general subsidies have been found to be less effective compared to specific provision of specific goods and social services (Special Rapporteur on the right to food 2010a, 17). Moreover, several measures can be taken which are not in themselves very costly. This includes the adoption of an adequate legislation for ending discrimination in ownership of land and other resources, or in the school system. Such measures will enhance participation and empowerment. One study finds that by ending discrimination in the school system, which is a central task for public authorities, malnutrition among children is reduced by 13.4 per cent in South Asia and 3 per cent in Africa (Smith et al. 2002), and another study finds that it leads to an increase average in economic growth with 0.4 to 0.9 per cent (Klausen 2002). This clearly shows that conscious social policies are integral to any food and food production policy.

3.3 Human rights principles

No decision by any intergovernmental body has defined or exemplified the term 'human rights principles'. One of the first appearances of the term is in General Comment 12 (CESCR

1999, paragraph 21), listing the following: dignity (ibid, paragraph 4), non-discrimination (ibid, paragraph 26), accountability, transparency, people's participation, decentralization, legislative capacity and the independence of the judiciary (ibid, paragraph 23). Subsequently, the so-called 'Common Understanding' among UN agencies gave a central role to human rights principles in the context of a human rights approach to development cooperation (United Nations Development Group, 2003; see also Haugen, 2011).

FAO has identified the following human rights principles: participation, accountability, nondiscrimination, transparency, human dignity, empowerment and the rule of law (FAO, 2007, 2). These are derived from human rights provisions and apply to all forms of conduct, and apply to development or investment actors, in addition to states (Haugen, 2011). By applying the term 'include' before this listing, FAO has open up for possible additions to this list. Both Eide (2009, 28) and Ssenyonjo (2009, 147) have applied FAO's seven human rights principles, but adding others, namely 'good governance' and 'monitoring', respectively. Good governance is the result of the careful observation of the human rights principles, while monitoring is the practical tools to ensure and improve compliance. Therefore, it is justified to keep the seven human rights principles as specified by FAO.

There will now be a brief outline of the seven human rights principles, specifying their application in the context of overall policies for food production, with examples taken from Brazil. As the dealing with indigenous peoples are likely to present the largest asymmetries, the quality of the conduct is even more crucial compared with dealing with actors and communities that are more equal in terms of power. The review will start with the most basic principles.

Dignity is the foundational basis for the constituting of both human beings and human rights. Dignity can also be applied in order to assess certain policies. As the life fulfillment of indigenous peoples in Brazil are so closely related to their territorial belonging, it cannot be considered controversial to identify the territorial expansion of agriculture which considerably reduces the living space of indigenous peoples to be a systematic challenges to human dignity of many in Brazil's indigenous peoples. During a research field trip to Brazil, including Mato Grosso do Sul in March and April 2011, the author was also informed by a representative of the agroindustry that the best way to promote the dignity of indigenous peoples was to get employment, and the author agrees that each person must be free to integrate in the larger society without being ostracized from one's indigenous community.

Four UN Special Rapporteurs, including the Special Rapporteur on the right to food, have addressed the situation for the Guarani Kaiowá in Mato Grosso do Sul (Special Rapporteur on the situation of human rights and fundamental freedoms of indigenous peoples, 2008, 19-21; see also Special Rapporteur on adequate housing as a component of the right to an adequate standard of living, 2010, paragraph 60). While acknowledging the stalemate these reports refer to, work is undertaken: six 'technical groups' have been mandated by Funai to identify Guarani Kaiowá territories in Southern Mato Grosso do Sul, and their reports are expected by the end of 2011. It must in this context be taken note of the Brazilian Constitution's emphasis on the maintenance of cultural diversity in Article 215 and 216, and the UNESCO Convention on the Protection and Promotion of the Diversity of Cultural Expressions, particularly Article 2.3, which links cultural diversity and human dignity.

Non-discrimination is about like treatment of like cases and unlike treatment of unlike case. Hence, to give preferential treatment to persons from certain categories which are substantively underrepresented or marginalized is not in violation of the non-discrimination principle – as long as this treatment ends as soon as substantive equality is achieved. Moreover, the non-discrimination prohibition applies generally to any "field regulated and protected by public authorities" and not only to the specific human rights which are recognized by the human rights treaties (HRC, 1989, paragraph 12). As persons from indigenous peoples are relatively less able to defend their rights, it is must be considered to be fully in line with the principle of non-discrimination that the Federal Prosecution Office takes a pro-active approach in accordance with the mandate given to it in Article 129.V of the Brazilian Constitution. The Funai mapping referred to above has been moving forward based on a 2007 agreement between Funai and the Federal Prosecution Office.

Rule of law is a comprehensive human rights principle, and encompasses the overall effectiveness of the judicial system and independence of judges and lawyers. It is also about applying the same standards irrespective of social, institutional and other weaknesses prevailing in certain regions. While it is difficult to single out Brazil from other states in South America, the relatively stronger ties between the various political and economical elites is a constant challenge to the quality of the court system's equal treatment of persons with and persons without formal and informal power. It must be acknowledged, however, that Brazilian courts are not necessarily acting in a restrictive manner. The Special Rapporteur on indigenous peoples observed that the Supreme Federal Tribunal's 2009 decision in the Raposa Serra do Sol case – which was exactly about the demarcation of an indigenous territory – found that the Tribunal went "far beyond the specific wording of the Constitution or of any applicable legislation..." (The Special Rapporteur on the situation of human rights and fundamental freedoms of indigenous peoples 2009, paragraph 39). This is an indication that rule of law can be operationalized strictly, as the requirement that any decision must only be based in the applicable legislation, or less strictly, where concerns beyond the mere wording of the legislation is influencing the interpretation.

Accountability is distinct from responsibility in that the former is based on external standards and external sanctions mechanisms to be applied in cases on non-compliance. Accountability is closely related to the quality of institutions, and how these institutions are able to operate independent of the prevailing power structures. The Special Representative on business and human rights, in a study of human rights impact assessment says that indigenous customary laws and traditions are among the legal and regulatory requirements to be catalogued before any intervention starts (Special Representative of the Secretary-General on the issue of human rights and transnational corporations and other business enterprises 2007, paragraph 23). In order to identify appropriate mechanism for finding and punishing violations of indigenous customary law, the (non-binding) UNDRIP contains certain relevant provisions (UN 2007, paragraph 27 and 28).

Transparency is about providing information about any relevant decision is a timely and appropriate manner. As an example, any process about investment decisions needs to be as transparent as possible. Not only documents and maps, but also models and physical places need to be made visible. The FPIC process (see section 3.2.2) is about giving consent based on a process free of compulsion or illegitimate rewarding, and where scope of the final project as well as both the negative and positive consequences are outlined to all the affected parties. The term 'process' indicates that one meeting is inadequate for the FPIC to have been undertaken appropriately.

Participation is facilitated by appropriate transparency. As noted in section 3.2.2, it can be asked whether the understanding of ILO 169 Article 6.1(a) ('in particular through their representative institutions') and UNDRIP Article 18 ('through representatives chosen by themselves in accordance with their own procedures') and the IACHR's emphasis on 'effective participation' is adequate in order to comply with the overall understanding of the human right to participation. ICCPR Article 25(a) specifies a human right "[t]o take part in the conduct of public affairs, directly or through freely chosen representatives." Hence, we see that participation through representatives and direct participation are made equal. On the other hand, the emphasis on the indigenous peoples' own procedures is fully legitimate, as there might be individuals and groups of individuals in any community who claims to speak on behalf of the community, when they are in fact speaking only on behalf of themselves. Still, the author is convinced that broad participation at an early stage of a decision-making process will reduce the number of conflicts at the later stages.

Empowerment is in many ways the outcome of gaining experiences with participation. Empowerment is also a precondition for the actual participation. The most elaborate understanding of empowerment in any human rights treaty is found in the ICESCR Article 13.1, recognizing the right to education, which

...shall be directed to the full development of the human personality and the sense of its dignity, and shall strengthen the respect for human rights and fundamental freedoms. They further agree that education shall enable all persons to participate effectively in a free society...

Hence, the development of the human personality and the sense of its dignity are interrelated, and are overall objectives of education. An inclusive educational system is not achieved in Brazil. Moreover, despite traditions for organizing among workers in general in Brazil, and recent efforts to improve the overall conditions for the agricultural workers – such as the 2005 National Pact for the Eradication of Slave Labour, which has been signed by 250 companies, associations and NGOs, and the 2009 National Commitment to Improve Labour Conditions in Sugarcane Production, which has been signed by 331 companies – it is fair to say that there is a lack of empowerment among large segments of Brazil's agricultural workers. Moreover, in order to protect the cultural diversity of Brazil's indigenous peoples, the authorities have adopted both a paternalistic approach and a form of segregation, by confining indigenous people to certain areas and not allowing non-indigenous persons to enter these areas. This can be partially explained by the relative weakness of many indigenous peoples in Brazil.

In summary, human rights principles are relevant and can serve to strengthen the implementation of the substantive human rights. Human rights principles should never operate alone, but only in conjunction with the substantive human rights. Attempts of introducing principles in isolation from human rights are deemed to provide weak and unaccountable results. Moreover, as the review has illustrated, it is necessary to have an appropriate understanding of the structure within which the human rights are to be exercised (Welling 2008), 950-953). This obviously includes an analysis of prevailing power structures. In an otherwise critical article about the right to food the Economist admits that a human rights approach can "force the state to perform where it has previously failed, and start to overturn the traditional power structures..." (Economist, 2010c; see also Economist, 2011a). Improving public conduct and constraining power abuse are elements of a human rights based approach.

4. Particular human rights challenges in Brazil

Certain challenges relating to food production are particularly relevant in Brazil. This section cannot do full justice to the magnitude of challenges facing Brazil's agriculture, but three issues have been identified, namely land reform, biofuel expansion, and the management and appropriation of genetic resources. These are also issues that are considered to be of most relevance for the overall theme of this chapter, identifying how human rights could guide food production in Brazil.

4.1 Land reforms

The human right to food was included in the Brazilian Constitution in 2010, as part of the social rights provision of Article 6. Moreover, Article 5.3 says that "international treaties and conventions on human rights ... have equal value as constitutional amendments." Hence, international human rights law must be presumed to influence public policies in Brazil. Based on CESCR's concerns, outlined in Section 2.1 and the review of the relevant forest legislation in Section 2.3, as well as the lack of legislation to limit the maximum property size, there are several issues within land reform that are contentious.

Brazil has a substantial land potential, as illustrated by it being ranked as number 14 among all states regarding 'land resource potential and constraints' (FAO, 2000, 112). The country ranking is based on seven variables: potential arable land as a percentage of total land area, deserts and drylands, steeplands, land degradation severity, land presently cultivated (actual arable land) per capita, cultivated land as a percentage of potential arable land, and population increase (percent per year). 10 per cent of the arable land of Brazil is currently under permanent crops, which indicates that there is a potential for agricultural expansion.

Among the most substantive land reforms undertaken during the last decades is the Cédula da Terra project, a World Bank-supported negotiated land reform program in four states in the North-East region in addition to Northern parts of Minas Gerais. It lasted from 1997 to 2002. According to a critical study (Sauer, 2006), the program suffered under substantial weaknesses, such as lack of participation in the process, and lack of relevant technical assistance, with the result that many became indebted. The disapproval of the overall directions of the Cédula da Terra is illustrated by the fact that a network 'National Forum for Agrarian Reform and Rural Justice' presented a case before the World Bank's Inspection Panel, which rejected the case (Sparovek & Maule, 2009, 297).

The Special Rapporteur on the right to food distinguishes between two models of security of tenure, one market-based titling process and one seeking to broaden entitlements of individuals and communities (Special Rapporteur, 2010b, paragraph 21). The former can result in confirmations of unequal distribution of lands (ibid, paragraph 17). Not surprisingly, the Special Rapporteur argues in favor of strengthening of customary land tenure systems (see also Commission for the Legal Empowerment of the Poor, 2008).

As Brazil have relatively better institutions operating in land management, the country is less likely to become subject to speculative land acquisition of the kinds reported in the Economist (2011c, 57-58). The report is partly based on Deininger and Byerlee (2010) which found that actual farming had only started on 21 per cent of the properties that had been sold or leased (ibid, xiv). Of great relevance is the fact that the Office of the Solicitor General

(AGU) on 19 October 2010 issued Informed Opinion LA-01, which reregulated the process of land appropriation by foreigners in Brazil (Sauer & Pereira Leite, 2011, 28). This was done by resuming Act 5709 of 1971, while affirming that this act must be accepted under the scope of the Constitution of 1988.

Negotiated land reforms were initiated by the successive Lula governments, and in 2008, Decree 6.672 was passed. It allowed agrarian reform to proceed in the absence of external funding. Moreover, the state-led agrarian reform based on expropriation of unproductive lands is continuing under different governments. While there are different views on the comprehensiveness of agrarian reforms, the clearest indication that the agrarian reforms have not had an adequate impact is the Gini coefficient for land distribution, being above 0.85 and unchanging.

4.2 Biofuels expansion

As we saw in Section 1, Brazil has emphasized food sovereignty in its 2006 law 11.346 on a 'National Food and Nutrition Security System (SISAN)'. The whole text of the relevant paragraph reads: "To attain the human right to adequate food, and food and nutrition security, sovereignty must be respected, and countries must be guaranteed the supremacy of their own decisions regarding food production and consumption." The policy area of biofuel expansion is exactly such an area where Brazil rejects the interference by external actors in its own policies, insisting that there is no conflict between biofuels expansion on the one hand and better food security on the other hand. There are several different definitions of food sovereignty contained in documents adopted by the food sovereignty movement itself (Haugen, 2009, 274-280). Rather than review these definitions, the analysis below builds on the relevant extract of the provision given in the Brazilian law, namely that 'countries must be guaranteed the supremacy of their own decisions regarding food production and consumption'.

Brazil is currently the only country in the world where biofuels represent more than 50 percent of all fuel, and the relevant actors say that its bioethanol production shall reach 70 billion liters by 2020, but the growth in production from 2008 to 2010 has been modest (Renewable Fuels Association, 2011). Most of this production is for the internal market. From 1 July 2007, there has been a 25 per cent mandatory blend of ethanol in fuel. As the biodiesel industry is of less magnitude and is more diverse concerning which crops that are available for processing, the emphasis of this section will be on sugarcane-based bioethanol. There will, however, also be a review of the policies to promote palm oil within the overall biodiesel policy.

Whether bioethanol expansion can be at the expense of the right to food is a contentious issue. Observing current global price and predicted future food availability impacts, the Economist – based on FAO estimates – noted: "Biofuels are an example of what not to do" (Economist 2011b). This general observation does not necessarily apply to each and every country where biofuels are produced. As noted in Section 4.1 above, there is land availability in Brazil, but the concern for vulnerable biomes, including the Atlantic Forest (Mata Atlantica) and the Cerrado, is most relevant to repeat in this context of biofuels expansion as part of the overall agricultural expansion (Deininger and Byerlee, 2010, xxix).

A study on the overall impacts of biofuels expansion in Brazil states: "The growth of sugarcane production for the production of ethanol could compete with the production of food" (Smeets et al., 2008, 793). After a further discussion of the current situation, the study, however, finds that "poverty is currently a major bottleneck for food security, much more than a lack of production means like land or labor due to the production of sugarcane" (ibid, 794). A similar finding is made by the NGO Reporter Brazil 2009: Brazil of Biofuels: Soybean, Castor Bean 2009, says on p. 8: "there are no conclusive elements to state that biofuel production is having an impact on food crops in Brazil…"

Hence, in principle, sugarcane expansion can take place without reducing the area for food production. As the expansion of sugarcane is planned to take place in those areas of Brazil which have the most fertile soil, this, however, should be a reason for concern in the longer term. Moreover, if the targets on biofuel blending in several countries – which are already very ambitious and relying on imports – are becoming even higher, this will necessitate new assessments. Finally, concerning food production, three general observations seem relevant.

First, acknowledging that Brazil does have considerable areas of unproductive land, degraded land and underutilized land on a national level does not imply that there will not be conflicts relating to land use on the local level. The focus for human rights is the vulnerable persons or communities that will be affected by the expansion of areas utilized by commercial and more powerful actors. Second, while Brazil has a diversity of agricultural production on the national level, this does not imply that monoculture in specific regions cannot be a problem, at least in the medium and long term. Third, notwithstanding the fact that Brazil can be considered as one national food market, food security in the various regions will be strengthened if there is an adequate diversity in food production in all regions. Adequate supply in all regions will also be an effective guarantee against rapid increases in food prices, which will affect vulnerable households.

The highest ambitions for the enhanced production of biodiesel are within palm oil. Presently, a relatively small area of Brazil is covered with palm oil, but the ambition is to have 1 million hectares of palm oil by 2020. As for all other regions, oil palm trees can only yield in a narrow belt along the equator, where there are rainforests. Hence, the only way to justify palm oil expansion is to establish strict prohibitions against deforestation to give way for plantations, allowing only palm oil in what is generally termed 'degraded land'. Three measures have recently been adopted for this purpose; an oil palm zoning, identifying 31.8 million hectares; the oil palm bill (Act 7.326 of 2010); and the Programme for the Sustainable Production of Palm Oil. Based on previous experiences with conduct and failed attempts of regulating this conduct by legislation, it is reasonable to expect that palm oil expansion will not only take place in degraded lands, but also affect natural forests.

4.3 Management and appropriation of genetic resources

When reviewing relevant legislation applying to the traditional agricultural model, in Section 2.4 above, the seed law's exemption for traditional cultivars was addressed. The fact that only modern cultivars are given acknowledgement through registration should, however, be criticized. In a report on seed policies and innovation, the Special Rapporteur on the right to food recommended that seed regulation and government-approved seed lists should not exclude farmers' or peasants varieties, and that community registers for such varieties should be developed, and that there should be incentives for wider use of the food produced from such varieties (Special Rapporteur on the right to food 2009c, 21). This emphasis is in accordance with the farmers' rights, which are based on a recognition of: the enormous contribution that the local and indigenous communities and farmers of all regions of the world, particularly those in the centres of origin and crop diversity, have made and will continue to make for the conservation and development of plant genetic resources which constitute the basis of food and agriculture production throughout the world (ITPGRFA, Article 9.1).

By stating that plant genetic resources is the basis for global food production and emphasizing crop diversity and conservation, the drafters gave a strong endorsement of the continued role of traditional varieties or peasants' varieties. Embrapa, which is the most important and advanced agricultural research institute in Brazil, provides the following information through its homepage: "Germplasm accessions of species conserved at Embrapa are available for exchange in accordance with ITPGRFA" (Embrapa, 2011a). Moreover, a web site has been constructed "where the collections of different species conserved by Embrapa can be accessed and are available for exchange ... for the multilateral system" (Embrapa, 2011b). As we saw in Section 3.2.3 above, this does not necessarily include plant genetic resources of another 'form' than these resources had when they were received, and which can be subject to patenting.

The UK Commission on Intellectual Property Rights, consisting of both patent attorneys and persons representing industry, noted after visiting one of the leading agricultural research institutes in the developing world: "We were struck both by the vigour with which intellectual property protection was being introduced, and by the conscious effort being made to change the traditionally open culture of research. " (UK Commission on Intellectual Property Rights, 2002, 125). The same institute that this observation applies to emphasizes in its guidelines that patent policy shall be implemented without sacrificing its social mission (ibid). One of the members of the Commission confirms that this leading research institute is Embrapa in Brazil. Without being able to go too deep into the impacts of the increased tendency to patent at Embrapa, and while acknowledging that it is fully possible to license patent product to the conditions that the patent holder sets to less commercial farmers, increased patenting will inevitably lead to an increased commercial orientation for the patent holder. As seen in Section 3.2.1, MDA's National Rural Extension Policy, which encompasses new seed varieties has grown considerably since its introduction in 2003.

Nevertheless, the promotion of monocropping that characterizes particularly the agroindustry in Brazil, the lack of formal recognition of traditional varieties, and the active promotion of patent protection in Embrapa, all point in a certain direction. While biological diversity is praised by Brazilian political authorities, the specific policies do not contribute to maintaining this diversity.

5. Conclusion

The Brazilian agricultural boom is more a result of conscious strategies and programmes than a result of 'miracles'. Brazil has comprehensive policies in both the scientific and the social realm – and in policy areas between these two. Several of Brazil's agricultural, science and social policies should be a model for other states.

Exporting the whole model, on the other hand, is not advisable. The Brazilian agricultural model is built on the colonial heritage with a highly unequal land distribution, and the

Brazilian society is still characterized by disempowerment and partial segregation. To a certain extent, this segregation is justified by the fact that many of the indigenous peoples need to be protected, but overprotection can imply patriarchal thinking and disempowerment. Moreover, the weak enforcement of laws and the adjustment of the laws to accommodate undesired conduct, which in effect facilitates the destruction of vulnerable biomes, implies that "agricultural investment and expansion pose serious environmental challenges" (Deininger and Byerlee, 2010, xxix).

Nevertheless, Brazil has shown impressive achievements in improving productivity in all three agricultural models, including the socio-biodiversity emphasis within the traditional farming model and the agroecological emphasis within family farming. Comprehensive policies applying to all three agricultural models has resulted in more efficient use and reuse of resources, which is the central characteristics of agroeceology (de Wit, 1992, see also Special Rapporteur, on the right to food 2010a). Such policies, however, risks being overshadowed by the continuing agricultural colonization both in the Amazon and in other vulnerable biomes, like the Atlantic Forest (Mata Atlantica) and the Cerrado.

6. Acknowledgement

The author wants to thank the Research Council Norway for the generous funding for the project Biofuels and human rights (2009-2013), within which a research field trip to Brazil was conducted in March and April 2011. This chapter builds on the information gathered during this field trip, and the author wants to express his sincere thanks to all those contributing with their time and insight. A report from the field trip is available upon request.

7. References

Alston, P. (1984). International Law and the Human Right to Food, In: *The Right to Food*, P. Alston & K. Tomasevski (Eds.), pp. 9-68, Martinus Nijhoff Publishers, ISBN 90-247-3087-2, Dordrecht.

Bogdanski, A., O. Dubois, C. Jamieson & R. Krell. (2011). *Making Integrated Food-Energy Systems Work for People and Climate. An Overview*, FAO, ISBN 978-92-5-106772-7, Rome.

Brazil. (2008). *Brazil's second periodic report to the Committee on Economic, Social and Cultural Rights, E/C.12/BRA/2.*

Brazilian Ministry of Agrarian Development. (2009). *Family Farming in Brazil and 2006's Agricultural and Livestock Census*, Brazilian Ministry of Agrarian Development, Brasília.

Collier, P. (2008). Politics of Hunger: How Illusion and Greed Fan the Food Crisis. *Foreign Affairs* Vol.87, No.1, pp. 67-80, ISSN 0015-7120.

Commission on the Legal Empowerment of the Poor. (2008). *Making the Law Work for Everyone Volume I*, UNDP, ISBN 978-92-1-126219-3, New York.

Committee on Economic, Social and Cultural Rights. (2009). *E/C.12/BRA/CO/2: Concluding observations of the Committee on Economic, Social and Cultural Right: Brazil.*

Committee on Economic, Social and Cultural Rights. (1999). *E/C.12/1999/5: General Comment 12. The right to adequate food (Art.11).*

Committee on Economic, Social and Cultural Rights. (1991). *E/1991/23, 83-87. General Comment No. 3, The nature of States parties obligations.*

de Wit, C.T. (1992). Resource Use Efficiency in Agriculture, *Agricultural Systems* Vol.40, No.1-3, pp. 125-151, ISSN 0308-521X.

Deininger, K. & D. Byerlee. (2010). *Rising Global Interest in Farmland. Can It Yield Sustainable and Equitable Benefits?* World Bank, ISBN 978-0-8213-8591-3, Washington DC.

Economist. (2011a). The Indian exception, Vol.399, No.8727 (2 April), p. 55, ISSN 0013-0613.

Economist. (2011b). Plagued by politics [special report on feeding the world], Vol.398, No.8722 (26 February), p. 6, ISSN 0013-0613.

Economist. (2011c). When others are grabbing your land, Vol.399, No.8732 (7 May), p. 55, ISSN 0013-0613.

Economist. (2010a). The Miracle of the Cerrado, Vol.396, No.8697 (28 August), pp. 46-48, ISSN 0013-0613.

Economist. (2010b). Affirmative anticipation, Vol.394, No.8673 (13 March), p. 53, ISSN 0013-0613.

Economist. (2010c). The rights approach, Vol.394, No.8674 (20 March), p. 56, ISSN 0013-0613.

Eide, A. (2009). *The Right to Food and the Impact of Liquid Biofuels (Agrofuels)*, FAO, ISBN 978-92-5-106174-9, Rome.

Eide, A. (2007). State Obligations Revisted, In: *Food and Human Rights in Development: Vol II: Evolving Issues and Emerging Applications*, W. Barth Eide and U. Kracht (Eds.), 137-158, Intersentia, ISBN 978-90-5095-459-4, Antwerpen.

Embrapa. (2011a). *Plant Accession Area Queries*, available from: <http://mwpin004.cenargen.embrapa.br/MCPDEnglish>

Embrapa. (2011b). *ITPGRFA » Brazil*, available from: <http://tirfaa.cenargen.embrapa.br/tirfaa/pg_english/brazil.htm>

Embrapa Agrienergy. (no date). *Brazilian Agriculture: Food and Energy for a Sustainable World.* Embrapa, Brasília.

Fundação Cultural Palmares. (2011). *Comunidades Quilombolas*, available from: <www.palmares.gov.br/?page_id=88>

FAO. (2011). *Land tenure and international investments in agriculture. A report by the High Level Panel of Experts on Food Security and Nutrition of the Committee on World Food Security*, HLPE Report 2, FAO, Rome

FAO. (2011). *International Treaty on Plant Genetic Resources for Food and Agriculture – Declarations*, available from: <www.fao.org/Legal/treaties/033s-e.htm>

FAO. (2009). *Low Greenhouse Gas Agriculture. Mitigation and Adaptation of Sustainable Farming Systems*, FAO, ISBN 978-92-5-106298-2, Rome.

FAO. (2007). *Right to food and indigenous peoples*, available from: <www.fao.org/righttofood/wfd/pdf2007/focus_indigenous_eng.pdf>

FAO. (2004). *Voluntary Guidelines to Support the Progressive Realization of the Right to Adequate Food in the Context of National Food Security, Adopted by the 127th Session of the FAO Council*, FAO, ISBN 92-5-105336-7, Rome.

FAO. (2001a). *International Treaty on Plant Genetic Resources for Food and Agriculture*, FAO Conference Resolution 3/01, entered into force 29 June 2004.

FAO. (2001b). *Land Resource Potential and Constraints at Regional and Country Levels, World Soil Resources Report 90*, FAO, Land and Water Development Division: Rome. Fearnside,

P. M. (2008). The roles and movements of actors in the deforestation of Brazilian Amazonia. *Ecology and Society* Vol.13, No.1, Article 23 [online], available from: <www.ecologyandsociety.org/vol13/iss1/art23>

Government of Brazil. (2010). *Brazil brings traditional people's land into compliance, benefiting 30,000 families in a year*, 29 August 2011, available from: <www.brasil.gov.br/news/history/2010/12/08/brazil-brings-traditional-peoples-land-into-compliance-benefiting-30-000-families-in-a-year>

Haugen, H.M. (2011). Human rights principles: Clarification and application to food and water', In: *Max Planck Yearbook of United Nations Law, Volume 15*, 419-444. Martinus Nijhoff Publishers, Leiden.

Haugen, H.M. (2009). Food sovereignty – an appropriate approach to ensure the right to food?, *Nordic Journal of International Law* Vol.78, No.3, pp. 263-292, ISSN 0902-7351.

Haugen, H.M. (2007). *The Right to Food and the TRIPS Agreement – With a Particular Emphasis on Developing Countries' Measures for Food Production and Distribution*, Martinus Nijhoff Publishers, ISBN 978-90-04-16184-9, Leiden.

HRC. (2009). *Poma v. Peru - CCPR/C/95/D/1457/2006* (24 April 2009).

HRC. (1989). *General Comment No. 18: Non-discrimination* (10 November 1989).

IAASTD (International Assessment of Agricultural Knowledge, Science and Technology for Development). (2009). *Agriculture at the Crossroads, Global Report*, Island Press, ISBN 978-1-59726-539-3, Washington DC.

Inter-American Court of Human Rights. (2008). *Saramaka People v. Suriname, Judgment of 12 August 2008 (Interpretation of the judgment on preliminary objections, merits, reparations, and costs) (Series C No. 185)*.

IFAD. (2001). *Rural Poverty Report 2001 - The Challenge of Ending Rural Poverty*, Oxford University Press, ISBN 0-19-924507-X, Oxford and New York.

ILO. (2011). *Application of International Labour Standards 2011 (1). Report III (Part 1a). Report of the Committee of Experts on the Application of Conventions and Recommendations. General Report and observations concerning particular countries. International Labour Conference 100th Session*, ILO, ISBN 978-92-2-123097-7, Geneva.

ILO. (1989). *ILO Convention 169 concerning Indigenous and Tribal Peoples in Independent Countries*, 1650 UNTS 28383, entered into force 5 September 1991.

INCRA (National Institute for Colonization and Agrarian Reform). (2008). *Instrução Especial Nº 20, de 28 de maio de 1980, Estabelece o Módulo Fiscal de cada Município, previsto no Decreto Nº 84.685 de 06 de maio de 1980*, 29 August 2011, available from: www.incra.gov.br/portal/index.php?option=com_docman&task=doc_download&gid=81&Itemid=136>

IPCC (Intergovernmental Panel on Climate Change). (2008). *Climate Change 2007: Working Group III: Mitigation of Climate Change* (B. Metz, O.R. Davidson, P.R. Bosch, R. Dave, L.A. Meyer. (Eds.)). Cambridge University Press, ISBN 978-05-2170-598-1, Cambridge.

IPSNews 2009, Agribusiness Driving Land Concentration (5 October), available from: <http://ipsnews.net/news.asp?idnews=48734>

Klausen, S. (2002). Low schooling for girls, slower growth for all? Cross-country evidence on the effect of gender inequality in education on economic development, *World Bank Economic Review* Vol.16, No.3, pp. 345-373, ISSN 0258-6770.

Nellemann, C., M. MacDevette, T. Manders, B. Eickhout, B. Svihus, A. Gerdien Prins & B.P. Kaltenborn. (2009). *The Environmental Food Crisis – The Environments's Role in*

Averting a Future Food Crisis. A UNEP Rapid Response Assessment, UNEP -GRID, ISBN 978-82-7701-054-0, Arendal.

Neumanna, K., P.H. Verburg, E. Stehfest & C. Müller. (2010). The Yield Gap of Global Grain Production: A Spatial Analysis, *Agricultural Systems* Vol.103, No.5, pp. 316-326, ISSN 0308-521X.

NGO Reporter Brazil. (2009). *Brazil of Biofuels: Impacts of Crops on Land, Environment and Society. Soybean, Castor Bean 2009*, NGO Reporter Brazil, ISBN 978-85-61252-12-0, São Paulo.Renewable Fuels Association. (2011). *World Fuel Ethanol Production*, available from:
<http://ethanolrfa.org/pages/World-Fuel-Ethanol-Production>

Santilli, J. (2009). Brazil's Experience in Implementing its ABS Regime: Suggestions for Reform and the Relationship with the International Treaty on Plant Genetic Resources for Food and Agriculture, In: *Genetic Resources, Traditional Knowledge and the Law: Solutions for Access and Benefit-Sharing*, E. C. Kamau & G. Winter (Eds.), 177-201, Earthscan, ISBN 978-1-84407-793-9, London.

Sauer, S. (2006). The World Bank's Market-Based Land Reform in Brazil, In: *Promised Land: Competing Visions of Agrarian Reform*, P. Rosset, R. Patel and M. Courville (Eds.) pp. 177-191, Institute for Food and Development Policy, ISBN 978-0-935028-28-7, New York.

Sauer, S. & S. Pereira Leite. (2011). *Agrarian structure, foreign land ownership, and land value in Brazil. Paper presented at the International Conference on Global Land Grabbing 6-8 April 2011*, available from:
<www.future-agricultures.org/index.php?option=com_docman&Itemid=971>

Smith, L.C., U. Ramakrishnan, A. Ndiaye, L. Haddad & R. Martorell. (2002). *The Importance of Women's Status for Child Nutrition in Developing Countries, IFPRI Report 131*, International Food Policy Researc Institute, ISBN 0-89629-134-0, Washington D.C.

E. Smeets, M. Junginger, A. Faaija, A. Walter, P. Dolzan & W. Turkenburg. (2008). The sustainability of Brazilian ethanol –An assessment of the possibilities of certified production, *Biomass and Bioenergy* Vol.32, No.8, pp. 781-813, ISSN 0961-9534.

Sparovek, G. & R.F. Maule. (2009). Negotiated Agrarian Reform in Brazil, In: *Agricultural Land Redistribution: Toward Greater Consensus*, H.P. Binswanger-Mkhize, C. Bourguignon & R. van den Brink (Eds.), 291-309, World Bank, ISBN 978-0-8213-7627-0, Washington D.C.

Special Rapporteur on adequate housing as a component of the right to an adequate standard of living. (2010). *A/HRC/13/20/Add.2: Follow-up to country recommendations: Brazil, Cambodia, Kenya.*

Special rapporteur on the right to adequate food. (2010a). *A/HRC/16/49: Report submitted by the Special Rapporteur on the right to food, Olivier De Schutter.*

Special rapporteur on the right to adequate food. (2010b). *A/65/281: Interim report of the Special Rapporteur on the right to food, Olivier De Schutter.*

Special rapporteur on the right to adequate food. (2009a). *A/HRC/13/33/Add.6: Report of the Special Rapporteur on the right to food, Olivier De Schutter. Addendum. Mission to Brazil.*

Special rapporteur on the right to adequate food. (2009b). *A/HRC/13/33/Add.2, Annex: Large-scale land acquisitions and leases: A set of minimum principles and measures to address the human rights challenge.*

Special rapporteur on the right to adequate food. (2009c). A/64/170: *Report of the Special Rapporteur on the right to food: Seed policies and the right to food: enhancing agrobiodiversity and encouraging innovation.*

Special Rapporteur on the situation of human rights and fundamental freedoms of indigenous peoples (2008). *A/HRC/12/34/Add.2: Report from Mission to Brazil.*

Special Rapporteur on the situation of human rights and fundamental freedoms of indigenous peoples (2008). *A/HRC/9/9/Add.1: Summary of cases transmitted to Governments and replies received.*

Special Representative of the Secretary-General on the issue of human rights and transnational corporations and other business enterprises. (2007). *A/HRC/4/74: Human rights impact assessments – resolving key methodological questions.*

Ssenyonjo, M. (2009). *Economic, Social and Cultural Rights in International Law,* Hart Publishing, ISBN 978-1-84113-915-9, Oxford.

Sua pesquisa. (no date). *Coeficiente de Gini,* 29 August 2011, available from: <www.suapesquisa.com/economia/coeficiente_gini.htm>*The Rio Times.* (2011). Amazon Deforestation Increases Alarming (24 May).

UK Commission on Intellectual Property Rights. (2002). *Integrating Intellectual Property Rights and Development Policy,* DFID, London.

United Nations. (2007). *A/61/295, Annex: Declaration of the Rights of Indigenous Peoples.*

United Nations. (1969). *Vienna Convention on the Law of Treaties,* 1155 U.N.T.S. 331, entered into force 27 January 1980.

United Nations. (1966a). *International Covenant on Economic, Social and Cultural Rights,* 993 U.N.T.S. 3, entered into force 3 January 1976.

United Nations. (1966b). *International Covenant on Civil and Political Rights,* 999 U.N.T.S. 171, entered into force 23 March 1976.

United Nations Development Group. (2003). *The Human Rights Based Approach to Development Cooperation: Towards a Common Understanding among UN Agencies,* available from:
<www.undg.org/archive_docs/6959-
The_Human_Rights_Based_Approach_to_Development_Cooperation_Towards_a_
Common_Understanding_among_UN.pdf>

von Braun, J. S. Fan, R. Meinzen-Dick, M.W. Rosegrant & A. Nin Pratt. (2008). *International Agricultural Research for Food Security, Poverty Reduction, and the Environment,* available from:
<www.ifpri.org/sites/default/files/publications/oc58.pdf>

Welling, J. (2008) International Indicators and Economic, Social and Cultural Rights, *Human Rights Quarterly* Vol.30, No.4, pp. 933-958, ISSN 0275-0392.

Wilkinson, J., B. Reydon & A. Sabbato. (2010). Dinâmica do mercado de terras na América Latina: o caso do Brasil. Santiago: FAO/Escritório Regional.

World Bank. (2007). *World Development Report 2008: Agriculture for Development,* World Bank, ISBN 978-0-8213-6807-7, Washington DC.

WTO. (1998). *WT/DS58/AB/R, United States - Import prohibition of certain shrimp and shrimp products.*

Part 2

Scientific Methods for Improving Food Safety and Quality

9

Physical Factors for Plant Growth Stimulation Improve Food Quality

Anna Aladjadjiyan
Agricultural University,
Plovdiv,
Bulgaria

1. Introduction

Fast growing of world population affected negatively the environmental conditions of our life. Increasing number of earth population resulted in growing consumption of food and energy. Both tendencies seriously exhaust the natural resources. The attempts to increase food and energy production for satisfying growing needs led to intensive development of plant production through the use of chemical additives, which in its turn caused more and more pollution of soil, water and air.

The environmental pollution cannot be limited in the region of the primary use of chemical additives. It spreads out horizontally - through the underground and surface waters it is passing out over the frontiers. Through the food chain including plants and animals it spreads out hierarchically and achieves our table. As a result the pollution negatively affects food quality.

The risk of food safety hazards occurring during on-farm production for fresh produce is closely related to the use of herbicides, fertilizers, and other chemicals. Soil and water contamination generates toxic compounds in plants that worsen the food quality. In order to raise food safety it is necessary to control and keep the concentration of harmful substances in foodstuff in reasonable limits. Even the biological agriculture hides danger of spreading infections like this one recently arisen with *Escherichia* Coli.

Chemical compounds are largely used for improving soil fertility; plant protection during growth and storage stage; for improving food quality at production, preservation and conservation stages; their application is partially allowed even in organic agriculture. Uncontrolled use of chemicals is hazardous for the contamination of raw materials for food production with toxins. Through the food chain the accumulation of different chemicals is dangerous for the health of consumers. Therefore, on-farm safety for fresh produce needs developing and implementing new methods for quality control and assurance. One possibility to reduce chemical contamination in raw materials is the substitution of chemical compounds with physical factors for plant growth stimulation.

2. Alternative technologies for increasing plant production

Nowadays, the diapason of agricultural practices lies between the swidden agriculture and high-level biotechnology. The use of physical factors takes interstitial position and represents interest for many scientists.

The use of physical factors for controlled influence on biological behaviour during development and storage of different cultures is a modern trend in combining the intensification of plant technologies with the ecological requirements. It could be important for biological and organic agriculture.

Physical methods for increasing the vegetable production are based on the use of different physical factors for plant treatment, particularly on the dill seeds. Most perspective factors include the treatment with electromagnetic waves, particularly optical emission, magnetic field as well as the ultrasound and ionizing radiation. All living processes are highly dependent on energy exchange between the cell and the environment. The core of physical methods is the energy supply through the treatment with physical factors.

Plants sensitivity to the influence of these physical factors has been elaborated during their evolutionary development since these physical factors are elements of their natural environment.

The influence of physical factors as microwave and laser radiation, magnetic field and ultrasound treatment is an alternative of soil additives and fertilizers. The substitution of chemical treatment by physical one can reduce toxins in raw materials and thus – raise the food safety.

Contemporary agriculture largely uses chemical compounds for improving soil fertility, plant production and protection against plant diseases and enemies.

The substitution of chemical amelioration with convenient physical methods of treatment has two advantages - one is reducing the use of fertilizers and thus decreasing pollution of on-farm produced raw materials for food production and the other – possibility for disinfection of seeds before sowing and during the storage through inactivation of microorganisms.

Physical factors might be good alternative of chemical products for the same purposes: raising the yield of agricultural production, improving plant protection and storage, as well as processing, preservation and conservation of food.

In the case of chemical amelioration the necessary substances are imported into the cell. Instead of this in the case of physical treatment the energy introduced in the cell creates conditions for molecular transformations and as a result, the necessary substances are provided for the cell.

2.1 Magnetic field

2.1.1 State of the art

One of most popular physical factors is magnetic field. Magnetic field is an attribute of the Earth. Its origin is the molten metallic core of our planet where circulating electric current appears. Magnetic field changes periodically its direction. Evidence for 171 magnetic field reversals during the past 71 million years has been reported. Its magnitude varies from Earth surface to the depth as well as over the surface of the Earth.

Numerous authors confirmed that a magnetic field of magnitude one or two orders above geomagnetic field strength (35 to 70 μT) could affect plant growth and metabolism (Galland & Pazur, 2005; Çelik et al., 2009; Racuciu 2011; Shine et al., 2011).

The impact of magnetic field on plants depends on its magnitude and its character – static or alternating. Different magnetic fields have been used for plant treatment – static (with constant magnetic induction), alternating (Aguilar et al., 2009), oscillating (Racuciu 2011), pulsed (Bilalis et al., 2011).

The positive influence of the stationary magnetic field on the plant seeds has been largely established (Galland & Pazur, 2005). The treatment fastens plants development (Florez et al. 2007; Gouda & Amer, 2009), improves germination and seedling growth (Carbonell et al., 2008; Martínez et al., 2009a; Odhiambo et al., 2009), activates protein formation and enzymes activity (Atak et al., 2007; Racuciu et al., 2007; Çelik et al., 2009). The investigations have shown that the treatment of the seeds with magnetic field increases the germination of non-standard seeds and improves their quality (Pietruszewski et al., 2007). Experiments have been made with large range of plants: grain - (Torres et al., 2008; Vashisth & Nagarajan, 2008; Vashisth & Nagarajan, 2010), leguminous (Podlesny et al., 2004; Podlesny et al., 2005; De Souza et al.; 2006, Odhiambo et al., 2009; Martínez et al., 2009b), and perennials (Çelik et al., 2008;, Dardeniz et al., 2006; Dhawi & Al-Khayri, 2009).

The effect of magnetic treatment on plants depends on the strength of magnetic field, expressed by magnetic induction, and the exposure time. There are significant differences in the induction B of investigated magnetic fields – from $B = 62\mu T$ (Odhiambo et al., 2009) to 250 mT (Vashisth & Nagarajan, 2010). Explored exposure times also vary largely - from 15 s (Muszinski et al., 2009) to 24 hours (Martinez et al., 2009a).

Some authors have explored the influence of irrigation with preliminary magnetized water and also have found positive impact on plant development (Abdul Qados & Hozayn (2010a, 2011b); Hozayn et al. (2010, 2011); Hozayn & Abdul Qados (2010); Vajdehfar et al. (2011).

2.1.2 Materials and methods

In our last experiments the influence of static magnetic field has been studied on seeds of lentils (*Lens Culinaris*, Med.). The induction of used magnetic field has been $B = 0,15$ T, measured with a digital Teslameter Systron - Donner. Seeds have been distributed in five variants and 5 replicate each one containing 10 seeds each. Seeds in each variant have been exposed to magnetic field treatment for different time: 0 min (control), 3 min, 6 min, 9 min, and 12 min.

Seeds have been preliminarily soaked in distilled water for 1 hour, presuming that the intra-cell water, due to its magnetic properties, plays role in the absorption of the energy of magnetic field. After the treatment the lentil seeds were cultured in small plastic pots (Ø =7.5 cm and h = 8.8 cm) on wet cotton. The natural light cycle was 9 h - light/15 h - darkness with daily temperature 21 ± 2°C, night temperature 15 ± 2°C.

In order to estimate the influence of the treatment on seeds in our experiments the next some criteria have been used hereafter:

- The germination energy (GE) of seeds in %, determined on the 4th day after the start of the experiment - as a ratio of the number of germinated to the total number of seeds for the corresponding variant;
- germination (G) of seeds in %, determined on 7th (8th) day as a ratio of the number of germinated to the total number of seeds;
- length of stems (SL) and main roots (RL) in mm determined on the 7th and 14th day;
- total mass (TM) in mg determined on the 14th day.

Data were statistically processed using the method of Fisher of dispersion analysis. The reported values are mean values for 50 measurements of each parameter.

2.1.3 Results

The germination energy (GE) and the germination (G) are shown at Fig.1. Both parameters – GE and G, have the highest values for non-treated samples. The values for all variants are enough high – not less than 80%. Differences between variances and control are not statistically significant. This observation seems to be in contradiction with our earlier results for soybean (Aladjadjiyan 2003c) and ornamental trees (Aladjadjiyan 2003a), where a significant rise of germination has been accounted for treated samples. This contradiction may be due to the different kind of seeds used in previous experiments.

The results for the length of stem and root of lentils seeds measured on the 7th day are similar - values for treated samples are lower than the control.

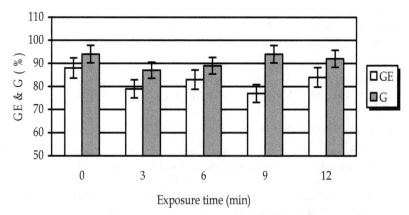

Fig. 1. Germination energy (GE) and germination (G) of lentils seeds exposed to static magnetic field.

The results for the measurements on the 14th day show more expressive differences. The stem lengths shown on fig. 2 and the total mass on fig. 3 have well expressed increase for the samples treated with magnetic field for 6 and 9 min. In the case of 6 min exposure the increase of SL and RL is 120% and 11%, respectively; in the case of 9 min exposure - 104%

and 12%, for SL and RL, respectively. Longer exposure of 12 min results in SL and RL shorter than those for exposure 3 min (fig.2). Similar conclusions are valid for the total mass (fig.3). Values of TM for exposure 6 and 9 min are the highest, the value for exposure 9 min is even less than the control one.

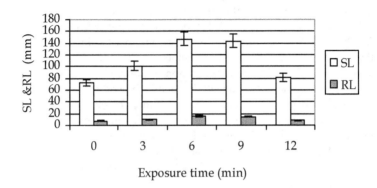

Fig. 2. Stem (SL) and root (RL) length of lentils seedlings measured at the 14th day.

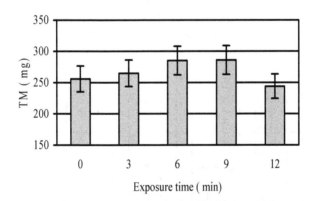

Fig. 3. Total mass (TM) of lentils seedlings exposed to static magnetic field at the 14th day.

The conclusion of our investigation is that for treatment of lentils seeds with stationary magnetic field with induction B=0,15 T the optimal exposure time is between 6 and 9 min. Our earlier experiments have shown that the best results for SL and TM in case of maize (Aladjadjiyan, 2002a) and soybean (Aladjadjiyan, 2003c) have been achieved for the same value of magnetic induction at 10 min exposure time, while in the case of *Nicotiana Tabacum* (Aladjadjiyan & Ylieva, 2003) a linear rise of G and GE in the interval 0 – 30 min has been registered. Similar results for maize have reported Hernandez et al., 2009. Their conclusion was that the impact of magnetic field treatment is most effective for the combination B=60mT and t=7,5 min, but depends also on seeds genotype. For higher intensity B=100mT the longer exposure time (30 min) leads to worse results. We have accounted the same for B=150mT and t=12 min.

2.1.4 Discussion

We tried to compare the effect of magnetic treatment in our investigation with those reported by different authors. We used the data of authors investigated the influence of static magnetic fields with value close to 0,15 T and exposure times ranging between 3 and 60 min. We calculated the maximum growth of germination (G), stem length (SL), and total mass (TM) in percentage towards the control. Data of Shine et al., 2011, and Flórez et al., 2007, have been used for producing additional data along with our own (Aladjadjiyan, 2002a, 2003a, b, c, 2010a) presented on fig.4. It seems that the influence of static magnetic field on the growth of investigated parameters G, SL, and TM reaches maximum for exposure time around 30 min. The investigation of Shine et al., 2011, on the treatment of soybean with magnetic field with intensity B=150mT showed the best results for all studied parameters at exposure 60 min. Compared to our results for soybean treated with stationary magnetic field with the same intensity it shows a difference – our best results for G have been achieved for exposure 15 min, and for stem length and total mass at 10 min (Aladjadjiyan 2003c), but we have investigated the interval of exposure times between 0 and 30 min. Muszynski et al., 2009, concluded that short exposure time (15 and 30s) does not influence wheat seedling growth.

The authors that have investigated longer exposure time (longer than 30 min) have accounted influence of the treatment on the concentration of chlorophyll *a* and *b* (Vashist & Nagarajan, 2008); mitotic activity (Racuciu, 2011); activities of superoxide dismutase and catalase (Celic et al., 2009). All these parameters have been significantly increased.

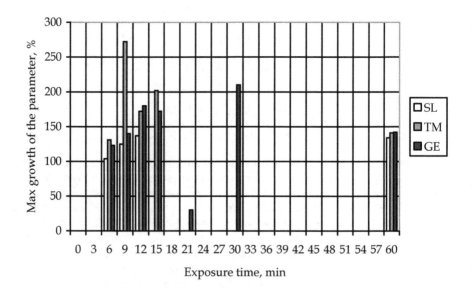

Fig. 4. The effect of static magnetic field on the growth of plant parameters.

This selective effect of different doses of magnetic field treatment may be explained with ions properties. Ions in the cell have the ability to absorb magnetic energy corresponding to specific parameters related to their vibration and rotation energy sublevels. This phenomenon represents a kind of resonance absorption and could explain the stronger effect of applying definite values of magnetic field induction, observed as well in the works of Martinez et al., 2009.

One of the possible explanation of observed positive effect of magnetic treatment could be found in paramagnetic properties of some atoms in plant cells and pigments, i.e. chloroplasts. In outer magnetic field magnetic moments of these atoms turn align the field. Magnetic properties of molecules determine their ability to absorb the energy of magnetic field, then transform it in other kind of energy and transfer this energy later to other structures in plant cells, thus activating them. Magnetic effects on plants can be explained in the framework of the ion cyclotron-resonance and the radical pair models, two mechanisms that also play an essential role in the magnetoreception of other organisms (Galland & Pazur 2005). The hypothesis proposed by Shine et al., 2011, underlines the role of Ca^{2+} ions. The increased concentration of Ca^{2+} ions after the treatment possibly plays the role of signal for cells to enter earlier in mitotic cycle. Magnetic field affects also the electroconductivity (Szczez et al., 2011) thus changing cell status.

2.2 Laser

2.2.1 State of the art

Light is most important attribute of natural environment for life. The sensibility of plants to light is well known. Light is necessary condition for photosynthesis – most important process in plants' life. Plants' sensor for light is chlorophyll. In chlorophyll centrums the light energy is being transformed and used for producing carbohydrates. Plants are sensitive to light intensity, its spectral composition and alternating of light and dark periods.

Laser (Light Amplification by Stimulated Emission of Radiation) is specific light source. The light emitted by laser is notable for its high degree of spatial and temporal coherence, unattainable using other technologies. Laser light is uniformly polarized and monochromatic.

A detailed review of the use of laser treatment for plant stimulation has been published recently by Hernandez et al., 2010. As it is underlined there, laser light is used in agriculture for biostimulation of seeds, seedlings and plants on the basis of the synergism between the polarized monochromatic laser beam and the photoreceptors absorbing it, which activate numerous biological reactions.

Lasers used in agriculture belong to different types – solid, gaseous, semiconductor (depending on the state of emitting medium). Their emission also differs – from red light emitted by ruby laser (694 nm), and helium-neon (632,8 nm); green light emitted by YAG:Nd laser (532 nm); blue - by argon (514,5 nm); ultraviolet emission by nitrogen (337,1 nm); visible and infrared diapason covered by semiconductors (510, 632, 650, 670, 810, 940, and 980 nm), and carbon dioxide in far infrared (10 600 nm). A laser can be classified as operating in either continuous or pulsed mode, depending on whether the power output is essentially continuous over time or whether its output takes the form of pulses of light on one or another time scale.

The effect of laser treatment depends on the wave length of its emission, but also on the output power of the laser and the exposure time. The investigated output powers vary in large diapason between 200 kW (Govil et al., 1991, as cited in Hernandez et al., 2010) and 5 mW (Grygierzec, 2008 as cited in Hernandez et al., 2010). The exposure times as well vary largely – between 30 s used by Michtchenko & Hernández, 2011, and 120 min in the work of Khalifa &. El Ghandoor, 2011.

In Bulgaria the treatment with helium-neon laser was most spread-out (Aladjadjiyan 2008) probably because its relatively low cost but highly coherent emission. Due to the impact of helium-neon laser, an acceleration of germination and development at the early phases for dill seeds of different cultures was established. It was found out that the effect of stimulation depended on the laser wave length, the exposition on irradiation, the reiteration and the pre-history of the samples (i.e. preliminary soaking of seeds in water).

2.2.2 Materials and methods

We have investigated the influence of impulse and continuous helium-neon laser on the development of some plants. An impulse laser with wave length 632,8 *nm* and intensity 100 Wm^{-2} was used for irradiation of seeds of carrot (*Daucus carrota* L., cv.Nantes). The light-impulse's durability was 1 min. Three variants of treatment have been experienced: 5-, 7- and 9-fold irradiation. To assess the effect of the treatment, the germination energy GE, the germination G, and the total mass TM on the 7th day had been measured as described in previous part 2.1.2.

The influence of laser treatment has been investigated also on bean (*Phaseolus vulgaris* L., cv. Plovdiv) seeds. In this case continuously emitting He-Ne laser with intensity 176 Wm^{-2} has been used. Seeds have been distributed in four variants. Each variant has been exposed to laser treatment for different time: 0 min (control), 5 min, 10 min, and 15 min. In this case the germination energy GE, the germination G, and the total mass TM on the 38th, 45th, and 52nd day had been measured according to description in part 2.1.2.

2.2.3 Results

The results of the treatment of carrot seeds with an impulse He-Ne laser are presented for G and GE on fig.5, and for TM – on fig.6, respectively. From fig. 5 it can be seen that the best results for GE of carrot seeds have been achieved in case of 5-fold (total accumulated exposure 5 min) treatment with He-Ne laser; for G - in case of 7-fold treatment. The differences for both values are not statistically significant.

The 9-fold treatment (total accumulated exposure 9 min) has shown an inhibitory impact on GE and G, which could be due to an excess of light energy absorbed by the seeds in this configuration.

Figure 6 shows the dependence of the total mass (TM) on the exposure time. It can be seen that the best result for TM is accounted for 9-fold treatment. In all cases for described experimental configuration the differences are not statistically significant, except of the 9-fold irradiation. Possible explanation of this observation is that the combination of laser light intensity and the exposure of seeds for more cases do not assure enough energy for demonstration of stimulating effect.

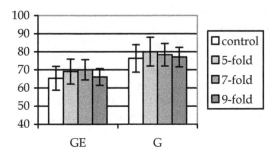

Fig. 5. Germination energy (GE) and germination (G) of carrot seeds treated with laser.

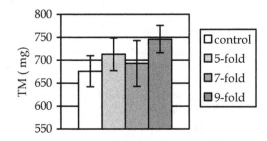

Fig. 6. Total mass (TM) of carrot seedlings after laser treatment, measured at the 7th day.

The results for the treatment of bean seeds with a continuously emitting He-Ne laser are presented on fig. 7. The stem length of 50 seedlings has been measured on different days

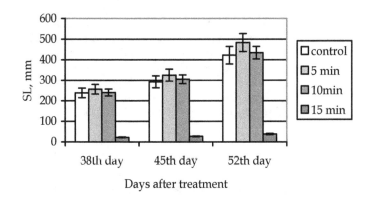

Fig. 7. Stem length (SL) of bean seedlings treated with He-Ne laser.

after treatment. It can be seen that the best result for this experimental configuration has been achieved for 5 min laser treatment. Longer laser treatment acts as an inhibitor. In the second configuration the intensity of laser emission is almost twice as strong as in the previous experiment. Exposure of seeds is equal (5 min) and longer than in the previous experiment. The combination of the higher light intensity with exposure time longer than 5 min probably introduce too much energy in the cell and instead of stimulation leads to inhibition of plant growth. We found that in this case the preliminary soaking of seeds in water showed worse results than those for dry seeds (Svetleva & Aladjadjiyan, 1996). In opposite - for magnetic field treatment the preliminary soaking improves the effect.

2.2.4 Discussion

Our results for the treatment with impulse He-Ne laser have shown the highest germination for the 5- and 7-fold irradiation. It can be compared with the results reported in the paper of Cwintal et al., 2010. Pre-sowing stimulation of seeds of alfalfa with He-Ne laser beam at intensity 6 mW/cm^2 (i.e. 60 Wm^{-2}) in their experiments has had a highest result for 3-fold and 5-fold exposure - it caused a significant increase in the content of specific protein, phosphorus and molybdenum in dry matter of the plants, and a decrease in the content of crude fibre.

Possible explanation of the inhibitory effect of preliminary soaking of seeds in water, observed in our investigation (Svetleva & Aladjadjiyan, 1996) could be related to re-distribution of energy emitted by laser. In the soaked seeds part of energy probably is absorbed by the imbibed water and the chlorophyll centrums received less energy compared to dry seeds.The explanation of these effects of laser treatment is not consistent. Some authors, as sited in Hernandez et al., 2010, accept that laser treatment can be regarded as a stress agent damaging cells and tissues. Salyaev et al., 2007, (as cited in Hernandez et al., 2010) suggest that a general cell response induced by laser light irradiation can be divided into two specific responses: the first one consists in a rapid stress effect resulting in an increase in the amount of lipid peroxidation products, and the second and longer one include the secondary reactions related to the adaptive metabolic changes and apparently accompanied by the stimulation of morphogenetic processes.

Numerous studies, cited in Hernandez et al., 2010, show a positive effects of pre-sowing laser irradiation, both on cereals (like rice, maize), and vegetables (tomato, radish, peas, cucumber, lettuce, onion, etc.). Several studies report that the seeds of vegetables are more sensitive and susceptible to laser stimulation than cereals (Drozd & Szasjener, 1999, and Gladyszewska, 2006, as cited in Hernandez et al., 2010). In opposite, the investigation of Yamazaki et al., 2002, as cited in Hernandez et al., 2010, showed 36% decrease in number of tiller spikes and 60% decrease in seed yield at final harvest of rice, treated with red-laser diode supplemented with blue light. The authors suggest the necessity of an optimization process when laser is applied as a pre-sowing treatment. Our experiments also support their opinion.

2.3 Ultrasound

2.3.1 State of the art

The sensibility of plants to ultrasound is related to mechanoperception (Telewski, 2006). Plants live in an environment including influence of mechanical forces – gravity, pressure of different flows (atmospheric, water), and have developed sensing of mechanical signals.

Ultrasound is a mechanical wave having frequency higher than 20 kHz. It was established that the treatment with ultrasound could change the state of the substances and even accelerate the reactions (Suslick, 1994). This fact motivated its application for stimulating the growth of different cultures. Sonification has been found favourable for the acceleration of early stages of plant development.

Ultrasound has been shown to have strong effect on plants, particularly on seed germination (Davidov, 1961; Timonin, 1966; Halstead & Vicario, 1969; Hageseth, 1974; Weinberger & Burton, 1981; Miyoshi & Mii, 1988 as cited in Telewski, 2006). Timonin, 1966, as cited in Telewski, 2006, reported that ultrasound treatment altered the viscosity of macromolecule solutions in seeds.

Ultrasound treatment of seeds often was used for industrial purposes like oil extraction and malts preparation (Kobus 2008; Tys et al, 2003), as well as for seeds' disinfection (Nagy et al. 1987; Nagy 1987). The obtained results are caused by cell destruction under the shock of the mechanical wave with high intensity. Cell destruction facilitates oil extraction; it also destroys infection transmitters.

Another application of ultrasound treatment for stimulating plant growth is also investigated. Different cultures were subjected to ultrasonic stimulation: pepper, tomatoes and cucumbers (Markov et al. 1987), fodder beans (Rubtsova 1967), radish (Shimomura, 1990), corn (Hebling et al. 1995), carrot (Aladjadjiyan 2002b), chickpea, wheat, pepper, and watermelon (Goussous et al. 2010), ornamental trees (Aladjadjiyan 2003a, 2003b), barley (Yaldagard et al. 2008a, 2008b, 2008c).

Obtained results have indicated that the effects of the treatment on seed germination depend on frequency of ultrasonic wave and exposure time as well as on plant species and cultivars. Most of the authors recommended the treatment with ultrasound of frequencies 15 – 100 kHz and exposition from 1 to 60 min, with radiation density between 1 and 10 Wcm^{-2} .

2.3.2 Materials and methods

In our previous works the effect of ultrasound treatment with a frequency of 22 kHz and a power of 150 W on the germinating energy and germination of carrot seeds (*Daucus carota* L.), cv. Nantes was studied (Aladjadjiyan 2002b). The maximum effect was established for 5 min treatment. Similar results have been obtained for some ornamental species (Aladjadjiyan 2003a, 2003b).

To explore the role of different parameters later the ultrasonic treatment of seeds has been implemented with an apparatus Carrera Sinus 2501 with frequency 42 kHz and power 100 W. In this investigation seeds of lentils (*Lens Culinaris*, Med.) and wheat (*Triticum aestivum*) have been used (Aladjadjiyan 2011).

From the theory of acoustics it is known that the intensity I of ultrasonic wave is related to its frequency ω:

$$I = \frac{\rho \omega^2 A^2}{2} v \qquad (1)$$

where ρ is the density of the medium, A is the amplitude of the ultrasonic wave, and v - the velocity of the sound.

The bigger frequency means bigger intensity of ultrasound wave and a stronger influence on the samples. That is because for treatment of lentils and wheat seeds shorter exposure times were chosen compared with those in the case of carrot seeds and ornamental trees (Aladjadjiyan, 2002b; Aladjadjiyan, 2003a, b).

Seeds have been distributed in four variants, respectively for lentils and wheat. Each variant has been repeated in 10 replicates containing 10 seeds each. Seeds in each variant have been exposed to ultrasound for different time: 0 min (control), 1 min, 2 min, and 3 min. Seeds have been soaked in plastic containers with tap water and placed in the centre of the ultrasonic apparatus. After the treatment the seeds were cultured in small plastic pots (∅ =7 cm and h = 7 cm) on wet filer paper.

To assess the effect of the treatment, the germination energy GE, the germination G, stem (SL) and root (RL) length and the total mass TM on the 7th and 14th day had been measured as described previously in part 2.1.2.

2.3.3 Results

The results for the SL and RL for seeds of lentils and wheat measured on the 7th day are illustrated on fig.8. It can be seen that the values for treated samples both for SL and RL rise with exposure time. The rise of SL for lentils at exposure time 1 min is 24%, for 2 min - 86%, and for 3 min – 100%. In case of wheat seeds the rise of SL is not so strong. It is respectively 9% for 1 min, 5% for 2 min and 16% for 3 min exposure. The rise of RL both for wheat and lentils seeds is well expressed. The values for lentils are 67% for 1 min exposure time, 130% for 2 min and 214% for 3 min. In the case of wheat it is 56%, 122 and 148%, respectively.

The dependence of SL and RL *vs.* exposure time in the case of lentils is approximately linear, while in the case of SL for wheat it is weak and the differences are not statistically significant.

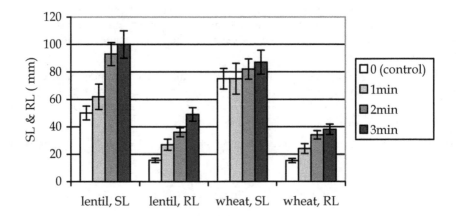

Fig. 8. Stem (SL) and root (RL) length of sonified lentils and wheat seeds at 7th day

The results for the measurements on the 14th day show some differences between the behaviour of lentils and wheat seeds. In the case of lentils SL (fig.9) for treated samples are longer than the control one. The change is respectively 35% for 1 min, 18% for 2 min and 18,5% for 3 min. A well expressed maximum is accounted for exposure time 1 min. In the case of wheat the longest SL is measured for untreated (control) samples. The values of SL for treated samples are less than control one. The values of RL for samples treated at exposure time 1 min are bigger than the control, both for lentils and wheat. The rises are 7% and 10% at exposure time 1 min, for lentils and wheat, respectively. In the case of wheat the difference is statistically significant. For exposure time 3 min the rise 6% has been accounted only for lentils.

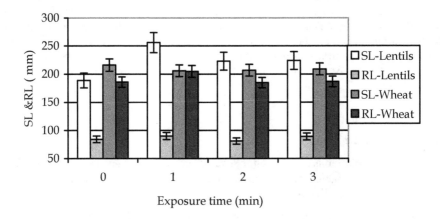

Fig. 9. Stem (SL) and root (RL) length of sonified lentils and wheat seeds at 14th day.

It have to be pointed out that only the length of the main root has been measured without taking into account the lateral roots. This can partially explain the differences in the rise of TM for the samples, for which a rise of SL and RL was not accounted.

The dependence of the total mass of lentils and wheat seeds *vs.* exposure time is presented on fig. 10. It can be noticed that both for lentils and wheat the TM of all the treated samples

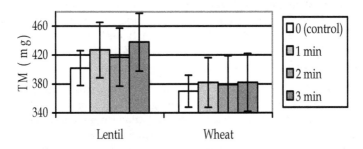

Fig. 10. Total mass (TM) of sonified lentils and wheat seeds at 14th day.

are bigger than control ones. The comparison shows that the TM of the samples of lentils are bigger than those of wheat, but in the case of wheat the difference between TM of treated and control samples is smaller than that for lentils. For all the variants the differences are not statistically significant. The rise of TM is 6%, 4%, and 9% for exposure times 1min, 2min, and 3min, respectively, in the case of lentils. In the case of wheat it is 3%, 2%, and 4%, respectively. The differences are not statistically significant.

2.3.4 Discussion

Suslick (1994) mentioned that the chemical effects of ultrasound are diverse and include substantial improvements in chemical reactions. In some cases, ultrasonic irradiation can increase reactivity by million times. The rise in plant growth characteristics for both lentils and wheat, better expressed for the measurements on the 7th day, may be explained with the increased reactivity of biological substances in the seeds under the influence of ultrasound. Later measurements on the 14th day show that the differences between the characteristics of treated and control samples decrease. This fact may be attributed to the kinetics of the effect of ultrasonic treatment on chemical reactions - possibly occurs some attenuation of this effect in the time.

2.4 Microwave radiation

2.4.1 State of the art

The microwave radiation is electromagnetic radiation with frequencies between 0,3GHz and 300GHz. Most investigated is the radiation 2,45 GHz because it is absorbed by water molecules, present in all live cells.

Banik et al., (2003) reviewed the bioeffects of microwave, mostly on animal and human health. In their paper the most popular opinion has been outlined, that the effect of microwave is attributed mainly to the heating. Nevertheless it has been mentioned that there are also non-thermal microwave effects in terms of energy required to produce molecular transformations.

It has been accepted (Buffler, 1993), that the thermal effect of microwave is related to the interaction with charged particles and polar molecules. Microwave fields are a form of electromagnetic energy and its interaction with charged particles and polar molecules leads to their agitation which is defined as heat. Biological material placed in such radiation absorbs an amount of energy which depends on the dielectric characteristics of the material. The thermal effect of electromagnetic fields from radiofrequency diapason on biological objects is evaluated by Specific Absorption Rate (SAR), defined as the power absorbed per mass of tissue and measured in Wkg^{-1}. The use of SAR for assessment of microwave impact is reported in literature for different biological objects but not for seeds.

In most of the published investigations concerning agriculture the microwave treatment has been used for disinfection of seeds before sowing. Bhaskara Reddy et al. 1995, 1998 used successfully the treatment with electromagnetic radiation from the radio- (10 - 40 MHz) and microwave diapason (2,45 GHz) on seeds of mustard, wheat, soybean, peas and rice seeking to eliminate the microorganisms (Fusarium graminearum) before seed storage. Similar aims have been described in the PhD thesis of V. Rajagopal, 2009. He has treated grain seeds with microwave radiation aiming disinfection, too. In his work a pilot-scale industrial microwave dryer operating at 2,45 GHz was used to determine the mortality of life stages of *Tribolium castaneum* (Herbst), *Sitophilus granarius* (L.) and *Cryptolestes ferrugineus* (Stephens) adults in wheat, barley, and rye. In the listed works there were no data about SAR evaluation. Tylkowska et al., 2010 have investigated the influence of treatment with microwave radiation (2,45 GHz) with output power 650 W and exposure times between 15 and 120 s on bean seeds infected with 13 fungi species and have found decreasing of infection and increasing of seed germination.

Some authors have investigated the influence of microwave treatment on different properties of seeds. Yoshida et al., 2000 treated soybean seeds with microwave radiation (2,45 GHz) for 6 to 12 min with the aim to improve the distribution of triglycerides in the seed coat. Oprică, 2008 has studied microwave treatment with power density under 1 mW/cm^3 on rapeseeds (*Brassica Napus*) and concluded that the microwaves determined variations of catalase and peroxidase activities depending on the age of the plants, time of exposure and state of seeds (germinated and non germinated) exposed to microwave. In all above-mentioned studies the microwave treatment was oriented to produce effects not related to plant stimulation.

Ponomarev et al., 1996 have investigated the influence of low intensity microwave radiation on the germination of cereals (winter and spring wheat, spring barley, and oats). Radiation with wavelength λ = 1 cm at exposition up to 40 min was used. An increasing of germination for all the treated seeds was observed, the optimum effect of stimulation being accounted at the exposition for 20 min.

Jakubowski (2010) has examined the impact of microwave radiation at frequencies ranging within (2.45-54) GHz on selected potato plant life processes and has found positive impact of microwave radiation at frequency 2.45 GHz on the weight of irradiated seed potato germs, and tubers. For the other investigated frequencies no positive results were accounted.

The treatment with microwave radiation as a stimulation agent in agriculture is not enough investigated yet. The stimulation effect of microwave treatment has been investigated by Aladjadjiyan & Svetleva, 1997 on bean (*Phaseolus vulgaris*) and on some ornamental perennial species *Caragana arborescens Lam.*, *Robinia pseudoacacia L.*, *Gleditsia triacanthos* and *Laburnum anagyroides* Med. (Aladjadjiyan, 2002a) and some encouraging results have been established.

The stimulation effect of microwave treatment for longer exposure time and higher irradiation power by investigating its influence on the early stage development of lentil seeds (*Lens culinaris*, Med.) has been performed by Aladjadjiyan, 2010.

2.4.2 Materials and methods

In our experiments the influence of microwave irradiation with wave-length 12 cm on seeds of lentil (*Lens culinaris*, Med) has been investigated. A magnetron OM75P(31) with frequency of radiation 2,45 GHz and maximum output power 900 W according to supplier's data has been used as microwave source. Maximum density of irradiation has been estimated at 45 kW/m³. The estimation has been obtained by dividing the output power of the device (900 W) to the working volume having dimensions 0,19x0,33x0,32 m³.

In earlier investigation (Aladjadjiyan & Svetleva, 1997) we have found that preliminary soaking of seeds in distilled water increased the effect of stimulation by more than 25 % due to the specific absorption of microwave radiation with wavelength of $\lambda=12$ cm by water molecules. That is because lentils seeds have been preliminarily soaked in distilled water for 1 hour, presuming that the imbibed water plays an important role in the absorption of the energy of microwave radiation.

Seeds for the experiment have been distributed in five variants and 5 replicates each containing 10 seeds. The variants differ by the time of exposure to the microwave radiation. Seeds have been exposed to the microwave radiation for 0 s (control), 30 s, 60 s, 90 s, 120 s. Two modifications of output powers of magnetron – 450 W and 730 W, corresponding to intensities - 22,5 kW/m³ and 36,5 kW/m³ respectively, have been applied.

2.4.3 Results

The effect of microwave treatment of lentils seeds on the germination energy GE and germination G s presented on fig.11, on SL and RL measured at the 7th and 14th day of sowing– on fig.11 and fig.12, respectively. Total mass of seedlings measured at the 14th day is presented on fig.13.

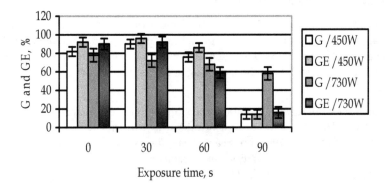

Fig. 11. Germination G and germination energy GE for lentils seeds treated with microwave for different exposure time (0, 30, 60, and 90 s) and output power (450 & 730W).

It can be noticed from fig.11 that for microwave treatment with output power 450 W the highest results for GE and G have been obtained for the exposure time 30 s. This exposure time has shown stimulation effect. All data were significantly different from control. For irradiated samples GE has risen with 9,8 %, while G – with 4,3 %.

The microwave treatment with output power 730 W shows that as well as in the case of treatment with 450 W, the values of G also demonstrate an effect of stimulation for the

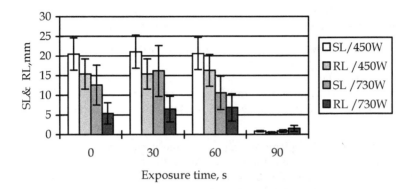

Fig. 12. Length of stems (SL) and roots (RL) for lentils seeds on the 7th day.

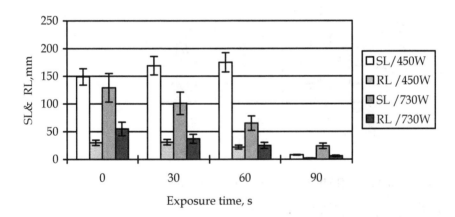

Fig. 13. Length of stems (SL) and roots (RL) for lentils measured on the 14th day.

exposure time 30 s. The differences for GE at exposure 30 s and G at exposure 60 s from the control are not significant. An inhibition of GE can be accounted for longer exposure time (60 and 90 s) as well as for G at exposure 90 s.

The comparison of data for 450 W and 730 W allows concluding that the positive effect of treatment generally is stronger for the lower output power of microwave irradiation – 450 W. Shorter exposure time (30 s) demonstrates higher stimulation effect than longer ones. Exposure time 120 s causes total inhibition.

It have to be pointed out that only the length of the main root has been measured without taking into account the lateral roots. This can partially explain the differences in the rise of TM for the samples, for which a rise of SL and RL was not accounted.

It have to be pointed out that only the length of the main root has been measured without taking into account the lateral roots. This can partially explain the differences in the rise of TM for the samples, for which a rise of SL and RL was not accounted.

The image on fig.13 shows that SL has higher values for the plants treated with microwaves with power 450 W than those for 730 W. The positive effect is accounted for the exposure times of 30 and 60 s. For the treatment with power 450 W at exposure 30 s the value of SL is 12,5% longer than the control one and for exposure 60 s the SL is 13,7% longer. For the treatment with power 730 W the values of SL are shorter than the control. All the differences are statistically significant.

The total mass (Fig.14) of plants *vs.* exposure time rises linearly for the treatment with microwave power 450 W from 0 to 60s, while for the one treated with 730 W there is a maximum at exposure time 30 s. Longer exposure times for the configuration with power 730 W demonstrate an inhibitory impact on total mass values.

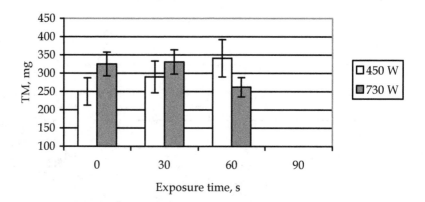

Fig. 14. The total mass (TM) of lentils seeds measured on the 14th day.

2.4.4 Discussion

Total mass for the samples treated with 430 W at 30 s is 16% higher, and for those at 60 s TM is 36,4 % higher than the control. One can conclude that for 450 W the exposure time 60 s is more effective in later stages of development than the exposure at 30 s. A controversy with the data about RL for the same configuration could be noticed. The results on fig.12 show that at exposure 30 s root length is 3% longer, but at exposure 60 s it is 30 % shorter than the control. This controversy could be attributed to the fact that the RL only of the main root is measured; but there are lateral roots that contribute to the weight and are not accounted for root length. This explanation refers also for the accounted rise of TM with 5% for the samples, treated with 730 W at exposure 30 s.

Compared to other examined methods microwave is considered as most harmful one. There are investigations contending the negative influence of microwave treatment on plant development - Jangid et al. 2010 have found that the treatment with microwave oven 2,45

GHz, 800 Wcm⁻² for 7 s induces mutations in seedlings of moth bean [*Vigna aconitifolia* (Jacq.) Marechal]. The authors have found positive influence for shorter exposure time (1, 3, and 5s). Our observations are similar – longer exposure time inhibits the development of seedlings. On the basis of these experiments the suggestion maybe formulated to use cautiously the treatment with microwave and shorter exposure times to be chosen.

3. Conclusion

The treatment with physical factors like different kind electromagnetic fields as well as ultrasound improves germination and early stages of development of plant seeds.

Correct application of physical methods of stimulation requires preliminary experimental investigation and establishment of convenient regimes, which for all the studied cases strongly depends on plant characteristics, intensity of physical factor and exposure time.

Experimental investigations of the physical factors' influence on plant development may help to clarify the mechanisms of energy exchange in molecules and thus stimulation of plant development.

The substitution of chemical methods of plant growth stimulation with physical ones can help avoiding the pollution of food raw materials with toxic substances.

4. References

Abdul Qados A.M.S. & Hozayn M. (2010a). Magnetic Water Technology, a Novel Tool to Increase Growth, Yield and Chemical Constituents of Lentil (Lens esculenta) under Greenhouse Condition. *Am-Euras. J. Agric. & Environ. Sci.*, vol.7, No.4, pp. 457-462.

Abdul Qados A.M.S. & Hozayn M. (2010b). Response of Growth, Yield, Yield Components and Some Chemical Constituents of Flax for Irrigation with Magnetized and Tap Water. *World Applied Sciences Journal*, vol.8, No.5, pp. 630-634.

Aladjadjian A.& Svetleva D. (1997). Influence of Magnetron Irradiation on Common Bean *(Phaseolus vulgaris L.)* Seeds. *Bulgarian Journal of Agricultural Science*, vol.3, pp. 741-747.

Aladjadjiyan A. (2002a). Study of the influence of magnetic field on some biological characteristics of Zea mais. *Journal of Central European Agriculture*, vol.3, No.2, pp. 89-94.

Aladjadjiyan A., (2002b). Increasing carrot seeds (Daucus carota L.), cv. Nantes viability through ultrasound treatment. *Bulgarian Journal of Agricultural Science*, vol.8, No.5-6, pp.469-472.

Aladjadjiyan A. (2002c). Influence of microwave irradiation on some vitality indices and electroconductivity of perennial crops. *Journal of Central European Agriculture*, vol. 3, No.4, pp.271-276

Aladjadjiyan A. (2003a). The effect of pre-sowing treatment by physical methods on seed germination in some ornamental tree species. *Rastenievudni Nauki*, vol.30, No.2, pp.176-179.

Aladjadjiyan A. (2003b). The effect of pre-sowing treatment by physical methods on seedlings length and fresh mass in some ornamental tree species. *Rastenievudni Nauki,* vol.30, No.3, pp. 278 - 282.

Aladjadjiyan A. (2003c). Use of physical factors as an alternative to chemical amelioration. *Journal of Environmental Protection and Ecology (JEPE)*, vol.4, No. 1, pp. 662-667.

Aladjadjiyan A. (2007). The Use of Physical Methods for Plant Growing Stimulation in Bulgaria. *Journal of Central European Agriculture*, vol.8, No.4, pp. 369-380.

Aladjadjiyan A.& Ylieva T. (2003). Influence of Stationary Magnetic Field on the Early Stages of the Development of Tobacco Seeds (*Nicotiana Tabacum* L.). *Journal of Central European Agriculture*, vol.4, No.2, pp.131-138.

Aladjadjiyan A. (2010a). Influence of stationary magnetic field on lentil seeds. *Int. Agrophysics*, vol. 24, No.3, pp. 321-324

Aladjadjiyan A. (2010b). Effect of Microwave Irradiation on Seeds of Lentils (*Lens Culinaris*, Med.). *Romanian Journal of Biophysics*, vol.20, No.3, pp. 213-221

Aladjadjiyan A. (2011). Ultrasonic stimulation of the development of lentils and wheat seedlings. *Romanian Journal of Biophysics*, vol.21, No.3, pp. 179-188

Atak, C., Celik, O. , Olgun, A., Alikamanoglu, S. & Rzakoulieva, A. (2007), Effect of magnetic field on peroxidase activities of soybean tissue culture. *Biotechnology and Biotechnological Equipment*, vol.21, No.2, pp.166-171.

Banik S., Bandyopadhyay S. & Ganguly S. (2003). Bioeffects of microwave – a brief review. *Bioresource Technology*, vol.87, pp. 155-159

Bhaskara Reddy M.V., Kushalappa A.C., Raghavan G.S.V. & Stevenson M.M.P. (1995). Eradication of seedborne *Diaporthe phaseolorum* in soybean by microwave treatment. *Journal of Microwave Power and Electromagnetic Energy* vol.30, No.4, pp. 199-204

Bhaskara Reddy M.V., Raghavan G.S.V., Kushalappa A.C. & Paulitz T.C. (1998). Effect of microwave treatment on quality of Wheat seeds infected with *Fusarium graminearum*. *Journal of Agricultural Engineering Research*, vol. 71, No.2, pp. 113-117.

Bilalis D., Katsenios N., Efthimiadou A., Efthimiadis P. & Karkanis A. (2011) Pulsed electromagnetic fields effect in oregano rooting and vegetative propagation: A potential new organic method. *Acta Agriculturae Scandinavica, Section B - Plant Soil Science*. Available from:
http://www.informaworld.com/smpp/title~content=t713394126

Buffler, C.R. (1993). *Microwave Cooking and Processing: Engineering fundamentals for food scientists*. New York, NY: Van Nostrand Reinhold

Cakmak T, Dumlupinar R. & Erdal S. (2010). Acceleration of germination and early growth of wheat and bean seedlings grown under various magnetic field and osmotic conditions. *Bioelectromagnetics*, vol. 31, No.2, pp.120-129.

Carbonell, M.V., Martinez, E., Flórez, M., Maqueda, R., López – Pintor A. & Amaya, J.M. (2008). Magnetic field treatments improve germination and seedling growth in *Festuca arundinacea* Schreb. and *Lolium perenne* L. *Seed Science and Technology*, vol.36 No.1, pp.31-37.

Çelik, O., C, Atak, & A. Rzakoulieva, (2008). Stimulation of rapid regeneration by a magnetic field in Paulownia node cultures. *Journal of Central European Agriculture* vol.9, No.2, pp.297-304.

Çelik, Ö., Büyükuslu, N., Atak, Ç., & Rzakoulieva, A. (2009) Effects of magnetic field on activity of superoxide dismutase and catalase in *Glycine max* (L.) Merr. roots. *Polish Journal of Environmental Studies*, vol.18, No.2, pp.175-182.

Cwintal M., Dziwulska-Hunek A., & Wilczek M., (2010). Laser stimulation effect of seeds on quality of alfalfa. *International Agrophysics*, vol. 24, No.1, 15-19

Dardeniz A., Tayyar S., & Yalcin S., (2006). Influence of Low-Frequency Electromagnetic Field on the Vegetative Growth of Grape cv. Uslu. *Journal of Central European Agriculture*, vol.7, No.3, pp. 389-395.

De Souza , A., D. García , L. Sueiro , F. Gilart, Porras E., & Licea L. (2006) Pre-sowing magnetic treatments of tomato seeds increase the growth and yield of plants. *Bioelectromagnetics*, vol.27, No.4, pp.247 – 257.

Dhawi, F.& Al-Khayri, J.M. (2009). The effect of magnetic resonance imaging on date palm (Phoenix dactylifera L.) elemental composition. *Communications in Biometry and Crop Science* vol.4, No.1, pp.14–20.

El-Naggar S.M. & Mikhaiel A.A. (2011). Disinfestation of stored wheat grain and flour using gamma rays and microwave heating. *Journal of Stored Products Research*, vol.47, No.3

Flórez, M., Carbonell, V. & Martínez, E. (2007) Exposure of maize seeds to stationary magnetic fields: Effects of germination and early growth. *Environmental and Experimental Botany* vol.59, No.1, pp.68-75

Fraczek J., Hebda T., Slipek Z. & Kurpaska S. (2005). Effect of seed coat thickness on seed hardness. *Canadian Biosystems Engineering*, vol.47, pp.4.1-4.5

Galland, P.& Pazur A., (2005). Magnetoreception in plants. *Journal of Plant Research*, vol.118, No.6, pp. 371-389.

Gouda, O.E. & Amer, G.M. (2009). Performance of crops growth under low frequency electric and magnetic fields. *2009 6th International Multi-Conference on Systems, Signals and Devices*, SSD 2009, art. No. 4956688

Goussous, S.J., Samarah N. H., Alqudah A. M. & Othman M. O. (2010). Enhancing seed germination of four crop species using an ultrasonic technique. *Experimental Agriculture*, vol. 46, No.2, pp. 231-242.

Hebling, S.A. & W.R.da Silva (1995). Effects of low intensity ultrasound on the germination of corn seeds (Zea mays L.) under different water availabilities. *Scientia Agricola*, vol.52, No.3, pp. 514-520.

Hernandez A. C., Dominguez-Pacheco A, Carballo C.A., Cruz-Orea A., Ivanov R., López Bonilla J. L.   & Valcarcel Montañez J. P.(2009) Alternating Magnetic Field Irradiation Effects on Three Genotype Maize Seed Field Performance, *Acta Agrophysica*, vol.14, No.1, pp. 7-17

Hernandez A.C., Dominguez P.A., Cruz O.A., Ivanov R., Carballo C.A. & Zepeda B.R. (2010). Laser in agriculture. *International Agrophysics*, vol. 24, No.4, pp.407-422

Hozayn M., Abdul Qados A.M.S. &Amany Abdel-Monem A. (2010). Utilization of Magnetic Water Technologies in Agriculture: Response of Growth, Some Chemical Constituents and Yield and Yield Components of Some Crops for Irrigation with Magnetized Water. *Int.Journ. Water Resources and Arid Environment,*1-7

Hozayn M.& Abdul Qados A. M. S. (2010). Magnetic water application for improving wheat (Triticum aestivum L.) crop production. *Agriculture and Biology Journal of North America*, vol.1, No.4, pp. 677-682

Hozayn M., Amany Abd El-Monem A. & Abdul Qados A.M.S. (2011). Irrigation with Magnetized Water, a Novel Tool for Improving Crop Production in Egypt. *25th ICID European Regional Conference: Integrated water management for multiple land use in flat coastal areas*. Paper II-15. Available from: http://www.icid2011.nl/

Jakubowski T. (2010). The impact of microwave radiation at different frequencies on weight of seed potato germs and crop of potato tubers. *Agricultural Engineering* vol.6, (124), pp.57-54.

Jangid R. K., Sharma R., Sudarsan Y., Eapen S., Singh G. & Purohit A. K. (2010) Microwave treatment induced mutations and altered gene expression in Vigna aconitifolia. *Biologia Plantarum* vol. 54, No. 4, pp.703-706.

Khalifa N. S. & H. El Ghandoor (2011). Investigate the Effect of Nd-Yag Laser Beam on Soybean (*Glycin max*) Leaves at the Protein Level. *International Journal of Biology*, vol.3, No.2, pp.135-144

Kobus Z. (2008). Dry matter extraction from valerian roots (*Valeriana officinalis* L.) with the help of pulsed acoustic field. *International Agrophysics*, vol.22, pp.133-137

Markov G., Krastev G., & Gogadzhev A. (1987). Influence of ultrasound on the productivity of pepper and cucumbers. *Rastenievudni nauki*, vol.24, No.4

Martínez, E., Flórez, M., Maqueda, R., Carbonell, M.V. & Amaya, J.M.(2009a). Pea (Pisum sativum, L.) and lentil (Lens culinaris, Med.) growth stimulation due to exposure to 125 and 250 mT stationary fields. *Polish Journal of Environmental Studies* , vol.18, No.4, pp.657-663

Martínez E., Carbonell M.V., Flórez M., Amaya J.M., & Maqueda R. (2009b) Germination of tomato seeds (Lycopersicon esculentum L.) under magnetic field. *International Agrophysics* vol.23, pp.45-49

Muszynski S., Gagos M. & Pietruszewski S.(2009) Short-term pre-germiination exposure to ELF magnetic fields does not Influence seedling growth in durum wheat (*Triticum durum*). *Polish Journal of Environmental Studies*, vol.18, No.6, pp1065-1072.

Nagy J. & Ratkos J. (1987). Effect of ultrasound and fungicides on the infectedness of sunflower with *Peronospora*. *International Agrophysics*, vol.3, pp.301-306

Nagy J. (1987). Increased efficacy of seed-dressing agents by ultrasound exposure of rice nematodes. *International Agrophysics*, vol.3, pp. 291-300.

Odhiambo, J.O., Ndiritu F.G. & Wagara I.N., (2009) Effects of Static Electromagnetic fields at 24 hours incubation on the germination of Rose Coco Beans (*Phaseolus vulgaris*) *Romanian Journal of Biophysics*, vol.19, No.2, pp. 135–147.

Oprică L., (2008). Effect of Microwave on the Dynamics of some Oxidoreductase Enzymes in *Brassica Napus* Germination Seeds. *Analele Ştiinţifice ale Universităţii „Alexandru Ioan Cuza", Secţiunea Genetică şi Biologie Moleculară*, TOM IX

Pietruszewski S., Muszynski S., & Dziwulska A., (2007) Electromagnetic field and electromagnetic radiation as non-invasive external stimulants for seeds (selected methods and responses). *International Agrophysics*, vol. 21, pp.95-100.

Podlesny J., Pietruszewski S., & Podlesna A., (2004) Efficiency of the magnetic treatment of broad bean seeds cultivated under experimental plot conditions. *International Agrophysics*, vol.18, pp.65-71.

Podlesny J., Pietruszewski S., & Podlesna A. (2005) Influence of magnetic stimulation of seeds on the formation of morphological features and yielding of the pea. *International Agrophysics*, vol.19, pp.61-68.

Ponomarev L. I., V. E. Dolgodvorov, Popov V.V. , Rodin S.V., & Roman O.A. Roman. 1996. The effect of low-intensity electromagnetic microwave field on seed germination. *Proceedings of Timiryazev Agricultural Academy*, vol.2, pp. 42-46

Racuciu M., D.Creanga, & I.Horga (2007). Plant growth under static magnetic field influence. *Romanian Journal of Physics*, vol.53, No.1-2, pp. 331-336

Racuciu M. (2011). 50 Hz Frequency Magnetic Field Effects on Mitotic Activity in the Maize Root. *Romanian Journal of Biophysics*, vol.21, No. 1, pp.53-62

Rubtsova, I. D., (1967) Effect of Ultrasound on the Germination of the Seeds and on Productivity of Fodder Beans, *Biofizika*, vol.12, No.3, pp. 489-492.

Rajagopal V. (2009). Disinfestation of Stored Grain Insects Using Microwave Energy. *PhD Thesis*, University of Manitoba, USA

Shimomura, S. (1990). The effects of ultrasonic irradiation on sprouting radish seed, Ultrasonics Symposium, in the *Proceedings, IEEE*, vol.3, pp. 1665-1667.

Shine M.B., K.N.Guruprasad, & A. Anand (2011). Enhancement of germination, growth and photosynthesis in soybean by pre-treatment of seeds with magnetic field. *Bioelectromagnetics* . v.32, No.6. pp. 474-484.

Suslick K.S. (1994). The chemistry of ultrasound. *The Yearbook of Science & the Future*; Encyclopaedia Britannica: Chicago, 1994; pp.138-155.

Svetleva D. & Aladjadjian A., (1996). The effect of helium-neon laser during irradiation of dry bean seeds. *Bulgarian J. Agric. Sci.*, vol.2, No.5, pp.587-593

Szczes A., Chibowski E., Holysz L. & Rafalski P. (2011). Effects of static magnetic field on electrolyte solutions under kinetic conditions. *The Journal of Physical Chemistry A*, vol .115, pp.5449-5452.

Telewski F. W. (2006). A unified hypothesis of mechanoperception in plants. *American Journal of Botany* vol.93. No.10, pp. 1466–1476.

Torres, C., Diaz, J. E. & Cabal, P. A. (2008). Magnetic fields effect over seeds germination of rice (*Oryza sativa* L.) and tomato (*Solanum lycopersicum* L.). *Agronomia Colombeana*, vol.26, No.2, pp.177-185.

Tylkowska K., M.Turek & R.Blanco Prieto (2010). Health, germination and vigour of common bean seeds in relation to microwave irradiation. *Pytopathologia*, vol. 55, pp.5-12

Rybacki R. & Malczyk P.(2003). Sources for contamination of rapeseed with benzo (a) pyrene. *International Agrophysics*, vol.17, pp. 131–135

Vajdehfar T. S., M. R.Ardakani, F. Paknejad, M. M.A. Boojar & S.Mafakheri (2011). Phytohormonal Responses of Sunflower (*Helianthus Annuus* L) to Magnetized Water and Seed Under Water Deficit Conditions. *Middle-East Journal of Scientific Research*, vol.7, No.4, pp.467-472.

Vashisth, A., & Nagarajan, S. (2008) Exposure of seeds to static magnetic field enhances germination and early growth characteristics in chickpea (Cicer arietinum L.) *Bioelectromagnetics*, vol.29, No. 7, pp. 571-578.

Vashisth, A., & S. Nagarajan, (2010) Effect on germination and early growth characteristics in sunflower (Helianthus annuus) seeds exposed to static magnetic field. *Journal of Plant Physiology*, vol.167, No.2, pp. 149-156.

Yaldagard M., S.Mortavani, & F.Tabatabaie, (2008a). Effect of Ultrasonic Power on the Activity of Barley's Alpha-amylase from Post-sowing Treatment of Seeds. *World Applied Sciences Journal*, vol. 3, No.1, pp. 91-95.

Yaldagard M., S.Mortavani, & F.Tabatabaie, (2008b). Application of Ultrasonic Waves as a Priming Technique for Accelerating and Enhancing the Germination of Barley

Seed: Optimization of Method by the Taguchi Approach. *Journal Inst. Brewing,* vol.114, No. 1, pp. 14-21.

Yaldagard M., S.Mortavani, & F.Tabatabaie, (2008c) Influence of ultrasonic stimulation on the germination of barley seed and its alpha-amylase activity. *African Journal of Biotechnology,* vol. 7, No.14, pp.2465-2471.

Yoshida H., S. Takagi, & Y. Hirakawa. (2000). Molecular species of triacylglycerols in the seed coats of soybeans following microwave treatment. *Food Chemistry,* No.70, pp.63-69.

Zdunek A., M. Gancarz, J. Cybulska, Z. Ranachowski, & K. Zgórska. (2008). Turgor and temperature effect on fracture properties of potato tuber (*Solanum tuberosum* cv. Irga) - *International Agrophysics,* vol. 22, pp.89-97.

10

Rapid Methods as Analytical Tools for Food and Feed Contaminant Evaluation: Methodological Implications for Mycotoxin Analysis in Cereals

Federica Cheli[1], Anna Campagnoli[2],
Luciano Pinotti[1] and Vittorio Dell'Orto[1]
[1]Dipartimento di Scienze e Tecnologie
Veterinarie per la Sicurezza Alimentare,
Università degli Studi di Milano, Milano,
[2]Università Telematica San Raffaele Roma, Roma,
Italy

1. Introduction

Over the past years, food quality is perceived to have improved and food safety has become an important food quality attribute (Röhr et al., 2005). This implies that all aspects of food production and therefore of the feed supply chain must be considered to ensure the safety of human food (Pinotti & Dell'Orto, 2011).

As a result, public authorities and regulatory agencies are pushing producers, manufacturers, and researchers to pay serious attention to food and feed production processes and to develop comprehensive quality policies and management systems to improve food safety and try to enhance consumer information to regain consumers trust in food.

From this point of view, the knowledge and control of the level and distribution of contaminants and undesirable substances in food and feed are become a worldwide topic of interest due to the high economic and sanitary impact on human/animal health. Since it is impossible to fully eliminate the presence of undesirable substances and contaminants, an adequate surveillance and frequent checks are fundamental to assure quality and safety of raw materials destined for direct consumption or industrial processes.

To guarantee food safety, the availability and the need for confirmatory methods of analysis with high sensitivity/accuracy to meet the regulatory requirements remain critical. However, the traditional methods have some typical drawbacks which include: high costs of implementation, long time of analysis and low samples throughput, and the need for high qualified manpower (Tang et al., 2009). The availability of fast, reliable and simple to use detecting tools for food feed products is therefore a target both for the safeguard of customer's health and production improvement (Tang et al., 2009) and it is undoubtedly one of the main challenges and an imperative for a modern feed and food industry.

In recent years, a number of cost-effective and fit-for-purpose approaches have been proposed to determine the effectiveness of the safety measures and to achieve logistical and operational targets. From this point of view, rapid analytical methods would keep commodities and products moving rapidly through the industrial processes, saving time and requiring less technical training. Analytical approaches that provide qualitative or semi-quantitative results for many chemical and microbiological applications are available and would reduce costs by operating a selection of samples to be submitted to more expensive, sensitive and specific analyses and can be recommended for use in sample screening. Among these, a group of rapid methods comprises some approach miming human/animal senses, for instance electronic nose. In many cases, these devices offer a particular kind of information, pointing on a general description of samples rather than providing a set of specific "discontinuous" analytical responses. This further aspect could result useful, under specific conditions, to give an evaluation regarding the "total quality" value of the matrices with a single analysis.

The aim of this chapter will be to evaluate the potentiality offered by rapid analytical approaches to food and feed evaluation, focusing on contaminants and undesirable substances. A critical overview, highlighting characteristics and applications of these techniques, will be offered with examples pointed on specific matrices and contaminants, cereals and mycotoxins, respectively.

2. Food and feed contaminants: Mycotoxins

Cereals are still by far the world's most important sources of food, both for direct human consumption and indirectly, as inputs to livestock production. FAO's latest forecast for world cereal production in 2011 stands at nearly 2 313 million tones, 3.3 percent higher than in 2010 (FAO, 2011). For the feed sector, cereals represent the main components of industrial feeds, which estimated production, worldwide, is more than 717 million tons (Best, 2011). These volumes make extremely complex the issue of the control and evaluation of quality and safety features and extremely high the amount of analysis that must be performed to meet the regulatory requirements or to give added value to products intended for human and animal consumption. In terms of food safety, cereals represent very heterogeneous materials characterized by a large set of undesirable substances and contaminants. Among the most important risks associated to cereals' consumption are mycotoxins (Codex Alimentarius , 1991).

Mycotoxins are metabolites of fungi capable of having acute toxic, carcinogenic, mutagenic, teratogenic, immunotoxic, and oestrogenic effects in man and animals (D'Mello et al., 1999; Wild & Gong, 2010). Since the discovery of aflatoxins in 1960 and subsequent recognition that mycotoxins are of significant health concern to both humans and animals, mycotoxins have received considerable attention as biotoxins in the food chain. Extensive mycotoxin contamination has been reported to occur in both developing and developed countries. It has been estimated that up to 25% of the world's crops grown for feed and food may be contaminated with mycotoxins (Fink-Gremmels, 1999; Hussein & Brasel, 2001). These data are in line with those reported by the Rapid Alert System for Food and Feed in the European Union (RASFF, 2009), for which of total 3 322 information notifications of possible risks to human health, 669 were related to mycotoxins. This also means that, if the estimated world production is about 2 300 million tonnes (2011), there are potentially about 500 million tonnes of mycotoxin contaminated grains entering the feed and food supply chain. Furthermore, according to the possible carry-over of mycotoxins, feed contamination can represent also a hazard for the safety of food of animal origin and can contribute to mycotoxin intake in human population (Monaci & Palmisano, 2004; Jorgensen, 2005). In this context, one of the latest surveys (Taylor-Pickard, 2009) confirms that feedstuffs are typically contaminated with more than one toxin, which may have a cumulative effect in terms of toxicity in the animals. This places a number of economic and food safety risks for growers, cereal food business operators and food and feed manufacturers. The risks of contamination are greater when raw materials are not traceable or derive from countries where adequate monitoring infrastructures are not in place (Pinotti et al., 2005;). In this field, the geographic origin of food and feed material is also important (Pinotti & Dell'Orto, 2011). Although it is known that mycotoxins are ubiquitous and not just limited to humid and hot countries, where the climate is more favourable to microbial and fungal contamination, it has been reported that some toxins can occur more frequently than other according to the producing area of the food/feed material. Thus zeralenone, fumonisin and aflatoxin were the most widespread toxins found in Asian commodities. By contrast, zeralenone and deoxynivalenol were the most prevalent toxins in continental Europe samples, even after adjusting for the seasonality of contamination for these different toxins (Taylor-Pickard, 2009). By-products typically contain higher levels of toxins' contamination compared to whole raw materials. From a safety perspective, it is well documented that milling and thermal processing such as baking, extrusion cooking and roasting are treatments that may affect redistribution, stability, change and removal of mycotoxins in the processed food (Brera et al., 2006; Bullerman & Bianchini, 2007; Castells et al., 2008; Cheli et al., 2010). Therefore, controls are needed at all stages of cereal production and processing in order to guarantee the quality and safety of the production.

The knowledge and control of the level and distribution of mycotoxins in food and feed are a worldwide objective of producers, manufacturers, regulatory agencies and researchers due to the high economic and sanitary impact on food and feed safety and human/animal health. As stated before, since it is impossible to fully eliminate the presence of undesirable substances and contaminants, maximum concentrations should be set at a strict level which is reasonably achievable considering the risk related to the consumption of the food and, consequently, an adequate surveillance and frequent checks are fundamental to assure quality and safety of raw materials destined for direct consumption or industrial processes. Communities fixed maximum levels for mycotoxins in foodstuffs through the Commission Regulation (EC) No 1881/2006 of 19 December 2006 and Commission Regulation (EC) 1126/2007 of 28 September 2007. In the field of animal nutrition, specific indications on mycotoxins and other undesirable substances in animal feed are considered in the Commission Directive 2003/100/EC of 31 October 2003 and in the Commission Recommendation 2006/576/EC of 17 August 2006.

3. Contaminated food and feed as analytical matrices. Approach to error reduction during sampling and analytical procedures

Ingredients for human foods as for animal feeds are typically very heterogeneous and complex matrices to be analyzed. On the other hand, food and feed contamination can be heterogeneous as well, including biological, chemical and physical contaminants. The biological contamination, comprising microorganism, natural occurring toxins (i.e. mycotoxins from fungi, phycotoxins from algae, toxins from cyanobacteria, histamine, vegetal alkaloids, etc.), and chemical contamination (i.e. agrochemicals as pesticides, plant growth regulators, veterinary drugs, and environmental contaminants as metals, dioxins, BCBs, etc.) get more concern for food and feed safety (Tang et al., 2009). When contaminants and undesirable substances have to be detected or quantified with reasonably confidence, a further critical aspect must be considered, such as their distribution, within a lot to be analyzed. This can be very different due to the characteristics of both food/feed matrices and undesirables molecules themselves. Usually contaminants are divided into two groups, substances uniformly distributed (pesticides, additives, heavy metals, PCBs, dioxins, medicine residues, etc) and non uniformly distributed (natural toxins, GMO, salmonellae, etc.). The type of distribution of contaminants in food and feed has major implications for attempting to precisely and accurately measure the level of contamination in a commodity bulk that is fundamental for products intended for food/feed uses in order to respect the final purposes, i.e. fixed maximum tolerable levels or other operational targets for food/feed industry. Once again a good example is provided by mould and mycotoxin distribution in food and feed commodities. It is well known that mycotoxin contamination is heterogeneously distributed in raw materials (Whitaker, 2004; Larsen et al., 2004). Bulk cereal moisture usually facilitates the development of localized clumps particularly rich in moulded kernels. These small percentages of extremely contaminated portions ("hot spots") are randomly distributed in a lot (average value usually registered about 0.1%) (Johansson et al., 2000a). This condition can lead to an underestimation of the real level of mycotoxin if a too small sample size without contaminated particles is analysed or, instead, to an overestimation of the true level in the case of a too small sample size featuring or more contaminated particles are analyses. Accordingly, when a quantification for a specific contaminant has to be performed in a specific food matrix, all the above mentioned aspects give a fundamental contribute to sampling variability, uncertainty of measurements and finally, to analytical results (Cheli et., 2007a). For these reasons, an analytical methodology to really be considered "fit-for-purpose" should be chosen taking into account not only the sensitivity / specificity, precision and accuracy of the measurement technique adopted, but also its compatibility with an adequate sampling method. In fact, under certain circumstances, as in the case of above described complex, coarse matrices and/or contaminants characterized by the tendency to heterogeneous distribution into the matrix, it appears intuitive that the sampling error could account for an important part of the total error of the final result. On the other hand this topic reveals further interesting implications. If is concrete the hypothesis that, in a specific condition, sampling uncertainty dominates in the uncertainty of the final result, then the choice of an expensive and effective analytical method could result an inefficient strategy. Otherwise, the adoption of a rapid, low cost and high sample throughput analytical approach able to test a high number of samples can represent a better option (Fearn, 2011). From this point of view some statistical approaches can represent helpful tools not only for results' analysis and final data interpretations but also to estimate the importance of the sampling error and in general to estimate the usefulness of a specific analytical application (French, 1989).

As a consequence, the definition of the concept of sampling procedure (also defined "Sampling plan"), and of sampling strategy, as a function of the final target of analysis, and, when possible, the selection of the opportune analytical technique, including rapid methods, represent topics that deserve further in-depth examination in order to achieve the optimization and the fitness of purpose of an analytical approach for contaminant evaluation in food and feed.

3.1 Plan a sampling procedure for mycotoxins

A sampling plan for mycotoxins may be defined as a "test procedure combined with a sample acceptance limit" (Johansson at al., 2000b). A sampling procedure is a multistage process and consists of a sampling phase and an analytical phase. The analytical phase can be further splitted into sample preparation and instrumental analysis (Whitaker, 2006). All the phases are associated to a variability which can impair the reliability of the final result. Each phase of a sampling plan is associated to a specific level of uncertainty and therefore, as mentioned above, in no circumstance is it possible to obtain a quantitative value for the contamination associated with 100% certainty (Whitaker, 2006). It is intuitive that each step of a sampling protocol specifically contributes to the final uncertainty of the procedure. The total variance of a specific sampling plan (TV) may be expressed by using statistic variance as a measure of variability and may be described as the sum of sampling variance (SV), and analytical variance (AV) as follows (1):

$$TV = SV + AV \tag{1}$$

(in which AV reassumes the sum of sample preparation variance (SPV) plus instrumental analysis variance (IV)). TV and variance distribution in the different steps of the sampling protocol give indications on the sampling plan efficiency and are also able to compare effectiveness of different sampling plans to the final purpose (Cheli et al., 2009a).

The contribution from SV has often been underestimate, though it is accountable for the largest source of variation associated to the quality of the final analytical result (Whitaker, 2003, Cheli et al., 2009a). There appears to be more substantial literature on food than feed (Cheli et al., 2009a).

Due to the frequently uneven contaminant and undesirable substance distribution in solid samples, such as grains and other alimentary commodities, raw material and matrices, obtaining a representative sample is a way of minimizing false results and increases the chances of accurate determination of mycotoxins in a batch or lot. When designing a specific sampling plan, all critical points have to be considered in order to reduce SV and increase the reliability of the final sample, such as collection of a sufficiently large number/size of incremental samples, choice of the sampling points, aggregate sample size properties, homogeneity of sample components in terms of size and specific weight. All these parameters must specifically consider the type of product and mycotoxin level of contamination. For mycotoxins, it becomes even more important than usual to consider the contribution of SV to the uncertainty of any measurement, and there are implications for the type of measurement technology that may be judged fit for purpose. The contribution of SV, SPV and IV to TV has been evaluated and quantified in several products (Table 1). In this context, quantitative data are available for foodstuffs, but are still lacking for the majority of feedstuffs.

Matrix, mycotoxin and test procedure	SV, %TV	SPV, %TV	IV, %TV	References
Shelled corn, 0.91 kg sample, Romer mill, 50 g subsample, 1 aliquot analysed, aflatoxin 20 ng/g	75.6	15.9	8.5	Whitaker, 2006
Shelled corn, 4.54 kg sample, Romer mill, 100 g subsample, 2 aliquots analysed, aflatoxin 20 ng/g	55.21	29.1	15.7	Whitaker, 2006
Shelled corn, 1.13 kg sample, Romer mill, 50 g subsample, 1 aliquot analysed, aflatoxin 20 ng/g	77.8	20.5	1.7	Johansson *et al.*, 2000c
Wheat, 0.454 kg sample, Romer mill, 25 g subsample, 1 aliquot analysed, Deoxynivalenol ppm	22	56	22	Whitaker *et al.*, 2002
Shelled corn, 5 kg sample, Romer mill, 100 g subsample, 1 aliquot analysed, aflatoxins 20 ng/g	59.8	34.5	5.7	Johansson *et al.*, 2000c
Peanut, 2.27 kg sample, 100 g subsample, aflatoxin 100 ppb	92.7	7.2	0.1	Whitaker *et al.*, 1994
Shelled corn, kg sample, 25g subsample, 1 aliquot analysed, fumonisin 2 mg/kg	61	18.2	20.8	Whitaker *et al.*, 1998

Table 1. Distribution of variability associated to each sampling step: sampling (SV), sample preparation (SPV) and instrumental analysis (IV) (modified from Cheli et al., 2009a).

The methods of sampling and analysis for the official control of the levels of mycotoxins, are reported in Commission Regulation (EC) No 401/2006 of 23 February 2006 and Commission Regulation (EC) No 152/2009 of 27 January 2009. These regulations provide different sampling plans according to the type of food and feed products, respectively. However, screening, monitoring, controlling, exposure studies or targeted purposes may require specific sampling and analytical approaches (Miraglia et al., 2005).

3.2 Toward optimization of sampling and analysis procedures

Some aspects related to sampling plan evaluation and the establishment of a decision strategy are more detailed by Fearn et al. (2002) in an interesting paper in which the authors describe a possible approach to the systematic optimization of the different phases during the entire sampling procedure. Later on, this approach enables an economic evaluation of the entire process, and, as a consequence, an objective comparison among different plans applicable to the same situation. Cost can be in fact defined as the measurement unit to take the optimal decision if it is considered that the optimal decision represents the choice of the most economic from different plans when quality of results are comparable. In a sampling procedure, total cost can be defined as the analytical cost plus the potential losses incurred in using the result.

To plan a sampling procedure, the analytical method, numbers of replicates samples, numbers of replicate measurements per samples and the sampling technique have to be selected. Thus, a systematic approach is first to optimize numbers of replicate samples and analyses separately for each combination of sampling technique and analytical methods. Then the optimised total costs of different methods may be compared.

As described in 3.1 paragraph, the uncertainty of the measurement can be expressed in terms of total measurement variance, calculated as the sum of sampling and analytical variance. Considering a measurement process in which n samples are taken and m replicate analyses per sample made, the uncertainty of the measurement is dependent on the number of samples and replicate analyses. Increasing the number of samples and/or analyses will reduce the uncertainty but will increase cost to obtaining the measurement. For a given cost, different allocations of resources between sampling and analysis may give different variances. As a consequence, for a fixed cost, a balance between sampling and analysis may be found with the aim to reach the best economic purpose and the minimum total measurement variance (besides usually there are few sampling or analytical methods available for a given problem so the choice can be simplified).

Thus, the total variance of the sampling plan can be more completely described as in (2)

$$TV^2 = (SV^2/n) + (AV^2/mn) \tag{2}$$

where n is the total number of samples taken and m is the number of analyses carried out on each sample; while the total cost of obtaining the measurement (cost of the entire sampling plan) (TC) including sampling (SC) and analysis cost (AC), can be defined as in (3)

$$TC = nSC + mnA \tag{3}$$

Either fixing the cost TC and minimizing the variance TV^2 or vice versa, the optimal number of replicate analyses can be shown to be (4)

$$m_{opt} = (SV/AV) \cdot \sqrt{(SC/AC)} \tag{4}$$

The value of m will need to be rounded to the nearest whole number. New rounded value for m give important information. If m does not seem sensible, this may indicate that the sampling and analytical methods are badly matched. Large values of m_{opt} will result if the analytical variance is large compared with sampling variance or if the sampling cost is large compared with the analytical cost. Then it may be better considering more precise analyses or less expensive and less precise sampling procedures to get a better balance. Of course not all choices can be permitted and each operational situations allow a specific range of possibilities, so some compromise value of m will need to be chosen. It will rarely be a good idea to make more than 4 or 5 replicate measurements on a sample. Values of much less than one for m_{opt} will occur if the sampling variance or analytical cost dominate. Again may be useful to consider alternative analytical procedures that are less precise and therefore less costly.

A practical example can be done. Starting from the assumption that the standard deviation and the cost for single sample of an analytical method are usually known and that frequently when a sampling methodology is consolidated the relative standard deviation and cost can be inferred, we can suppose the sampling has a SD=0.8 with a cost of 21.00 Euros, while the analysis has a SD=0.6 and a cost for single sample of 4.00 Euros. m_{opt} will be calculated as (5)

$$m_{opt} = (0.6/0.8) \cdot \sqrt{(21.00/4.00)} = 1.72 \tag{5}$$

so m_{opt} will be approximate to 2. Then each sample will cost 29.00 Euros (21.00+2*4.00) and results associated to a SD=0.91.

After having obtained cost and SD of result, the next step is to find the optimal level of sampling replications n, balancing measurement costs against possible losses. When choosing a value for n, then each optimized method can be compared with the other candidate methods. If the optimal n is less than one in situations where an m of greater than one has been used it may be reasonable trying smaller values of m.

When optimising each method separately, then they can be compared by comparing the total costs. In the absence of other operational or technical considerations the least cost option will be chosen.

As general consideration the use of a decision strategy like those described allows a rational approach to the problem of choosing analytical methods, a sampling scheme and how to mach efficiently these two phases of the sampling procedure. Under certain circumstances, there is no doubt that some parameters may be difficult to quantify. Probably for instance, the most problematic of the inputs will usually be the losses arising from measurement errors. In situations where the potential losses are very large, it may be necessary to take account of a nonlinear utility for money. Despite these aspects, it can be state that is still possible to get useful results from this approach.

4. Rapid methods for mycotoxin analysis

The use of so called "Rapid Methods" is highly relevant for improving the knowledge on the presence and distribution of mycotoxins in food and feed and for creating a reliable database (Stroka et al., 2004). These low cost, simple, rapid and reliable methods may be applied in laboratory and non-laboratory environment and combine effective sampling with analysis of a large number of samples for a screening approach. As a general rule, rapid methods that provide qualitative or semi-quantitative results are recommended in sample screening. An analytical method is usually referred to as "rapid" when it requires, at most, a few minutes to obtain a result (van Amerongen et al., 2007). Currently, there are three main tendencies to develop rapid methods for mycotoxin analysis in order to reduce the quantity of assays and, therefore, to shorten time and to lower costs for feed and food quality control: 1) improvement of speed, user-friendliness, reliability, non-destructiveness, 2) use in a non-laboratory environment, 3) simultaneous determination of multiple mycotoxins (Maragos,

2004). In recent years, a number of rapid, cost-effective and fit-for-purpose approaches have
been proposed to determine the effectiveness of the safety measures, to determine legal
compliance, to achieve logistical and operational targets, to keep commodities and products
moving rapidly through marketing channels, to save time and investments in complex
instruments. Some are advanced enough for field studies and have already reached the
stage of commercialization, some are at a transition phase between research and application
to analysis of food/feed samples, other still have to face the challenge of validation by
multiple laboratories. A list of the emerging rapid methods for mycotoxin analysis is
reported in Table 2.

Methods	Advantages	Disadvantages	References
LFD (lateral flow device)	Rapid No expensive equipment Easy to use	Semi-quantitative Validation required for each matrix	Maragos, 2004; Zeng et al., 2006; Goryacheva et al., 2007.
FPI (fluorescence polarization immunoassay)	High sensitivity Low matrix interference	Not usable for simultaneous detection of several individual mycotoxins	Maragos, 2004; Goryacheva et al., 2007.
CE (capillary electrophoresis)	High sensitivity Non polluting technology Possible simultaneous multi-component analysis	Expensive equipment Expensive Clean-up may be required	Maragos, 2004; Maragos & Appel, 2007.
SPR (surface plasmon resonance)	Rapid No clean up	Cross reactivity	Tudos et al., 2003; Van der Gaag et al., 2003; Maragos, 2004.
MIP (molecularly imprinted polymers)	Low cost Stable Reusable	Poor selectivity	Maragos, 2004; Logrieco et al., 2005; Krska & Welzig, 2006.
IR spectroscopy (NIR, FR-NIR)	Rapid Non destructive measurements No clean up Easy to use	Expensive equipment Calibration model must be validated Good for classification	Kos et al., 2002, 2003; Petterson & Aberg, 2003; Berardo et al., 2005; De Girolamo et al., 2009.
EN (electronic nose)	Rapid Non destructive measurements No clean up	Calibration model must be validated Good for classification	Keshri & Magan, 2000; Olsson et al., 2002; Presicce et al., 2006; Cheli et al., 2009b; Campagnoli et al., 2011.

Table 2. Examples of emerging rapid methods for mycotoxin analysis.

Emerging technologies and their potential application in rapid mycotoxin detection have been recently reviewed (Maragos, 2004; Krska & Welzig, 2006; Zeng et al., 2006; Goryacheva et al., 2007; Cheli et al., 2008; Maragos & Busnam, 2010). The most known rapid screening methods for mycotoxin detection, especially for the screening of raw materials, are antibody-based methods, ELISA test. The ELISA methods have been commercially available since many years and are extensively used as rapid screening methods. Kits are available in quantitative, semi-quantitative or qualitative formats (Zeng et al., 2006). These methods are easy to use, fast and suitable for testing mycotoxin in the field too. Within the concept of flexible out of laboratory testing, non instrumental (visual) membrane based immunoassays (dipstick, lateral flow and flow-through tests) have been developed and are commercially available for several mycotoxins and matrices. The main advantages of non instrumental ELISA methods are field portability, not requirement of any specialized equipment and simple sample preparation procedures, while the main disadvantages are subjective interpretation, lower sensitivity and higher cost/test compared with instrumental ELISA methods (Zeng et al., 2006; Goryacheva et al., 2007). Although immunochemical methods have become one of the most useful tools for mycotoxin rapid screening, the price for simplification may be usually lower sensitivity. The main problems with antibody-based methods are related to the characteristics of the antibody, test specificity (cross-reactivity), matrix interference and interpretation of the result, if the method is semi-quantitative, when the mycotoxin concentration is close to the method cut-off level. Still insufficient validation studies of ELISA methods for all commodities limit their use to those matrices for which they were validated.

Apart from ELISA, the more recent and best candidates as mycotoxin analytical methods for further developments in terms of rapid methods, multi-mycotoxin assays, easy to use and to be validated by multiple laboratories are capillary electrophoresis (CE), fluorescence polarization immunoassay (FPI) and surface plasmon resonance (SPR). CE methods are laboratory-based methods because of the size and required automation of the instrumentation, while FPI and SPR methods may be much more portable and therefore may be used outside the laboratory and have reached the stage of commercialization. CE methods for aflatoxins, fumonisins, ochratoxin A, deoxynivalenol, moniliformin and zearalenone have been reviewed by Maragos (1998). The main advantage of CE is the possibility to reach a sensitivity comparable to that of established HPLC methods. Combination of CE with immunoassay makes it possible a simultaneous multi-component analysis due to the high resolving power of CE.

FPI are solution based-assays in which a mycotoxin-fluorophore conjugate (tracer) is used. Applications of FPI assays have been described for detection of deoxynivalenol, fumonisins, aflatoxins, zearalenone and ochratoxin A in cereals, semolina and pasta (Maragos, 2004; Goryacheva et al., 2007). Good correlation have been found between comparative analyses performed by FPI and HPLC. The main advantages of FPI are a high sensitivity and a low matrix interference. The potential speed of FPI assays combined with the portability of commercially available devices, suggests this to be a promising technology for mycotoxin detection. A limit of FPI is that it cannot be used for simultaneous detection of several individual mycotoxins.

SPR is a measure of mass changes that occur in a sensor surface. Applications of SPR assay for detection of DON, fumonisins, aflatoxins, zearalenone and ochratoxin A have been developed and optimized (Daly et al., 2000; Schnerr et al., 2002; Tudos et al., 2003; van der Gaag et al., 2003). SPR sensitivity for aflatoxin B1 has been demonstrated to be higher than ELISA assay. Studies on naturally contaminated samples showed that SPR results are in agreement with liquid chromatography mass spectrometry (LC-MS) measurements (Tudos et al., 2003; van der Gaag et al., 2003). A technique for the simultaneous detection of four different mycotoxins in a single measurement using SPR commercially available portable equipment was recently reported (van der Gaag et al., 2003).

Emerging challenge of sensors for mycotoxins is represented by the development of non-biologically based binding, such as molecularly imprinted polymers (MIPs) (Maragos, 2004; Logrieco et al., 2005; Krska et al., 2005). Rapid future applications of MIPs are expected if affinity problems are overcome. Mimicking antibodies is the basic idea of MIPs technology. The preliminary results of MIPs technology in zearalenone, deoxynivalenol, and ochratoxin A analysis has been reported (Visconti & De Gerolamo, 2005; Krska & Welzig, 2006). Although the affinity of MIPs are not yet competitive with those of antibodies, this technique offers a good potential for further developments.

Near Infrared (NIR) Spectroscopy, micro system technology tools based on DNA arrays, electronic noses and tongues, biosensors and chemical sensors for the detection of fungal contaminants in feed and food are other emerging, available and promising methods (Larsen et al., 2004; Maragos, 2004; Logrieco et al., 2005; Zeng et al., 2006; Cheli et al., 2008). Infrared (IR) spectroscopy has been continuously evolving, as can be deduced comparing the old mid-IR equipment manufactured in the 1950s and based on dispersive monochromators with the present customized near infrared (NIR) instrumentation. The incorporation of the Fourier transform technique (FT) together with the interferometric spectrometers into the mid-IR instruments has increased the use of this technique in food analysis (Ibañez & Cifuentes, 2001). Although NIR spectroscopy has been used routinely since many years as a rapid method in feed and food industry for determination of constituents such as humidity, proteins, lipids with a precision comparable with that of the official methods of analysis, a limited number of publications concerning mycotoxins and NIR spectroscopy have been reported. This is because the concentration of mycotoxins normally found in feed and food has been considered low for this technique. Recently NIR and mid-infrared (MI) spectroscopy with attenuated total reflection (IR/ATR and FT-IR/ATR) have been used in order to rapidly detect the presence of fungal infection and estimation of fungal metabolites and mycotoxins in naturally and artificially contaminated products (Kos et al., 2002, 2003; Petterson & Aberg, 2003; Berardo et al., 2005; De Girolamo et al., 2009). Multivariate analysis for the extraction of additional information from the recorded spectra gave promising results on the capability of these techniques as tools and models not only for the detection of mould presence, but also for the prediction of the presence of mycotoxins. Chemometric models applied to FT-IR/ATR analysis enabled correct classification of non contaminated and contaminated maize and wheat with deoxynivalenol (Kos et al., 2003; De Girolamo et al., 2009). The developed method enabled the separation of samples with a cut off level for DON of 300 µg/kg, a value below the maximum level and guidance value proposed by the EU for maize and wheat intended for human and animal consumption. Improvements of the classification performance of FT-IR/ATR analysis can be achieved optimising sample preparation procedure and applying particle size analysis to samples (Kos et al., 2007). The use of NIR spectroscopy for the determination of DON in wheat and fumonisin B1 in maize has been investigated (Petterson & Aberg, 2003; Berardo et al., 2005). It has been shown that it is possible to predict DON concentration in wheat kernels by NIR at levels higher than ca. 400 µg/kg (Petterson & Aberg, 2003), indicating the high potential of IR spectroscopy for accurately predicting the presence or absence of mycotoxins in cereals.

4.1 The analytical approaches miming senses: The example of electronic nose

Further example of rapid methods are those based on electronic senses, which represent an evolution of sensory evaluation traditionally entrusted to the human/animal senses. The evaluation of food and feed in terms of smell, taste, morphology and colour is often overlooked, but contains a lot of information directly related to quality and safety. In particular, the smell and aroma of a food, due to the presence of many volatile chemicals, are sensory parameters of great interest, which can be used as indicators of food quality (Cheli et al., 2007b). Fungal spoilage induces nutritional losses, off-flavours, organoleptic deterioration often associated to mycotoxins formation. Research studies correlated fungal activity with the production of volatile metabolites characterized by gas chromatography mass spectrometry (GC-MS) (Magan & Evans, 2000). These authors conclude that accumulation and pattern of fungal volatiles can be used as indicators of fungal activity and as taxonomic markers in order to differentiate between fungal species and between toxigenic and non toxigenic fungal strains. Since volatile headspace analysis can be evaluated as a whole by the use of electronic nose (EN), this technique is becoming widespread in order to evaluate mould spoilage, quality and safety of food and feed. An EN is an instrument which comprises an array of electronic chemical sensors with partial specificity and an appropriate pattern recognition system, capable of recognizing simple or complex odours (Gardner & Bartlett, 1994)(Fig. 1). The array of non-specific chemical detectors interacts with different volatile compounds and provide signals that can be utilised effectively as a fingerprint of the volatile molecules rising from the samples analysed. After the achievement of a fingerprint, the identification and/or quantification of the odours by means of a pattern recognition system become possible.

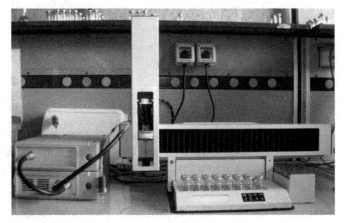

Fig. 1. An example of electronic nose.

The electronic nose does not distinguish each volatile substance, but express the global odour of a product (Gardner & Bartlett, 1994). This ability, as in the case of other devices as electronic tongue or certain applications of computer image analysis, can enable a general evaluation regarding the "total quality" value of the food and feed analyzed. The process is completed with the aid of appropriate mathematical and statistical methods. As previously cited, the use of EN for evaluating the quality of stored grain has been reported. Sensor technology has been shown to enable to determine the mycological quality of grains. The first type of study carried out with EN technology has been made in order to differentiate between non-infected and infected samples with different species or strain of fungi, through the variation of the metabolic pathway due to the contamination of grains. The ability of EN to differentiate grains and bakery products clean or contaminated (naturally or artificially infected) with different mould species have been demonstrated (Magan & Evans, 2000; Olsson et al., 2000; Balasubramanian et al., 2007; Paolesse et al., 2006). Detection and differentiation between mycotoxigenic and non-mycotoxigenic strains of *Fusarium* spp. using volatile production profiles evaluated by EN has been also reported (Keshri & Magan, 2000; Magan & Evans, 2000; Falasconi et al., 2005; Presicce et al., 2006; Sahgal et al., 2007). Further developments of studies carried out with EN technology have been made in order to evaluate the possibility of using fungal volatile metabolites as indicators of mycotoxin presence (Campagnoli et al., 2009b). Results from a study carried out on naturally contaminated barley samples showed that it was possible to use volatile compounds to predict whether the OTA level in samples was below or above 5 µg/kg; seven of 37 samples were misclassified (Olsson et al., 2002). EN analysis enabled correct classification of naturally contaminated maize with aflatoxins (Campagnoli et al., 2009a, 2009b; Cheli et al., 2009b). EN analysis was applied to wheat in the case of naturally DON contaminated samples (Tognon et al., 2005; Dell'Orto et al., 2007; Campagnoli et al., 2009b). A simple analytical protocol, combined with the application of the CART (Classification and Regression Tree) model and PCA (Principal Component Analysis) for the selection of variables and the classification of samples was used in another paper (Campagnoli et al., 2011). Results obtained indicated that the EN equipped with ten MOS (Metal Oxide Semiconductor) sensors array allows the classification of naturally contaminated samples on the basis of DON content into three classes on the basis of the European Union limits for DON in unprocessed durum wheat: (a) non-contaminated; (b) contaminated below the limit (DON < 1,750 µg/kg); (c) contaminated above the limit (DON > 1,750 µg/kg); with a validated prediction error rate of 0% when a 20-sample dataset was considered. (Campagnoli et al., 2011). The same model was used with a 122-sample dataset, 9 contaminated and 113 non-contaminated samples, more faithfully reproducing a real-life situation characterised by unbalanced classes. Although, classifying performance was lower than in the 20-sample dataset case, reasonable results were achieved, with a validated prediction error rate of 3.28% (Table 3). Four errors were computed in prediction; however, none of the contaminated samples were misclassified as non-contaminated, avoiding the worst eventuality under in-field conditions.

Less information is available regarding quantification capability of electronic nose in order to predict mycotoxins concentration in cereals. Tests were conducted on DON levels in barley and wheat. Positive correlation was found between electronic nose data and reference concentration of DON (Olsson et al., 2002). However the performance of the regression model on prediction was quite low (PRESS =0.65, R^2 =0.63, adjR^2 =0.63) (Tognon et al., 2005; Dell'Orto et al., 2007).

	Total true samples		Misclassification matrix (Samples fitted assignment)			Validated misclassification matrix (Samples predicted assignment)		
			Assigned class			Assigned class		
			a	b	c	a	b	c
Class a	113		112	0	1	111	1	1
		rate	0.991	0.000	0.009	0.982	0.009	0.009
Class b	6		0	6	0	0	5	1
		rate	0.000	1.000	0.000	0.000	0.833	0.167
Class c	3		0	0	3	0	1	2
		rate	0.000	0.000	1.000	0.000	0.333	0.667

Table 3. EN use for DON analysis in wheat: performances of classification for a 122-samples dataset. Class a) samples non-contaminated; Class b) samples below the legal limit; Class c) samples above the legal limit (modified from Campagnoli et al., 2011).

5. Conclusion

The plan of an effective sampling procedure for food and feed contaminants' detection or quantification represents a complex challenge for operators. Special attention has to be paid when matrices are coarse and contaminants are characterized by a non uniform distribution, as in the case of mycotoxins in cereal commodities, that represent the most important worldwide human and animal food and feed resources. Under these conditions, sampling uncertainty dominates in the final uncertainty result, then the choice of expensive, precise, sensible, specific analytical method could result an inefficient strategy. Instead, the adoption of a rapid, low cost but high sample throughput analytical approach able to test a high number of samples can represent a better option. This is one of the most important reason for which R&D regarding these analytical approaches and statistical data analysis specifically dedicated merits further implementation. Fearn (2009) states that "*The safest policy is to use the simplest method you can, and within that the simplest model you can, avoiding the temptation to add a lot of extra complexity for a small gain in performance*". Therefore, some analytical methods reveal further useful characteristics for screening purposes. For example, methods miming senses, i.e electronic nose, that, by means of rapid and simple analytical protocols, can provide a general description regarding the quality of complex matrices of interest. Then, samples could be classified and a limited selected number submitted to more expensive and time-consuming quantitative analyses with useful costs reduction.

6. References

Balasubramanian, S., Panigrahi, S., Kottapalli, B. & Wolf-Hall, C.E. (2007). Evaluation of an artificial olfactory system for grain quality discrimination. *LWT-Food Science and techonology*, Vol.40, No.10, (December 2007), pp. 1815-1825, ISSN 0023-643

Berardo, N., Pisacane, V., Battilani, P., Scandolara, A.,Pietri, A. & Marocco, A. (2005). Rapid detection of kernel rots and mycotoxins in maize by near-infrared reflectance spectroscopy. *Journal of Agricultural and Food Chemistry*, Vol.53, No.21, (October 2005), pp. 8128-8134, ISSN 0021-8561

Best, P. (2011). World Feed Panorama: Expansive grains slows industry expansion. *Feed International*, Vol.32, No.1, (January/February 2011), pp. 10-12, ISSN 0274-5770

Brera, C., Catalano, C., De Santis, B., Debegnach, F., De Giacomo, M., Pannunzi, E. & Miraglia, M. (2006). Effects of industrial processing on the distribution of aflatoxins and zearalenone in corn-milling fractions. *Journal of Agricultural and Food Chemistry*, Vol.54, No.14, (July 2006), pp. 5014–5019, ISSN 0021-8561

Bullerman, L. B. & Bianchini, A. (2007). Stability of mycotoxins during food processing. *International Journal of Food Microbiology*, Vol.119, No.1-2, (October 2007), pp. 140-146, ISSN 0168-1605

Campagnoli, A., Cheli, F., Polidori, C., Zaninelli, M., Zecca, O., Savoini, G., Pinotti, L. & Dell'Orto, V. (2011). Use of the electronic nose as a screening tool for the recognition of durum wheat naturally contaminated by deoxynivalenol: a preliminary approach. *Sensors*, Vol.11, No.5, (May 2011), pp. 4899-4916, ISSN 1424-8220

Campagnoli, A., Cheli, F., Savoini, G., Crotti, A., Pastori, AGM. & Dell'Orto, V. (2009)a. Application of an electronic nose to detection of aflatoxins in corn. *Veterinary Research Communications*, Vol.33, Suppl.1, (September 2009), pp. 273-275, ISSN 0165-7380

Campagnoli, A., Dell'Orto, V., Savoini, G. & Cheli, F. (2009)b. Screening cereals quality by electronic nose: the example of mycotoxins naturally contaminated maize and durum wheat, *Proceedings of the 13 International Symposium on Olfaction and Electronic Nose*, pp. 507-510, ISBN 978-0-7354-0674-2, Brescia, Italy, April 15-17, 2009

Castells, M., Marin, S., Sanchis, V. & Ramos, A.J. (2008). Distribution of fumonisins and aflatoxins in corn fractions during industrial corn flakes processing. *International Journal of Food Microbiology*, Vol.123, No.1-2, (March 2008), pp. 81-87, ISSN 0168-1605

Cheli, F., Campagnoli, A., Pinotti, L., Fusi, E. & Dell'Orto, V. (2009)a. Review article - Sampling feed for mycotoxins: acquiring knowledge from food. *Italian Journal of Animal Science*, Vol.8, No.1, (March 2008), pp. 5-22, ISSN 1828-051X

Cheli, F., Campagnoli, A., Pinotti, L., Fusi, E. & Dell'Orto, V. (2007)a. Map a plan for estimating mycotoxin risks. *Feed Management*, (September/October 2007), pp. 10-11, ISSN 0014-956X

Cheli, F., Campagnoli, A., Pinotti, L., Maggioni, L., Savoini, G. & Dell'Orto, V. (2007)b. Testing feed quality: the "artificial senses". *Feed International*, (May/June 2007), pp. 24-26, ISSN 0274-5771

Cheli, F., Campagnoli, A., Pinotti, L., Savoini, G. & Dell'Orto, V. (2009)b. Electronic nose for determination of aflatoxins in maize. *Biotechnology, Agronomy, Society and Environment*, Vol.13, No.S, (2009), pp. 39-43, ISSN 1370-6233

Cheli, F., Campagnoli, A., Ventura, V.; Brera, C., Berdini, C., Palmaccio, E. & Dell'Orto, V. (2010). Effect of industrial processing on the distribution of deoxynivalenol, cadmium and lead in wheat milling fractions. *LWT-Food Science and Technology*, Vol.43, No.7, (September 2010), pp. 1050–1057, ISSN 0023-6438

Cheli, F., Pinotti, L., Campagnoli, A., Fusi, E., Rebucci, R. & Baldi, A. (2008). Mycotoxin analysis, mycotoxin producing fungi assays and mycotoxin toxicity bioassays in food mycotoxin monitoring and surveillance. *Italian Journal of Food Science*, Vol.20, No.4, (December 2008), pp. 447-462, ISSN 11201770

Codex Alimentarius. (1991). Codex standard for durum wheat semolina and durum wheat flour 178-1991 (Rev. 1-1995). Rome: FAO/WHO

Commission Directive 2003/100/EC of 31 October 2003 amending Annex I to Directive 2002/32/EC of the European Parliament and of the Council on undesirable substances in animal feed. *Official Journal of the European Union* 1.11.2003, L 285, pp. 33-37

Commission Recommendation 2006/576/EC of 17 August 2006 on the presence of deoxynivalenol, zearalenone, ochratoxin A, T-2 and HT 2 and fumonisins in products intended for animal feeding. *Official Journal of the European Union* 23.8.2006, L 229, pp. 7-9

Commission Regulation (EC) No 1126/2007 of 28 September 2007 amending Regulation (EC) No 1881/2006 setting maximum levels for certain contaminants in foodstuffs as regards *Fusarium* toxins in maize and maize product. *Official Journal of the European Union* 29.9.2007, L 255, pp. 14-17

Commission Regulation (EC) No 152/2009 of 27 January 2009 laying down the methods of sampling and analysis for the official control of feed. *Official Journal of the European Union* 26.2.2009, L 54, pp. 1-130

Commission Regulation (EC) No 1881/2006 of 19 December 2006 setting maximum levels for certain contaminants in foodstuffs. *Official Journal of the European Union* 21.12.2006, L 364, pp. 5-24

Commission Regulation (EC) No 401/2006 of 23 February 2006 laying down the methods of sampling and analysis for the official control of the levels of mycotoxins in foodstuffs. *Official Journal of the European Union* 9.3.2006, L 70, pp. 12-34

D'Mello, J.P.F., Placinta, C.M. & Macdonald, A.M.C. (1999). *Fusarium* mycotoxins: a review of global implications for animal health, welfare and productivity. *Animal Feed Science and Technology*, Vol.80, No.3-4, (August 1999), pp. 183-205, ISSN 0377-8401

Daly, S.J., Keating, G.J., Dillon, P.P., Manning, B.M,; O'Kennedy, R., Lee, H.A. & Morgan, M.R.A. (2000). Development of surface plasmon resonance-based immunoassay for aflatoxin B-1. *Journal of Agricultural and Food Chemistry*, Vol.48, No.11, (November 2000), pp. 5097-5104, ISSN 0021-8561

De Girolamo, A., Lippolis, V., Nordkvist, E. & Visconti, A. (2009). Rapid and non invasive analysis of deoxynivalenol in durum and common wheat by Fourier-Transform Near Infrared (FT-NIR) spectroscopy. *Food Additives and Contaminants*, Vol.26, No.6, (June 2009), pp. 907-917, ISSN 0265-203X

Dell'Orto, V., Savoini, G., Nichilo, A., Campagnoli, A. & Cheli F. (2007). Impiego del naso elettronico abbinato a modelli chemometrici. *Rapporti ISTISAN*, Vol.7, No.37, (September 2007), pp. 207-212, ISSN 1123-3117

Falasconi, M., Gobbi E., Pardo, M., Della Torre, M., Bresciani, A. & Sberveglieri, G. (2005). Detection of toxigenic strains of Fusarium verticilloides in corn by electronic olfactory system. *Sensors Actuators B: Chemical*, Vol.108, No.1-2, (July 2005), pp 250-257, ISSN 0925-4005

FAO (Food and Agriculture Organization of the United Nations). (2011). Cereal Supply and Demand Brief. In: *FAO Cereal Supply and Demand Situation*, (July 2011), Available from: http://www.fao.org/worldfoodsituation/wfs-home/csdb/en/

Fearn, T. (2009). Making stable and robust calibrations in NIR analysis. *Feed Tech*, Vol.13,
No.3, (April 2009), ISSN 1387-1978 Available from:
http://www.alalboutfeed.net/background/making-stable-and-robust calibrations-
in-nir-analysis-11479.html

Fearn, T. (2011). Sampling errors: problems and opportunities. *Proceedings of Cost Action
FA0802 International Workshop WG2 & WG3 Feed for Health: "Feed Quality and Safety:
technology, traceability and labeling"*, Gijon, Spain, April 2011

Fearn, T., Fisher, S.A., Thomson, M. & Ellison, S. L. R. (2002). A decision theory approach to
fitness for purpose in analytical measurement. *Analyst*, Vol.127, No.6, (June 2002),
pp. 818-824, ISSN 0003-2654

Fink-Gremmels, J. (1999). Mycotoxins: Their implications for human and animal health. *The
Veterinary quarterly*, Vol.21, No.4, (November 1999), pp. 115-120, ISSN 0165-2176

French, S. (1989). *Readings in Decision Analysis*, Chapman and Hall/CRC, ISBN 10:
041232170X, London, Uk

Gardner, J.W., Bartlett, P.N. (1994). A brief history of electronic noses. *Sensors Actuators B:
Chemical*, Vol.18, No.1-3, (April 1994), pp 210-211, ISSN 0925-4005

Goryacheva, I.Y., De Saeger, S., Eremin, S.A. & Van Peteghem, C. (2007). Immunochemical
methods for rapid mycotoxin detection: Evolution from single to multiple analyte
screening. A review. *Food Additives and Contaminants*, Vol.24, No.10, (October 2007),
pp. 1169-1183, ISSN 0265-203X

Hussein, H.S. & Brasel, J.M. (2001). Toxicity, metabolism, and impact of mycotoxins on
human and animals. *Toxicology*, Vol.167, No.2, (October 2001), pp. 101-134, ISSN
0300-483X

Ibañez, E. & Cifuentes, A. (2001). New Analytical Techniques in Food Science. *Critical
Reviews in Food Science and Nutrition*, Vol.41, No.6, (September 2001), pp. 413-450,
ISSN 1040-8398

Johansson, A.S., Whitaker, T.B., Giesbrecht, F.G., Hagler J., W.M. & Young, J.H. (2000)a.
Testing shelled corn for aflatoxin, Part III: evaluating the performance of aflatoxins
sampling plans. *Journal of the Association of Official Analytical Chemists International*,
Vol.83, No.5, (September 2000), pp. 1279- 1284, ISSN 1060-3271

Johansson, A.S., Whitaker, T.B., Giesbrecht, F.G., Hagler Jr, W.M., Young, J.H. (2000)b.
Testing shelled corn for aflatoxin, Part II: modelling the observed distribution of
aflatoxins test results. *Journal of the Association of Official Analytical Chemists
International*, Vol.83, No.5, (September 2000), pp. 1270- 1278, ISSN 1060-3271

Johansson, A.S.; Whitaker, T.B.; Hagler Jr, W.M.; Giesbrecht, F.G.; Young, J.H. & Bowman
D.T. (2000)c. Testing shelled corn for aflatoxin, Part I: estimation of variance
components. *Journal of the Association of Official Analytical Chemists International*,
Vol.83, No.5, (September 2000), pp. 1264-1269, ISSN 1060-3271

Jorgensen, K. (2005). Occurrence of ochratoxin A in commodities and processed food: a
review of EU occurrence data. *Food Additives & Contaminants*, Vol.22, No.S1, (2005),
pp. 562-567, ISSN 0265–203X

Keshri, G., & Magan, N. (2000). Detection and differentiation between mycotoxigenic and
non-mycotoxigenic strains of two Fusarium spp. using volatile production profiles
and hydrolytic enzymes. *Journal of Applied Microbiology*, Vol.89, No. 5, (December
2001), pp. 825-833, ISSN 1365-2672

Kos, G., Lohninger, H. & Krska, R. (2002). Fourier-transform mid-infrared spectroscopy with attenuated total reflection (FT-IR/ATR) as a tool for the detection of *Fusarium* fungi on maize. *Vibrational Spectroscopy*, Vol.29, No.1-2, (July 2002), pp. 115-119, ISSN 0924-2031

Kos, G., Lohninger, H. & Krska, R. (2003). Development of a method for the determination of Fusarium fungi on corn using mid-infrared spectroscopy with attenuated total reflection and chemometrics. *Analytical Chemistry*, Vol.75, No.5, (March 2003), pp. 1211-1217, ISSN 0003-2700

Kos, G., Lohninger, H., Mizaikoff, B. & Krska, R. (2007). Optimization of a sample preparation procedure for the screening of fungal infection and assessment of deoxynivalenol content in maize using mid-infrared attenuated total reflection spectroscopy. *Food Additives & Contamiants*, Vol.24, No.7, (June 2007), pp 721-729, ISSN ISSN 0265-203X

Krska, R. & Welzig, E. (2006). Mycotoxin analysis: an overview of classical, rapid and emerging techniques. In: *The mycotoxin factbook*, D. Barug, D. Bhatnagar, H.P. van Egmond, J.W. van der Kamp, W.A. van Osenbruggen, & A. Visconti. (Eds.), pp. 223-247, Wageningen Academic Publisher, ISBN 908686-006-0, Wageningen, The Netherland

Krska, R., Welzig, E., Berthiller, F., Molinelli, A. & Mizaikoff, B. (2005). Advances in the analysis of mycotoxins and its quality assurance. *Food Additives and Contaminants*, Vol.22, No.4, (April 2005), pp. 345-353, ISSN 0265–203X

Larsen, J.C., Hunt, J., Perrin, I. & Ruckenbauer, P. (2004). Workshop on trichothecenes with a focus on DON: summary report. *Toxicology Letters*, Vol.153, No.1, (October 2004), pp. 1-22, ISSN 0378-4274

Logrieco, A., Arrigan, D.W.M., Brengel-Pesce, K., Siciliano, P. & Tothill, I. (2005). DNA arrays, electronic noses and tongues, biosensors and receptors for rapid detection of toxigenic fungi and myctoxins: A review. *Food Additives and Contaminants*, Vol.22, No.4, (April 2005), pp. 335-344, ISSN 0265–203X

Magan, N. & Evans, P. (2000). Volatiles as an indicator of fungal activity and differentiation between species, and the potential use of electronic nose technology for early detection of grain spoilage. *Journal of Stored Products*, Vol.3, No.4, (April 2000), pp. 319-340, ISSN 0022-474X

Maragos C.M. (1998). Analysis of mycotoxins with capillary electrophoresis. *Seminars in Food Analysis*, Vol.3, No.2, (March 1998), pp. 353, ISSN 1084-2071

Maragos, C.M. & Appel, M. (2007). Capillary electrophoresis of the mycotoxin zearalenone using cyclodextrin-enhanced fluorescence. *Journal of Chromatography A*, Vol.1143, No. 1-2, (March 2007), pp. 252-257, ISSN 0021-9673

Maragos, C.M. & Busman, M. (2010). Rapid and advanced tools for mycotoxin analysis: a review. *Food Additives & Contaminants* Vol.27, No.5, (May 2010), pp. 688-700, ISSN 0265-203X

Maragos, C.M. (2004). Emerging technologies for mycotoxin detection. *Journal of Toxicology-Toxin Review*, Vol.23, No.2-3, (January 2004), pp. 317-344, ISSN 1556-9543

Miraglia, M., De Santis, B., Minardi, V., Debengnach, F. & Brera, C. (2005). The role of sampling in mycotoxin contamination: An holistic view. *Food Additives & Contaminants*, Vol.22, No.S1, (2005), pp. 31-36, ISSN 0265–203X

Monaci, L. & Palmisano, F. (2004). Determination of ochratoxin A in foods: state-of-the-art
and analytical challenges. *Analytical and Bioanalytical Chemistry*, Vol.378, No.1,
(January 2004), pp. 96-103, ISSN 1618-2642

Olsson, J., Borjesson, T., Lundstedt, T. & Schnuerer, J. (2000). Volatiles for microbiological
quality grading of barley grains: determination using gas chromatography-mass
spectrometry and and electronic nose. *International Journal of Food Microbiology*,
Vol.59, No.3, (September 2000), pp.167-178, ISSN 0168-1605

Olsson, J., Borjesson, T., Lundstedt, T.& Schnuerer, J. (2002). Detection and quantification of
ochratoxin A and deoxinivalenol in barley grain by GC-MS and electronic nose.
International Journal of Food Microbiology, Vol.72, No.3, (February 2002), pp.203-214,
ISSN 0168-1605

Paolesse, R., Alimelli, A., Martinelli, E., Di Natale, C., D'Amico, A., D'Egidio, M.G., Aureli,
G., Ricelli A. & Fanelli, C. (2006). Detection of fungal contamination of cereal grain
samples by an electronic nose. *Sensors and Actuatuators B: Chemical*, Vol.119, No.2,
(December 2006), pp.425-430, ISSN 0925-4005

Pettersson, H. & Aberg, L. (2003). Near infrared spectroscopy for determination of
mycotoxins in cereals. *Food Control*, Vol.14, No.4, (June 2003), pp. 229-232, ISSN
0956-7135

Pinotti, L. & Dell'Orto, V. (2011). Feed safety in the feed supply chain. *Biotechnology,
Agronomy, Society and Environment*, Vol.15, No.S1, (January 2011), pp. 9-14, ISSN
1370-6233

Pinotti, L., Moretti, V.M., Baldi, A., Bellagamba, F., Campagnoli, A., Savoini, G., Cantoni, C.
& Dell'Orto, V. (2005). Feed authentication as an essential component of food safety
and control. *Outlook on Agriculture*, Vol.34, No.4, (December 2004), pp. 243-248,
ISSN 0030-7270

Presicce, D.S., Forleo, A., Taurino, A.M., Zuppa, M., Siciliano, P., Laddomadab, B., Logrieco,
A. & Visconti, A. (2006). Response evaluation of an E-nose towards contaminated
wheat by Fusarium poae fungi. *Sensors and Actuatuators B: Chemical*. Vol.118, No.1-
2, (October 2006), pp.433-438, ISSN 0925-4005

RASFF (European Commission Rapid Alert System for Food and Feed). (2009). *RASFF
Annual Report 2009*. Available from:
http://ec.europa.eu/food/food/rapidalert/docs/report2009_en.pdf

Röhr, A., Lüddecke, K., Drusch, S., Müller, M.J. & Alvenslebenet, R.V. (2005). Food quality
and safety–consumer perception and public health concern. *Food Control*, Vol.16,
No.8, (October 2005), pp. 649–655, ISSN 0956-7135

Sahgal, N., Needeman, R., Cabanes, F.J. & Magan N. 2007. Potential for detection and
discrimination between mycotoxigenic and non-toxigenic moulds using volatile
production pattern: A review. *Food Additives & Contamiants*, Vol.24, No.10, (October
2007), pp 1161-1168, ISSN 0265-203X

Schnerr, H., Vogel, R.F. & Niessen, L. (2002). Correlation between DNA of trichotecene-
producing *Fusarium* species and deoxinilvalenol concentrations in wheat-samples.
Letters in Applied Microbiology, Vol.35, No.2, (August 2002), pp. 121-125, ISSN 1472-
765X

Stroka, J., Spanjer, M., Buechler, S., Barel, S., Kos, G. & Anklam, F. (2004). Novel sampling
methods for the analysis of mycotoxins and the combination with spectroscopic
methods for the rapid evaluation of deoxynivalenol contamination. *Toxicology
Letters*, Vol.153, No.1, (October 2004), pp. 99–107, ISSN 0378-4274

Tang, Y., Lu, L., Zhao, W. & Wang, J. (2009). Rapid detection techniques for biological and chemical contamination in food: a review. *International Journal of Food Engineering*, Vol.5, No.5, (November 2009), pp. article 2, ISSN 1556-3758

Taylor-Pickard, J. (2009). Mycotoxin contamination of feed: current global status. *Feed Tech*, Vol.13, No.7, (July 2009). Available from: http://www.allaboutfeed.net/background/mycotoxin-contamination-of-feed-current-global-status-11459.html

Tognon, G., Campagnoli, A., Pinotti, L., Dell'Orto, V. & Cheli, F. (2005). Implementation of the Electronic Nose for the Identification of Mycotoxins in Durum Wheat (*Triticum durum*). *Veterinary Research Communications*, Vol.29, Suppl.2, (August 2005), pp. 391–393, ISSN 0165-7380

Tudos, A.J., Lucas-van den Bos, E.R. & Stigter, E.C.A. (2003). Rapid surface plasmon resonance-based immunoassay of deoxynivalenol. *Journal of Agricultural and Food Chemistry*, Vol.51, No.20, (September 2003), pp. 5843-5848, ISSN 0021-8561

van Amerongen, A.; Barug, D.; Lauwaars, M. (2007). *Rapid Methods for Food and Feed Quality Determination*. Wageningen Academic Publishers., ISBN-9789076998930 Wageningen, The Netherlands

van der Gaag, B., Spath, S., Dietrich, H., Stigter, E., Boonzaaijer, G., van Osenbruggen, T. & Koopal, K. (2003). Biosensor and multiple mycotoxin analysis. *Food Control*, Vol.14, No.4, (June 2003), pp. 251-254, ISSN 0956-7135

Visconti, A. & De Girolamo, A. (2005). Fitness for purpose – Ochratoxin A analytical developments. *Food Additives & Contaminants*, Vol.22, No.S1, (2005), pp. 37-44, ISSN 0265–203X

Whitaker, T.B. (2003). Standardisation of mycotoxin sampling procedures: an urgent necessity. *Food Control*, Vol.14, No.4, (June 2003), pp. 233-237, ISSN 0956-7135

Whitaker, T.B. (2004). Sampling for mycotoxins. In: *Mycotoxins in food: detection and control*, N. Magan, M. Olsen (Eds), pp. 69-81, Woodhead Publishing Ltd., ISBN-10: 1855737337 Cambridge, UK

Whitaker, T.B. (2006). Sampling food for mycotoxins. *Food Additives & Contaminants*, Vol.23, No.1, (January 2006), pp. 50-61, ISSN 0265–203X

Whitaker, T.B., Dowell, F.E., Hagler Jr, W.M., Griesbrecht, F.G. & Wu, J. (1994). Variability associated with sampling, sample preparation, and chemical testing farmer's stock peanuts for aflatoxins. *Journal of the Association of Official Analytical Chemists International*, Vol.77, No.1, (January 1994), pp. 107-116, ISSN 1060-3271

Whitaker, T.B., Hagler Jr, W.M., Griesbrecht, F.G. & Johansson, A.S. (2002). Sampling wheat for deoxynivalenol. In: *Mycotoxin and food safety*, J.W. DeVries, M.W. Trucksess, and L.S. Jackson (Eds.), pp 73-83, Kluwer Academic/Plenum Publisher, ISBN 0306467801, New York, USA

Whitaker, T.B., Truckess, M.W., Johansson, A.S., Griesbecht, F.G., Hagler, Jr, W.M., & Bowman, D.T., (1998). Variability associated with testing shelled corn for fumonisin. *Journal of the Association of Official Analytical Chemists International*, Vol.81, No.6, (December 1998), pp.1162-1168, ISSN 1060-3271

Wild, C.P. & Gong, Y.Y. (2010). Mycotoxins and human disease: a largely ignored global health issue. *Carcinogenesis*, Vol.31, No.1 (January 2010), pp.71–82, ISSN 0143-3334

Zeng, M.Z., Richard, J.L. & Binder, J. (2006). A review of rapid methods for the analysis of mycotoxins. *Mycopathologia*, Vol.161, No.5, (May 2006), pp.261-273, ISSN 0301-486X

11

Natural Hormones in Food-Producing Animals: Legal Measurements and Analytical Implications

Patricia Regal, Alberto Cepeda and Cristina A. Fente
University of Santiago de Compostela,
Spain

1. Introduction

Hormones are chemicals that are naturally produced in the body of animals and human beings and have a number of important functions in life, such as reproduction or growth. They act as messengers through the different parts of the organism and trigger and modulate key reactions to support and promote life. However, and due to the important role of these chemicals in several body functions, they also have been exogenously applied to animals and humans in order to obtain some kind of benefit in health or even to improve physical and growth performance. Focusing on the veterinary field, the most desirable action of hormones has always been reducing costs and obtaining more products of animal origin in shorter productive times, increasing the benefit per unit head for farmers. As a matter of fact, anabolic steroid hormones have played a key role among veterinary products in farming history and they have been one the most used and controversial components among veterinary drugs.

Usually, hormones work in harmony in the body and this status must be maintained to avoid metabolic disequilibrium and the subsequent illness. Besides, it has been reported the influence of exogenous steroids (presence in the environment and food products) in the development of several important illness in humans. With regard to food safety when treating animals with exogenous hormones, consumers' concerns have led to a complete prohibition of the use of substances having a hormonal action in food producing animals in the EU. Even when several regulations and laws exist all over the world with regard to the use of natural and synthetic hormones in animal husbandry, natural hormones have arisen as a real weak point of residue monitoring plans due to their natural origin. The existence of high variability through animals in terms of natural hormonal levels has been reported. This latest fact makes almost impossible to establish legal thresholds to control any exogenous administration of natural hormones to animals. That is why no final legal solution has been found yet to control the misuse and abuse of natural hormones exogenously applied to farm animals, even though a number of promising analytical procedures have already been published.

2. Anabolic steroid hormones

Throughout history, a large number of natural and synthetic substances have been applied in stock farming to speed up and improve animal growth, and to decrease feed costs.

Anabolic agents or growth promoters are metabolic modifiers which improve efficiency and profitability of livestock production and improve carcass composition (Dikeman, 2007). Main physiologic effects of anabolic steroids include growth of muscle mass and strength, increased bone density, maturation of the sex organs, particularly important in the fetus, and at puberty the appearance of the secondary sex characteristics. The group of anabolic growth promotants includes compounds that naturally occur in an animal's body and synthetic chemicals that mimic the action of naturally occurring compounds. Meat industry have widely used anabolic hormones to quickly get larger quantities of meat and decrease inputs, reducing production costs, but also because they lead to a leaner carcass more in accordance to current consumer's preferences. Additionally, the zootechnical use of some sex hormones, such as estradiol or its esters (i.e., estradiol benzoate), which successfully regulate oestrus in cattle, has also led to important improvements and financial gain in stock farming (Cavalieri et al., 2005; Martínez et al., 2002).

Several illegal hormones have been used in the European Union (EU), as it has been reported in a series of European International Symposia and Conferences, such as EuroResidue Conferences on Residues of Veterinary Drugs in Food (Federation of European Chemical Societies, Division of Food Chemistry) and the Ghent Symposia on Hormone and Veterinary Drug Residue Analysis, amongst others. The number of active compounds is wide and continuously changing, as observed by the EU National Reference Laboratories (NRLs). Estrogenic, gestagenic and androgenic compounds (EGAs), as well as thyreostatic, corticosterois and β-agonist compounds, are also used alone or in growth promoting "cocktails" with low concentrations of several ones, that makes even more difficult their detection. There have been several European regulations regarding the use of EGAs as animal growth promoters because of their possible toxic effect on public health. In the Council Directive 96/22/EC (EC, 1996a) the EU prohibited the administration of substances having thyreostatic, oestrogenic, androgenic or gestagenic effects and of beta agonists in animal husbandry, while certain therapeutic applications of these drugs were still allowed. These anabolic steroids are included in group A substances according to Annex I of Directive 96/23/EC (EC, 1996b), which pertains to growth-promoting agents abused in animal fattening and unauthorized substances with no maximum residue limit (MRL). A zero-tolerance policy has been adopted, and especial analytical requirements have been stated in regard to these hormones (EC, 2002; European Commission, Directorate General for Health & Consumers, 2004). However, the possibility of widespread abuse of hormonal substances by unscrupulous farmers and veterinary professionals in some parts of Europe still exists, mainly due to the economic benefits these substances provide in animal husbandry. On the other hand, the use of hormones to promote growth is still a legal practice in some parts of the world, which facilitates the existence of a possible "black market" of substances from these areas.

2.1 Estrogenic drugs

Cattle are the main food-producing species in which estradiol products are used for therapy or growth promotion. Estradiol benzoate, one of the most applied steroids in animal husbandry, was authorized for the treatment of pyometra and endometritis, for dilation of the cervix in cases of abortion, to enhance the expression of estrous behaviour, and to provoke luteolysis incorporated into estrous synchronization drug devices (i.e. PRID,

CIDR), among other applications (Levy, 2010). In meat industry, it has been already reported that estrogenic implants (alone or in combination) increase carcass weight and longissimus muscle area and decrease intramuscular fat, compared with non-implanted steers (Boles et al., 2009; McPhee et al., 2006; Parr et al., 2011). Estrogenic implants also decrease kidney, pelvic and heart fat but apparently this fat increases for combination implants (McPhee et al., 2006). Cattle repeatedly treated with estradiol and trenbolone acetate implants have greater average daily gain and final weights than single-treated or non-treated steers, as well as more mature skeletons and higher protein content in their carcasses (Scheffler et al., 2003). However, hormonal treatments may have a negative effect on tenderness and meat quality of beef because they reduce marbling and advance skeletal or lean maturity (Dikeman, 2007; Hunter, 2010; Scheffler et al., 2003), this effect being more pronounced with combination implants than with estradiol alone. Beef flavour, juiciness and tenderness might be affected by trenbolone acetate implants but apparently this effect decreases with aging time (Igo et al., 2011).

On the other hand, the economic profitability of a dairy farm is based on the calving interval of the cows, in order to keep them as long as possible into lactating phase. To achieve this, the cow needs to get pregnant very quickly during postpartum, so the main step is the determination of the optimal time for insemination, basing on estrous behaviour. The expression of estrous behaviour is at a low level in modern dairy cows, resulting in low detection rates and longer calving intervals (Senger, 1994). Estradiol-based drugs, particularly those combined with progestins, appeared as a really effective and efficient solution to estrus detection problems in farm animals, allowing artificial insemination synchronization and high pregnancy rates to fixed-time artificial insemination in dairy cows, sheep and other farm animals (Burkea et al., 2001; Martínez et al., 2002). Although Directive 2003/74/EC, amending Directive 96/22/EC, permanently prohibited the use of estradiol-17β and its ester-like derivatives as growth promoters, a temporary exemption was given until 14 October 2006 for their use as an oestrous-induction tool in cows, horses, sheep or goats (EC, 2003). As alternative effective products exist and are implemented in the market (Lane et al., 2008; Vilariño et al., 2010), the European Parliament banned estradiol-17β and its ester-like derivatives, including those with a therapeutic purpose, in 2008 to ensure human health protection within the EU (EC, 2008). In the absence of estradiol-based products, alternatives for estrous synchronization are prostaglandin or the progesterone-releasing devices. Alternatives for the treatment of pyometra and endometritis could include the use of prostaglandins thanks to a combination of their direct ecbolic and luteolytic effects.

No estradiol-based drugs are in the European veterinary market anymore, except for its use in pets (EC, 2008). However, the possibility of widespread abuse of hormonal substances by unscrupulous farmers and veterinary professionals in some parts of Europe still exists, mainly due to the economic benefits that these substances provide in animal husbandry and the existence of authorized drugs in other non-European countries (Stephany, 2001). Limited research was found on the effects of anabolic implants in poultry, sheep, and pigs. Anabolic steroids are not approved for growth regulation in pigs in the United States (US) and numerous other countries. Even so, Lee et al., 2002 and Sheridan et al. 1999 studied the effect of anabolic steroids in pigs, concluding that they were not suitable agents to improve growth or carcass characteristics of pigs, but mid-back fat appeared reduced anyway (Lee et al., 2002; Sheridan et al., 1990).

2.2 Androgenic and gestagenic drugs

Androgenic and gestagenic growth promotants approved in the US include steroid hormone anabolic implants with testosterone, progesterone, trenbolone and melengestrol acetate, all of them banned in EU. With the exception of melengestrol acetate, the recommended administration of these drugs is by subcutaneous implantation of continuously releasing hormone pellets in the ear. Androgenic hormones (testosterone and trenbolone acetate) directly reduce fat content of the carcass (Hunter, 2010) and have proved to be also effective in chicken to increase muscle quality and quantity (Chen et al., 2010). Medroxyprogesterone, chlormadinone, megestrol and melengestrol are synthetic analogues of progesterone that are commonly administered orally as acetate derivatives. They are used for synchronization of estrous, but have also been used as growth promoters in cattle. Although forbidden within the EU, the misuse of these natural and synthetic hormones is well known. For this illegal purpose they are frequently injected into the animal body as 'hormone cocktails' including new compounds each day, such as gestagens delmadinone acetate and algestone acetophenide (Daeseleire et al., 1994).

3. Human health and hormones

Endogenously synthesized steroid hormones exert a wide range of biological effects on the body and not only in the reproductive organs, which is why they are vital in normal development and life. Possible effects vary according to a number of factors such as gender and age, ethnicity and even environment. However, exogenous steroidogenically active compounds may interfere in the hormonal endogenous equilibrium affecting health and natural body development. As any other chemicals of natural or synthetic nature, hormones can be "toxic" to living organisms under certain circumstances, due to an excessive exposure at an abnormal stage during development or adult life. The current increasing trends of cancer and reproductive disorders have been frequently related to exogenous steroids food intake and endocrine disrupters that are present in the environment. The major areas of concern expressed in the literature are related to cancer, mutagenicity and reproductive effects, in particular endocrine disruption. Generally, cancer and mutagenicity are well described and well understood but endocrine disruption has become, in recent years, an area where there has been concern about potential harmful outcomes for a wide range of chemicals previously unsuspected of causing such effects.

As a matter of fact, the possible impact of exogenous steroid hormones, such as natural and synthetic hormones present in food products, are more dangerous for certain groups of population which are considered to be more sensitive and vulnerable than the rest. As regard to naturally occurring sex hormones, such as estradiol or testosterone, daily endogenous production and exogenous intake (in food) seem to be key points to evaluate risk. Taking children as reference population with the lowest levels of endogenous synthesis of steroids, an assessment of their plasmatic levels and of the presence of these chemicals in food are crucial. For this purpose, highly sensible and accurate techniques based on chromatography and mass spectrometry are required. Additionally, circulating levels of hormones have resulted to be lower than previously reported for prepubertal children and fetuses. First assessments of estradiol levels in serum of prepubertal boys and girls were based on radioimmunoassay (RIA) and its concentration appeared in most cases in a range of difficult accurate measurement, very close or even below conventional detection limit, resulting in overestimated values (Aksglaede et al., 2006; Bay et al., 2004). Tandem mass spectrometry methods in combination with gas chromatography or liquid chromatography for sex steroid hormones have been developed and are the methods of choice for the accurate measurement of the low levels of testosterone found in children and females and even the low levels of estradiol in postmenopausal women, men and the prepubertal child (Kushnir et al., 2010; Moal et al., 2007; Nelson et al., 2004; Stanczyk et al., 2007). Actually, the current and more sensitive assays, mainly mass-spectrometry-based analysis, have revealed that previous RIA values were in fact overestimated and sex steroids in children are extremely low (Courant et al., 2010). There are no limits for hormones which assure children's safety under exposure to exogenous steroids and endocrine disruptors. Furthermore hormonal changes or disruption during fetal life or puberty may provoke serious subsequent problems in their adult life. Since no safe threshold has been established yet, it seems necessary to avoid unnecessary children's and fetuses' exposure to exogenous disruptors, natural or synthetic, present in food even at very low levels (Bay et al., 2004).

Both exogenous hormones and synthetic compounds mimicking their effects may change the endogenous balance of human body, provoking disturb in their natural functions. As a result of their low endogenous levels, children are extremely sensitive to exogenous steroid hormones and small variations in blood levels might trigger serious pubertal development effects and even future adult life problems (Aksglaede et al., 2006; Alves et al., 2007). Several epidemiological studies have proved the existence of a trend to earlier puberty in American girls during last decades, and incidence is on the rise. In 1997, Hermann-Giddens et al. reported an unexpected advance in timing of puberty in both African-American and white American girls (Herman-Giddens et al., 1997). An advance in timing of onset of puberty has not been noted yet in other countries, although it is likely to become more prevalent as other countries adopt American lifestyle and diets (Parent et al., 2003). Precocious puberty has health and social implications, it is complex and influenced by multiple factors such as ethnicity, gender, nutrition, endocrine disrupting chemicals, pollutants and exogenous sex steroids (Aksglaede et al., 2006; Cesario & Hughes, 2007; Daxenberger et al., 2001). However, there is a key difference between US and the rest of the world, since they still allow the use of some hormonal drugs in food producing animals. This fact might not be a bare coincidence and mean an increase on the exogenous intake of steroids for American children (Aksglaede et al., 2006; Partsch & Sippell, 2001).

On the other hand, a tendency to increasing incidence of certain cancer types, such as testicle, breast and prostate cancer, has not been fully clarified yet, though sex hormones are suspected to play a key role (Foster et al., 2008; Huyghe et al., 2003; Prins, 2008; Wigle et al., 2008). For instance lung cancer, which is the leading cause of cancer deaths in the United States and has surpassed breast cancer as the primary cause of cancer-related mortality in women, has been related to estradiol along with tobacco consumption by Meireles et al. (Meireles et al., 2010). Estrogens have also been linked to other types of cancer such as squamous cell carcinoma of the head and neck (HNSCC), which is the sixth most common type of cancer in the United States (Shatalova et al., 2011). Estrogen exposure is one of the established risk factors for breast cancer, the most commonly diagnosed cancer in women (Zhong et al., 2011). An association between the risk of breast cancer and persistently elevated blood levels of estrogen and androgen has been found in many studies (Kaaks et al., 2005; Yager & Davidson, 2006). Metabolites of zeranol, a non-steroidal anabolic growth promoter with potent estrogenic activity and widely used in the US, contained in meat produced from cattle after zeranol implantation, may be a risk factor for breast cancer (Zhong et al., 2011). Experimental and epidemiological data support a role for sex steroid hormones in the pathogenesis of endometrial cancer as well. As a matter of fact, circulating androgen levels were also related to endometrial cancer, although less strongly than circulating estrogen levels (Lukanova et al., 2004). However, the effect of elevated androgen (androstenedione and testosterone levels) on endometrial cancer risk seems to be mediated mainly through their conversion to estrogens.

As recently reported by Kvarnryd et al., progestogens exposure might have reproductive toxicity as well, in animals and humans, provoking defects on the development of female sex organs and subsequent infertility (Kvarnryd et al., 2011). The level of serum progesterone has not been associated with a risk of breast cancer in postmenopausal women, but in premenopausal women it appears to be inversely associated with the risk of breast cancer (Micheli et al., 2004). As for androgenic steroids, circulating concentrations of dehydroepiandrosterone (DHEA) and DHEA appear markedly decreased during aging, and thus this fact implicates the natural androgen in cognitive decline associated to age (Sorwell & Urbanski, 2010). On the other hand, increased blood levels of DHEA and its sulphate have been found in schizophrenia patients, and apparently these levels are strongly correlated to the severity of illness and aggressive behaviour of patients and to the pathophysiology of other stress-related psychiatric disorders (Garner et al., 2011; Strous et al., 2004).

As regard to hormonal content, all foodstuff of animal origin contains steroid hormones and metabolites, but their concentrations vary with the kind of food, species, gender, age and physiological stage of the animal (Daxenberger et al., 2001; Poelmans et al., 2005a, 2005b). As a matter of fact, meat is clearly one of the most naturally 'contaminated' foods (Maume et al., 2001; Maume et al., 2003; Poelmans et al., 2005a). Data published by Swan et al. in 2007 already suggested that maternal beef consumption may alter males' testicular development in utero and adversely affect his adult reproductive capacity (Swan et al., 2007). Even milk consumption, the hormone content of which is well known, has been associated with an increased risk of early menarche (Wiley, 2011). There are studies that find a relationship between milk and dairy products with human illnesses, such as teenagers' acne, prostate, breast, ovarian and corpus uteri cancers, many chronic diseases that are common in Western societies, as well as male reproductive disorders (Adebamowo et al., 2008; Ganmaa et al., 2011; Ganmaa et al., 2001; Givens et al., 2008; Melnik, 2009; Wiley, 2011). There are many possible contributory factors to these health problems, including steroid hormones which are well known as endocrine disruption agents. In this field, some studies have arisen regarding sex hormone levels in milk in relation to animals' pregnancy, most of them regarding estrogens and androgens (Courant et al., 2007; Farlow et al., 2009; Ganmaa & Sato, 2005; Maruyama et al., 2010; Pape-Zambito et al., 2010). Cow's milk contains considerable quantities of hormones and is therefore of particular concern (Courant et al., 2007). It is a fact that dairy milk consumption by humans started around 2000 years ago, but the milk which people drink today is quite different from traditional milk. As a result of modern farming and animal breeding, today's milk originates from genetically improved dairy cows such as Holstein, which are pregnant during most of their lactation period (Maruyama et al., 2010).

Regarding potential toxicological substances used in animal husbandry, for the endogenous sex steroids and their simple ester derivatives the US Food and Drug Administration (FDA) concluded that 'safety can be assured' because they are endogenous in both food-producing animals and people. Additionally, they stated that '...no additional physiological effect will occur in individuals chronically ingesting animal tissues that contain an increase of endogenous sex steroids from exogenous sources equal to 1% or less of the amount in micrograms produced by daily synthesis in the segment of the population with the lowest daily production. We believe that the 1% value is supported by scientific evidence, is reasonable, and reflects sound public health policy. For estradiol and progesterone, prepubertal boys provide the baseline benchmark. For testosterone, prepubertal girls provide the baseline benchmark...' (U.S. Department of Health and Human Services, 2006). The FDA stated that although not all sex steroids are demonstrated carcinogens, they should be regarded as suspect carcinogens. As a matter of fact, the FDA concluded that to establish the safety of a synthetic steroid animal testing is necessary. However, to show the safety of an endogenous sex steroid, the sponsor simply have to demonstrate that, under the proposed conditions of use, the concentration of residue of the endogenous steroid in treated food-producing animals is such that the increase will not exceed this 1% permitted increase. The Joint Food and Agricultural Organisation/World Health Organisation (FAO/WHO) Expert Committee on Food Additives (JECFA) and the US Food and Drug Administration (FDA) considered, in 1988, that the residues found in meat from treated animals were safe for the consumers. However, current recommendations might be overestimated and should be revised, altogether with hormonal levels in children. The lack of known proved hormonal thresholds, under which value no effects could be observed in humans, add uncertainty to this issue.

4. Legal implications

The use of hormonal growth promoters to increase the production of muscle meat has led to international disputes about the safety of meat originating from animals treated with such anabolics. Implants containing anabolic steroids are widely used in the US beef industry, among other countries, to fast growth and finish cattle and to improve feed efficiency. Growth promotants approved in the US include steroid hormone anabolic implants (17β-estradiol, testosterone, progesterone, trenbolone, zeranol, melengestrol acetate) and β-agonist feed additives (ractopamine) for finishing swine, cattle and turkeys, all of them banned in EU (U.S. FDA, 2010). With the exception of melengestrol acetate, the recommended administration of estradiol, progesterone, and testosterone (three natural hormones), and zeranol and trenbolone acetate (two synthetic hormones) is by subcutaneous implantation of continuously releasing hormone pellets in the ear. This ear would be then discarded during slaughtering but there is no withdrawal time for any of these legally approved implants. Melengestrol acetate is approved for its use as a feed additive. As a result of the existence of these legal drugs, a significant part of cattle raised in US feedlots are treated with growth promoting sex hormones. Over 97% of cattle weighing 700 lbs or more received at least one anabolic implant during the finishing period in 1999 (Salman et al., 2008). In general, a decrease on the use of growth promoting implants in US cattle over the past twenty years has been observed. More than one of four farms implanted some calves with growth promotants prior or at weaning in 1992, but fewer than one of eight did so in 2007 (USDA, 2009). The reason of this decline on the use of implanting, a profitable US management practice, could be the publicity surrounding hormonal implants and movement toward marketing cattle in natural or organic programs. The most used substances are estrogenic drugs, in the form of estradiol-17β, estradiol benzoate or the synthetic zeranol. Progesterone, testosterone and the two synthetic chemicals trenbolone acetate and melengestrol acetate are generally used in combination with estrogens. It is also a standard legal practice to use hormones to promote the growth of cattle in the meat industry in Australia (Hunter, 2010). These chemicals are approved, registered and regulated by the Australian Pesticides and Veterinary Medicines Authority (APVMA) which, as well as US Food and Drug Administration (FDA), keeps its position on saying that they are safe for consumers, not harmful to animals and effective when used according to label instructions. As a consequence, regulatory controls would differ sharply between the UE and the countries where hormonal active growth promoters are still legal.

The European ban of the use of hormones arose in the 70s due to the health consequences derived from the use of diethylstilbestrol (DES), a synthetic estrogen widely administered to women to prevent miscarriage and other pregnancy complications. This chemical led to reproductive problems in treated women, as well as reproductive alterations, gynecologic cancer and malformations in reproductive organs in their female children, above normal average values (Auclair, 1979; Cousins et al., 1980; Haney & Hammond, 1983; Rosenfeld & Bronson, 1980). In 1980, European consumer organizations called for a boycott of beef as a result of widespread publicity involving illegal use of diethylstilbestrol in European veal production. In response, EC agriculture ministers agreed to ban the use of hormones for raising livestock with the enactment of the first legal European ban of hormones in 1981 (EEC, 1981), with the adoption of restriction in livestock production prohibiting the use of synthetic hormones and substances having a hormonal activity and limiting the use of natural hormones to therapeutic purposes. The Directive 81/602/EEC prohibited the use of certain substances having hormonal action (estradiol-17β, progesterone, zeranol, trenbolone acetate, melegestrol acetate or MGA) and thyrostatic, as growth promoters in farm animals. However, the Council recognized that five of the hormones at issue here (all but MGA) were of a different status than the other banned hormones and directed the Commission to provide a report on the experience acquired and scientific developments, accompanied, if necessary, by proposals which take these developments into account. In the meantime, the individual Member State regulations would continue to apply to the use of these five hormones.

Seven years later, the Directive 88/146/EEC was enacted, aiming at banning the administration of synthetic hormones (zeranol and trenbolone acetate) with any purpose, and natural hormones (estradiol, progesterone and testosterone) to promote growth in cattle (EEC, 1988a). This Directive allowed State Members to authorize the administration of those natural hormones, under certain circumstances, with therapeutic and zootechnical purposes. Both intra community trade and importation from non-European countries of meat and meat products from animals treated with chemicals with estrogenic, progestogenic and androgenic or thyrostatic effects were specifically forbidden with Directive 88/146/EEC. Meat from animals treated with a therapeutic or zootechnical purpose was allowed under certain circumstances, established with Directive 88/299/EEC (EEC, 1988b). In 1996, Directive 96/22/EC, a revision and repealing of previous hormone Directives, established the ban of substances having thyrostatic, estrogenic, androgenic and gestagenic action in animal husbandry and aquaculture (EC, 1996a). The Directive 96/23/EC on measures to monitor certain substances and residues thereof in live animals and animal products, was released to establish that Member States should draft a national residue monitoring plan for the groups of substances detailed in its Annex I (EC, 1996b). These plans had to comply with the sampling rules in Annex IV to the Directive. It also established the frequencies and level of sampling and the groups of substances to be controlled for each food commodity. This Directive included the control of a wide range of veterinary drugs in food producing animals and goods derived from them, such as meat, eggs and honey. In Annex I, substances were classified in two groups: group A included substances having anabolic effect and unauthorized substances, and group B included authorized veterinary drugs, the MRL of which have been established, and contaminants. So that for residues of substances from group A, a 'zero tolerance' applied.

In 2003, the Council Directive 2003/74/EC amended Directive 96/22/EC and narrowed circumstances under which estradiol-17β and its ester-like derivatives could be administered, under strict veterinary prescription and for non-growth-promoting purposes (treatment of foetus maceration or mummification or treatment of pyometra in cattle, and in oestrus induction in cattle, horses, sheep or goats until 14 October 2006) (EC, 2003). Those authorized treatments had to be carried out by the veterinarian himself or herself on farm animals which have been clearly identified, and had to be registered by the veterinarian responsible. Lately, the Council Directive 2008/97/EC was enacted to take into account the European Protocol on protection and welfare of animals, limiting the scope of Directive 96/22/EC only to food-producing animals and withdrawing the prohibition for pet animals, as well as to adjust the definition of therapeutic treatment (EC , 2008). As a matter of fact, an efficient control of residues is an essential contribution to the maintenance of a high level of consumer protection in the EU and it was necessary to provide clear rules on how laboratory analysis had to be carried out and results interpreted. That was achieved with Commission Decision 2002/657/EC, implementing Council Directive 96/23/EC, which established criteria and procedures for the validation of analytical methods to ensure the quality and comparability of analytical results generated by official laboratories (EC, 2002). Moreover, the Decision established common criteria for the interpretation results and introduced a procedure to establish minimum required performance limits (MRPL) for analytical methods employed to detect substances for which no permitted limit (MRL) had been established. This is in particular important for substances whose use is not authorized or is specifically prohibited in the EU, such as hormonally active substances. For the first time, the concepts of decision limit (CCα) and detection capability (CCβ) were introduced, as quality parameters that must be established during the validation of an analytical method. Currently, the evolution in analytical equipment and progress in scientific research, accompanied by recent European regulatory changes, seems to demand an update or revision of the 2002/657/EC (Vanhaecke et al., 2011).

Unlike in European countries, a number of steroidogenic drugs, which are used as hormonal growth promoters, are registered for use in many countries including Australia, New Zealand, United States and Canada, among others. However, the EU has been constantly banning their use since early 80s with Directive 81/602/EEC and neither allows the importation of products from cattle given growth promoters. In 1998, the World Trade Organization (WTO) found the European ban not supported by scientific evidence and inconsistent with its WTO obligations, but Europe continues arguing consumers' concerns, animal welfare and meat quality so that its rule remains in place currently. Although the World Trade Organization has issued decisions that have questioned the validity of the European ban, the EU has repeatedly voted to maintain it, citing consumer worries, questions of animal welfare and meat quality.

5. Beef hormone: US versus EU and trade dispute

Growth hormones are used extensively around the world to enhance the performance of beef cattle. In 1981 the European Union adopted first restrictions on the use of hormones as growth promoters in beef production with Directive 81/602/EEC, the first hormone directive (EEC, 1981). This directive prohibited the utilization of stilbenes and thyrostatics, two hormonal substances presumed to have harmful effects. Later in 1989 the EU fully implemented ban on imports of meats treated with enhancing hormones, expanding their restrictions to other non-European countries wishing to export meat to EU and that would assume many of the rights and obligations of European single market. The ban of imported hormone-beef arose from European consumers' pressure more than from producers, and it meant great losses for the US meat industry. The EU justified the ban as needed to protect the health and safety of consumers from the illegal and unregulated use of hormones in livestock production in several European countries. During the 1980s, there were widespread press reports of black market sales of 'hormone cocktails' by a 'hormone mafia' as well as several reports of serious health effects from consuming meat from animals treated with enhancing hormones. Many European livestock producers support the hormone ban because of the possible existence of competition from cheaper imported beef from beef exporting countries using hormones to breed animals. Also consumers' increasing demand of hormone-free meat creates concerns among European farmers about maintaining the ban. Certain circumstances, such as the Italian hormone crisis (Loizzo et al., 1984; Loizzo, 1984) and the outbreak during the 1990s of bovine spongiform encephalopathy (BSE), so called 'mad cow disease', added consumer distrust about the safety of beef supply. Although the BSE problem had nothing to do with hormones, it also contributed further to an unfavourable politic-economic and social environment for resolving the beef hormone dispute between EU and US and Canada.

For the past 15 years, the United States and the European Union have been disputing the safety of growth promotants used in cattle. The disagreement over the use of hormones started when the EU banned the import of beef from cattle treated with hormones in 1989, cutting off exports of beef. Unlike EU, the use of natural hormones in farm animals keeps avoiding any legal ban as the US Food and Drug Administration (FDA) says use of hormones is 'safe and scientifically backed up with research'. Since the 1950s, the FDA has approved a number of steroid hormone drugs for use in farm animals, including estradiol, progesterone, testosterone, and their synthetic versions zeranol, melengestrol acetate and trenbolone acetate. FDA claims that people are not at risk eating food from animals treated with these drugs because the amount of additional hormone following drug treatment is very small compared with the amount of natural hormones that are normally found in the meat of untreated animals and that are naturally produced in the human body. Consequently, hormones have continued to be used to promote growth in beef cattle both legally in the US and elsewhere in the world and illegally within the EU.

In response to US threats to challenge the ban in early 1996, the European Parliament voted unanimously to keep it, citing consumers' concerns, animal welfare and meat quality, among other reasons. European farm ministers from different EU countries also supported the ban, only the Minister of Agriculture of UK voted to end it arguing that there was no scientific basis to maintain it. The trade dispute took turn in 1996 when US presented its case against EU hormone ban to the World Trade Organization (WTO). The WTO, a dispute-settlement mechanism born in 1995, found that the European ban was not based on evidence in 1997. However, the European Commission (EC) appealed against this statement and sponsored research studies to clarify the risk for consumers of hormonally active substances applied in food producing animals. Altogether six substances were at issue in the dispute, three naturally occurring hormones (estradiol-17β, testosterone, and progesterone) whose level in animals can vary significantly, depending on age, sex, and sexual development of the animal, among other factors, and three synthetic substances (trenbolone, zeranol, and melengestrol acetate) that are produced synthetically to mimic the effect of the three naturally occurring hormones. The EC's Scientific Committee on Veterinary Measures relating to Public Health (SCVPH) concluded that the risk from hormone-treated food was higher than previously thought and proposed that there was a significant body of scientific evidence suggesting that 17β-estradiol should be considered a complete carcinogen. It also concluded that there were risks to consumers from the other five hormones examined and no threshold concentrations could be defined. The EU invoked the precautionary principle as a rationale for its banning the import of beef produced using hormones. The Agreement on Application of Sanitary and Phyto-sanitary Measures (SPS) permits precautionary measures when a government considers the scientific evidence insufficient to permit a final decision on the safety of a product, as is the case of hormonally produced food. The WTO Panel upheld US position and the EU was given until May 13 1999 to bring its measure into compliance. However, the EU Commission voted again unanimously to continue the ban. In maintaining its unscientific ban, the EU does nothing to further the objective of protecting public health, but instead undermines the WTO Sanitary and Phytosanitary Agreement (SPS) and invites other countries to renege on their international obligations.

Despite the attempts of US to solve this dispute, the EU reaffirmed its position that there is a possible risk to human health associated with hormone-treated meat, basing on available scientific data. To date, the EU continues to ban import of meat from animals treated with hormones and only imports high-quality beef certified as produced without the use of hormones. However, on May 13 in 2009, following a series of negotiations, the United States and the EU signed in Geneva a memorandum of understanding (MOU) implementing an agreement that could resolve this longstanding dispute. Under MOU the EU expanded the market access of US beef, at zero duty, from cattle raised under control measures specified in USDA's Non-Hormone Treated Cattle (NHTC) program, from cattle grown in approved farms/feedlots. To become eligible to export non-treated beef, producers must obtain certification for their cattle through the NHTC program. Meanwhile, the US and Canada continue retaliating against the EU hormone ban based on the additional costs of producing non-hormone treated beef for the European Union and the lack of evidence of its harmful effects in humans. Despite all the US controversy, in a survey of US consumers it was found that most respondents desired the existence of mandatory labelling of food produced with growth hormones, even when labelling costs causing beef prices increase up to 17% (Lusk & Fox, 2002). While the dispute is between Canada and the US and the EU, other important beef-producing countries have approved the use of growth-promoting hormones in beef production such as Canada, New Zealand, South Africa, Mexico, Chile, and Japan, among others. Like for US meat, thigh controls are in place to ensure all beef exported to EU comes from non-hormone treated cattle.

6. Progress on analytical methodology

During the past few years, many authors have described the application of LC-MS/MS methods for the analysis of anabolic steroids in various biological samples, including urine, serum, hair, kidney and fat (Draisci et al., 2000; G. Kaklamanos et al., 2009a; Kaklamanos et al., 2009b; Kaklamanos et al., 2011; Shao et al., 2005), all validated according to the criteria set out in Decision 2002/657/CE for banned substances. Although normally the levels of steroids that accumulate in animal tissues are lower than in other matrices, many effective methods are known currently for the determination of these compounds in muscle tissue, sometimes monitoring a wide range of anabolic compounds (Courant et al., 2008; Kaklamanos et al., 2007; Vanhaecke et al., 2011; Yang et al., 2009). Since the number of growth promoters is high and includes natural and synthetic compounds, the use of multianalyte techniques is becoming more interesting (Vanhaecke et al., 2011; Xu et al., 2006; Yang et al., 2009). The use of ultra-resolution liquid chromatography techniques (UPLC), coupled to mass spectrometry devices, provides a rapid separation of analytes, shortening analytical times and improving the simultaneous detection of multiple steroids (Stolker et al., 2008; Vanhaecke et al., 2011).

6.1 Synthetic and semi-synthetic steroids

Synthetic hormones are xenobiotic substances that do not naturally occur in animal organisms. These exogenous drugs mimic the effects of natural endogenous hormones, such as the case of synthetic versions of estradiol, progesterone and testosterone: zeranol, melengestrol acetate (MGA) and trenbolone acetate, respectively. In general and due to their entirely exogenous character, since these compounds do not exist naturally, there are no major difficulties in determining analytic synthetic steroids. Thus, their mere presence in the animal organism is a clear evidence their administration. With regard to the confirmation of use of xenobiotic analogues of natural sex steroids and non-steroidal compounds, such as stilbenes and zeranol, there is an extensive range of successful methods that has been performed on different analytic matrices. These analytical procedures have made the confirmation of illicit administrations of anabolics in cattle feasible (De Brabander et al., 2007; Duffy et al., 2009; Kaklamanos et al., 2011; Noppe et al., 2008).

On the other hand, many veterinary hormonal preparations, although not all, consist on ester derivatives of the corresponding endogenous steroid, such as testosterone decanoate or estradiol benzoate. As hormonal esters do not naturally occur in the animal organism, the detection of these synthetic substances in the body of an animal provides irrefutable evidence of the abuse of these promoters. Although the administration of esters of natural hormones can be detected through hair analysis (Duffy et al., 2008; Gratacós-Cubarsí et al., 2006; Pedreira et al., 2007; Rambaud et al., 2005), it has been very difficult to detect intact steroid esters in body fluids or tissues. It is likely that esters quickly hydrolyze in the body of the animal, releasing the corresponding natural hormone (Stolker et al., 2009). So, the simple detection and confirmation of the presence of these semisynthetic versions are not always possible. That is particularly true for estradiol benzoate, which has only been detected in hair up to date (Duffy et al., 2009; Hooijerink et al., 2005; Rambaud et al., 2005; Stolker et al., 2009), because estradiol benzoate undergoes the loss of the ester group once it reaches the bloodstream. In order to detect the administration of these esterified substances, new analytical approaches are required, in the same manner that they are required to detect the administration of exogenous hormones in their natural chemical form (estradiol, testosterone or progesterone).

6.2 Exogenous natural steroid hormones

Hormones of natural origin, such as estradiol-17β, testosterone or progesterone, are still a weak area in residue-monitoring plans due to their endogenous origin, as the target compound is always present. In such a case, the confirmation of an exogenous administration involves logical difficulties associated with distinguishing an exogenous origin from an endogenous (naturally occurring) presence of these hormones. The demonstration of an exogenous administration of natural steroids, such as testosterone, estradiol or cortisol, remains problematic. No official threshold has been stated for natural hormone concentrations, mainly due to the fact that concentrations of naturally occurring hormones are highly variable and depend on the type of animal product, breed, gender, age, disease, medication and physiological condition (Angeletti et al., 2006; Hartmann et al., 1998; Le Bizec et al., 2009; Pleadin et al., 2011). The development of methods to provide unequivocal discrimination between the natural presence of an endogenous hormone and its presence as a consequence of an illegal exogenous administration remains a challenge. Some promising analytical approaches have been published in the past few years regarding this critical point of controlling residues in food of animal origin.

7. Monitoring exogenous natural hormones: An unresolved problem

The confirmation of an exogenous administration of natural hormones involves logical difficulties associated with the distinction between an exogenous origin and the natural presence of these endogenous hormones (synthesis in the body). In fact, it has been found that treatments with testosterone or estradiol in bovines lead to equal or lower plasma concentrations of these compounds (Scippo et al., 1994; Simontacchi et al., 2004). On the other hand, exogenous natural hormones are usually administered as simple semi-synthetic esters (i.e., 17β-estradiol benzoate and testosterone decanoate) to increase their effective half-life. Subsequently, a rapid hydrolysis of these compounds takes place as soon as they reach the bloodstream, where they generate non-esterified forms that are indistinguishable from naturally occurring forms (Stolker et al., 2009). These exogenous natural compounds (or even hormonal esters) follow the same pathways as the natural compounds biosynthesized by the animal, making the detection and confirmation of their exogenous administration difficult, if not impossible. These circumstances have led to the lack of success in detecting hormone esters such as estradiol benzoate in serum or plasma, which has only been confirmed in hair from animals treated with this ester (Duffy et al., 2009; Regal et al., 2008). Some promising approaches have been published in recent years in regard to the development of methods that allow unambiguous discrimination between the presence of a natural endogenous hormone and its presence as a result of an illegal exogenous administration.

7.1 Approaches based on hormonal metabolism

Up to date, it is unknown to what extent the administration of exogenous hormones interferes with the hormonal metabolism of the animal. When administered exogenously, natural steroid hormones must be metabolized in the same way and using the same metabolic pathways that endogenous compounds, so that the ratios of these hormones against their precursors and/or their metabolites may be somehow altered by the hormonal treatment (Pinel et al., 2010). This analytical approach is already a common practice in the supervision of doping in sports and horse racing (Sottas et al., 2008; Strahm et al., 2009; Torrado et al., 2008). However, the known approaches based on metabolic parameters that have proved to be useful in humans, as the ratio testosterone/epitestosterone, not seem to succeed in cattle (Angeletti et al., 2006). Early attempts to detect illegal hormone treatments in cattle assessing hormonal levels and levels of their precursors and metabolites (and hormonal ratios) appeared in plasma and urine of cattle (Becue et al., 2010; Fritsche et al., 1999; Simontacchi et al., 2004). In 1999, Fritsch et al. concluded that implants of estradiol benzoate and progesterone did not to affect the total concentration of steroids (precursors and metabolites) in meat from treated cattle. However, they encountered significant differences in 17β-estradiol/17α-estradiol+Estrone ratio and in cortisone/hydrocortisone ratio; so that the administration of natural hormones did seem to influence the hormonal metabolism (Fritsche et al., 1999). On the other hand, Maume et al. concluded that estradiol implants increased the concentration of 17β-estradiol (E_2) in beef, following a dose-dependent relationship. Besides, the concentrations of E_2 and αE_2 glucuronides were also increased in tissues of treated animals, an effect also being dose-dependent (Maume et al., 2003). As for the administration of steroid precursors or prohormones, Becue et al. recently published results in urine for cattle treated with DHEA orally or intramuscularly. They concluded that the administration of DHEA causes significant differences in overall levels of DHEA sulphate and 5-androstenediol, and also in the levels of free DHEA and 17α-testosterone in animals treated orally. DHEA levels, both free and sulphate, 5-androstenediol and 17α-testosterone were modified for animals treated via intramuscular (Becue et al., 2010).

7.2 $^{13}C/^{12}C$ isotopic ratio

The measurement of $^{13}C/^{12}C$ ratios by gas chromatography/combustion/isotope ratio mass spectrometry (GC-C-IRMS) can be a powerful tool to trace the true origin of the steroids, and it is one of the most promising approaches to control of exogenous administration of natural hormones. The administration of natural hormones to cattle will lead to an alteration of the $^{13}C/^{12}C$ ratio of these compounds and their metabolites, whereas the isotopic composition remains constant for their precursors. This methodology has been applied successfully in urine to trace the administration of testosterone and estradiol esters in cattle (Buisson et al., 2005; Hebestreit et al., 2006). However, no thresholds of reference for these changes in the 13C/12C ratios of steroid hormones have been established yet, since their variations depend largely on the diet of the animal and also other factors such as age, sex or breed (Cawley et al., 2009; Ferchaud et al., 2000). In addition, GC-C-IRMS methods imply long and complicated steps of extraction and purification of the sample and also the use of semi-preparation HPLC and derivatization procedures (Buisson et al., 2005; Hebestreit et al., 2006).

7.3 Omic technologies: Metabolomics

In recent years, there is a clear trend towards more intelligent procedures for data evaluation. New methodological approaches based on untargeted and global measurements are emerging strongly as analytical tools for the analysis of residues in food-producing animals (Pinel et al., 2010; Riedmaier et al., 2009). Traditionally, samples are analyzed searching for the presence of specific target analytes. Metabolomics is an emerging field within the *omic* methodologies (proteomics, transcriptomics, metabolomics) that focuses, in an untargeted and global scale, in the high yield measurement of small molecules (called metabolites) in biological matrices. These techniques are based on the physiological changes that are expected to appear in the animal due to the administration of an anabolic agent, and they are largely used to search potential biomarkers of such administration (Mooney et al., 2008; Mooney et al., 2009; Riedmaier et al., 2009). Metabolomics is based on detecting small molecules and excluding big biopolymers such as proteins, generating this way a large set of descriptors characteristic of the biological matrix under investigation in different experimental groups. Metabolomic approaches have already shown to be affective in relation to different biochemical processes, such as drug toxicity and diseases (Brindle et al., 2002; Coen et al., 2003; Lindon et al., 2004; Vallejo et al., 2009), prediction of gender (Lutz et al., 2006), nutritional effects studies (Wang et al., 2005) and even doping control in horses (Kieken et al., 2011) or the clenbuterol use in cattle (Courant et al., 2009). Basically, the general principle of metabolomic studies is the characterization of the biological system or sample in question through the generation of metabolomic fingerprints.

There are not many metabolomics studies regarding the detection of sex steroid abuse in livestock production, neither useful biomarkers to be used in targeted analysis. However, Rijk et al. developed a metabolomic strategy of screening in urine for bovine animals treated with DHEA and pregnenolone. This methodology is an useful tool to track the abuse of these prohormones, but their reference levels remain unknown, making difficult their use in targeted analysis, at least until the metabolites of this non-focused approach are elucidated (Rijk et al., 2009). On the other hand, Dervilly-Pinel et al. recently published a metabolomic study in urine for cattle treated with a combination of estradiol benzoate and nandrolone laureate. These authors proved the existence of changes in urinary metabolomic profiles of animals treated with steroids, when comparing them with control animals, opening the door to possible screening strategies. Dervilly-Pinel et al. also stated the importance of several potential biomarkers in the discrimination of hormonally treated animals from control cattle (Dervilly-Pinel et al., 2011).

There is a range of analytical platforms which could be employed in metabolomics to collect data, including gas chromatography-mass spectrometry (GC-MS), liquid chromatography-mass spectrometry (LC-MS) and nuclear magnetic resonance (NMR) spectrometry, just to extract as much information as possible from the matrix of interest. Traditionally, NMR has played a key role in the development of metabolomics and was the preferred analytical tool. However, due to the higher sensitivity compared to NMR and the increasing reproducibility of today's GC-MS and LC-MS systems, the have become important analytical platforms for all types of metabolomic applications (Brown et al., 2009; Lutz et al., 2006). By coupling high chromatography separation with different mass spectrometric techniques, the partial or complete separation of metabolites prior to detection is achieved. To this end, gas chromatography was first used, however, the necessary thermal stability of the investigated compounds, the required derivatization of non-volatile compounds prior to analysis and the frequent absence of molecular ions have made liquid chromatography coupled to atmospheric-pressure ionization a more and more used tool for metabolomics. Last generation of high resolution devices including time of flight and *orbitrap* instruments are becoming more and more used in this field due to their high performance in terms of sensitivity and specificity (Dunn et al., 2008; Scheltema et al., 2008).

Moreover, due to the large amount of data generated during metabolomic fingerprinting, bioinformatics tools are required for processing and analysing this huge volume of complex data. Several software solutions are today available either with free access (XCMS, MZmine, Metalign, MathDAMP, COMSPARI and METIDEA) or commercially (SIEVE from Thermo Fisher Scientific, Markerlynx from Waters, GeneSpring from Agilent and MarkerView from MSDSciex). Data processing is composed of four major steps; background noise correction, peak alignment, peak deconvolution and peak sorting. Then, resulting data are of multivariate character, since metabolomic procedures handle several hundreds to thousands of variables/metabolites. Thus the use of multivariate statistical techniques are required to analyse such data set and finally to point out potential signals (i.e. biomarkers) of interest. Principal component analysis (PCA) has become a popular tool for visualizing datasets and for extracting relevant information, as well as partial least-squares projection to latent structures (PLS) (Eriksson et al., 2004). Recently, the Orthogonal PLS methodology (OPLS) has shown to be useful for elucidating differences between many samples and many variables (Pohjanen et al., 2007; Werner et al., 2008).

8. References

Adebamowo, C. A.; Spiegelman, D.; Berkey, C. S.; Danby, F. W.; Rockett, H. H.; Colditz, G. A.; Willett, W.C. & Holmes, M.D. (2008). Milk consumption and acne in teenaged boys. *Journal of the American Academy of Dermatology*, Vol.58, No.5, pp. 787-793.

Aksglaede, L.; Juul, A.; Leffers, H.; Skakkebæk, N. E. & Andersson, A. (2006). The sensitivity of the child to sex steroids: Possible impact of exogenous estrogens. *Human Reproduction Update*, Vol.12, No.4, pp. 341-349.

Alves, C.; Flores, L. C.; Cerqueira, T. S. & Toralles, M. B. P. (2007). Environmental exposure to endocrine disruptors with estrogenic activity and the association with pubertal disorders in children. *Cadernos Saude Publica*, Vol.23, No.5, pp. 1005-1014.

Angeletti, R.; Contiero, L.; Gallina, G. & Montesissa, C. (2006). The urinary ratio of testosterone to epitetosterone: A good marker of illegal treatment also in cattle? *Veterinary Research Communications*, Vol.30, Suppl.1, pp. 127-131.

Auclair, C. A. (1979). Consequences of prenatal exposure to diethylstilbestrol. *Journal of Obstetric, Gynecologic, and Neonatal Nursing*, Vo.8, No.1, pp. 35-39.

Bay, K.; Andersson, A. & Skakkebaek, N. E. (2004). Estradiol levels in prepubertal boys and girls - analytical challenges. *International Journal of Andrology*, Vol.27, No.5, pp. 266-273.

Becue, I.; Van Poucke, C.; Rijk, J. C. W.; Bovee, T. F. H.; Nielen, M. & Van Peteghem, C. (2010). Investigation of urinary steroid metabolites in calf urine after oral and intramuscular administration of DHEA. *Analytical and Bioanalytical Chemistry*, Vol.396, pp. 799-808.

Boles, J. A.; Boss, D. L.; Neary, K. I.; Davis, K. C. & Tess, M. W. (2009). Growth implants reduced tenderness of steaks from steers and heifers with different genetic potentials for growth and marbling. *Journal of Animal Science*, Vol.87, pp. 269-274.

Brindle, J. T.; Antti, H.; Holmes, E.; Tranter, G.; Nicholson, J. K.; Bethell, H. W. L.; Clarke, S.; Schofield, P.M.; McKilligan, E.; Mosedale, D.E. & Grainger, D.J. (2002). Rapid and non-invasive diagnosis of the presence and severity of coronary heart disease using 1H NMR -based metabonomics. *Nature Medicine*, Vol.8, No.12, pp. 1439-1444.

Brown, M.; Dunn, W. B.; Dobson, P.; Patel, Y.; Winder, C. L.; Francis-McIntyre, S.; Begley, P.; Carroll, K.; Broadhurst, D.; Tseng, A.; Swainston, N.; Spasic, I.; Goodacre, R. & Kell, D.B. (2009). Mass spectrometry tools and metabolite-specific databases for molecular identification in metabolomics. *Analyst*, Vol.134, No.7, pp. 1322-1332.

Buisson, C.; Hebestreit, M.; Weigert, A. P.; Heinrich, K.; Fry, H.; Flenker, U.; Banneke, S.; Prevost, S.; Andre, F.; Schaenzer, W.; Houghton, E. & Le Bizec, B. (2005). Application of stable carbon isotope analysis to the detection of 17 beta-estradiol administration to cattle. *Journal of Chromatography A*, Vol.1093, No.1-2, pp. 69-80.

Burkea, C. R.; Mussardb, M. L.; Grumb, D. E. & Day, M. L. (2001). Effects of maturity of the potential ovulatory follicle on induction of oestrus and ovulation in cattle with oestradiol benzoate. *Animal Reproduction Science*, Vol.66, No.3-4, pp. 161-174.

Cavalieri, J.; Rabiee, A. R.; Hepworth, G. & Macmillan, K. L. (2005). Effect of artificial insemination on submission rates of lactating dairy cows synchronised and resynchronised with intravaginal progesterone releasing devices and oestradiol benzoate. *Animal Reproduction Science,* Vol.90, No.1-2, pp. 39-55.

Cawley, A. T.; Trout, G. J.; Kazlauskas, R.; Howe, C. J. & George, A. V. (2009). Carbon isotope ratio (δ13C) values of urinary steroids for doping control in sport. *Steroids,* Vol.74, pp. 379-392.

Cesario, S. K. & Hughes, L. A. (2007). Precocious puberty: A comprehensive review of literature. *Journal of Obstetric, Gynecologic, & Neonatal Nursing*, Vol.36, No.3, pp. 263-274.

Chen, T. T.; Huang, C. C.; Lee, T. Y.; Lin, K. J.; Chang, C. C. & Chen, K. L. (2010). Effect of caponization and exogenous androgen implantation on muscle characteristics of male chickens. *Poultry Science*, Vol.89, No.3, pp. 558-563.

Coen, M.; Lenz, E. M.; Nicholson, J. K.; Wilson, I. D.; Pognan, F. & Lindon, J. C. (2003). An integrated metabonomic investigation of acetaminophen toxicity in the mouse using NMR spectroscopy. *Chemical Research in Toxicology*, Vol.16, No.3, pp. 295-303.

Courant, F.; Aksglaede, L.; Antignac, J.P.; Monteau, F.; Sorensen, K.; Andersson, A.; Skakkebaek, N.; Juul, A. & Le Bizec, B. (2010). Assessment of circulating sex steroid levels in prepubertal and pubertal boys and girls by a novel ultrasensitive gas chromatography-tandem mass spectrometry method. *Journal of Clinical Endocrinology and Metabolism*, Vol.95, No.1, pp. 82-92.

Courant, F.; Antignac, J.P.; Laille, J.; Monteau, F.; Andre, F. & Le Bizec, B. (2008). Exposure assessment of prepubertal children to steroid endocrine disruptors, 2: Determination of steroid hormones in milk, egg, and meat samples. *Journal of Agricultural and Food Chemistry*, Vol.56, No.9, pp. 3176-3184.

Courant, F.; Antignac, J.P.; Maume, D.; Monteau, F.; Andre, F. & Le Bizec, B. (2007). Determination of naturally occurring oestrogens and androgens in retail samples of milk and eggs. *Food Additives and Contaminants*, Vol.24, No.12, pp. 1358-1366.

Courant, F.; Pinel, G.; Bichon, E.; Monteau, F.; Antignac, J.P. & Le Bizec, B. (2009). Development of a metabolomic approach based on liquid chromatography-high resolution mass spectrometry to screen for clenbuterol abuse in calves. *Analyst*, Vol.134, pp. 1637-1646.

Cousins, L.; Karp, W.; Lacey, C. & Lucas, W. E. (1980). Reproductive outcome of women exposed to diethylstilbestrol in utero. *Obstetrics and Gynecology*, Vol.56, No.1, pp. 70-76.

Daeseleire, E.; Vanoosthuyze, K. & Van Peteghem, C. (1994). Application of high-performance thin-layer chromatography and gas chromatography-mass spectrometry to the detection of new anabolic steroids used as growth promoters in cattle fattening. *Journal of Chromatography A*, Vol.674, No.1-2, pp. 247-253.

Daxenberger, A.; Ibarreta, D. & Meyer, H. H. D. (2001). Possible health impact of animal oestrogens in food. *Human Reproduction Update*, Vol.7, No.3, pp. 340-355.

De Brabander, H. F.; Le Bizec, B.; Pinel, G.; Antignac, J.P.; Verheyden, K.; Mortier, V.; Courtheyn, D. & Noppe, H. (2007). Past, present and future of mass spectrometry in the analysis of residues of banned substances in meat-producing animals. *Journal of Mass Spectrometry*, Vol.42, No.8, pp. 983-998.

Dervilly-Pinel, G.; Weigel, S.; Lommen, A.; Chereau, S.; Rambaud, L.; Essers, M.; Antignac, J.P.; Nielen, M.W.F. & Le Bizec, B. (2011). Assessment of two complementary liquid chromatography coupled to high resolution mass spectrometry metabolomics strategies for the screening of anabolic steroid treatment in calves. *Analytica Chimica Acta*, Vol.700, No.1-2, pp. 144-154.

Dikeman, M. E. (2007). Effects of metabolic modifiers on carcass traits and meat quality. *Meat Science*, Vol.77, No.1, pp. 121-135.

Draisci, R.; Palleschi, L.; Ferretti, E.; Lucentini, L. & Cammarata, P. (2000). Quantitation of anabolic hormones and their metabolites in bovine serum and urine by liquid chromatography–tandem mass spectrometry. *Journal of Chromatography A*, Vol.870, No.1-2, pp. 511-522.

Duffy, E.; Rambaud, L.; Le Bizec, B. & O'Keeffe, M. (2009). Determination of hormonal growth promoters in bovine hair: Comparison of liquid chromatography-mass spectrometry and gas chromatography-mass spectrometry methods for estradiol benzoate and nortestosterone decanoate. *Analytica Chimica Acta*, Vol.637, No.1-2, pp. 165-172.

Duffy, E. F.; Danaher, M. & O'Keeffe, M. (2008). Determination of steroid esters in bovine hair using LC-MS/MS. *Proceedings of the EuroResidue VI (Conference on Residues of Veterinary Drugs in Food)*, Vol. 3, pp. 1025-1030.

Dunn, W. B.; Broadhurst, D.; Brown, M.; Baker, P. N.; Redman, C. W. G.; Kenny, L. C. & Kell, D.B. (2008). Metabolic profiling of serum using ultra performance liquid chromatography and the LTQ-orbitrap mass spectrometry system. *Journal of Chromatography B*, Vol.871, No.2, pp. 288-298.

EEC (1988a). Directive 81/602/EEC. *Official Journal of the European Communities*, No.L222, pp. 32-33.

EEC (1988b). Directive 88/299/EEC. *Official Journal of the European Communities*, No.L128, pp. 36-38.

EC (1996a). Directive 96/22/EC. *Official Journal of the European Communities*, No.L125, pp. 3-9.

EC (1996b). Directive 96/23/EC. *Official Journal of the European Communities*, No.L125, pp. 10-32.

EC (2002). Decision 2002/657/EC. *Official Journal of the European Union*, No.L221, pp. 8-36.

EC (2003). Directive 2003/74/EC. *Official Journal of the European Union*, No.L262, pp. 17-21.

EC (2008). Directive 2008/97/EC. *Official Journal of the European Union*, No.L318, pp. 9-11.

Eriksson, L.; Antti, H.; Gottfries, J.; Holmes, H.; Johansson, E.; Lindgren, F.; Long, I.; Lundstedt, T.; Trygg, J. & Wold, S. (2004). Using chemometrics for navigating in the large data sets of genomics, proteomics and metabolomics (gpm). *Analytical and Bioanalytical Chemistry*, Vol.380, pp. 419-429.

European Commission, Directorate General for Health & Consumers (2004). SANCO/2004/2726rev1: Guidelines for the implementation of Decision 2002/657/EC.

Farlow, D. W.; Xu, X. & Veenstra, T. D. (2009). Quantitative measurement of endogenous estrogen metabolites, risk-factors for developement of breast cancer, in commercial milk products by LC-MS/MS. *Journal of Chromatography B*, Vol.877, pp. 1327-1334.

Ferchaud, V.; Le Bizec, B.; Monteau, F. & Andre, F. (2000). Characterization of exogenous testosterone in livestock by gas chromatography/combustion/isotope ratio mass spectrometry: Influence of feeding and age. *Rapid Communications in Mass Spectrometry*, Vol.14, No.8, pp. 652-656.

Foster, P. A.; Chander, S. K.; Parsons, M. F. C.; Newman, S. P.; Woo, L. W. L.; Potter, B. V. L.; Reed, M.J. & Purohit, A. (2008). Efficacy of three potent steroid sulfatase inhibitors: Pre-clinical investigations for their use in the treatment of hormone-dependent breast cancer. *Breast Cancer Research and Treatment*, Vol.111, No.1, pp. 129-138.

Fritsche, S.; Rumsey, T.; Meyer, H.; Schmidt, G. & Steinhart, H. (1999). Profiles of steroid hormones in beef from steers implanted with synovex-S (estradiol benzoate and progesterone) in comparison to control steers. *Zeitschrift Für Lebensmitteluntersuchung Und -Forschung A*, Vol.208, No.5-6, pp. 328-331.

Ganmaa, D.; Cui, X., Feskanich, D., Hankinson, S. E., & Willett, W. C. (2011). Milk, dairy intake and risk of endometrial cancer: A twenty six-year follow-up. *International Journal of Cancer*, (DOI: 10.1002/ijc.26265).

Ganmaa, D. & Sato, A. (2005). The possible role of female sex hormones in milk from pregnant cows in the development of breast, ovarian and corpus uteri cancers. *Medical Hypotheses*, Vol.65, No.6, pp. 1028-1037.

Ganmaa, D.; Wang, P. Y.; Qin, L. Q.; Hoshi, K. & Sato, A. (2001). Is milk responsible for male reproductive disorders? *Medical Hypotheses*, Vol.57, No.4, pp. 510-514.

Garner, B.; Phassouliotis, C.; Phillips, L. J.; Markulev, C.; Butselaar, F.; Bendall, S.; Yung, Y. & McGorry, P.D. (2011). Cortisol and dehydroepiandrosterone-sulphate levels correlate with symptom severity in first-episode psychosis. *Journal of Psychiatric Research*, Vol.45, No.2, pp. 249-255.

Givens, D. I.; Morgan, R. & Elwood, P. C. (2008). Relationship between milk consumption and prostate cancer: A short review. *Nutrition Bulletin*, Vol.33, pp. 279-286.

Gratacós-Cubarsí, M.; Castellari, M.; Valero, A. & García-Regueiro, J. A. (2006). Hair analysis for veterinary drug monitoring in livestock production. *Journal of Chromatography B*, Vol.834, No.1-2, pp. 14-25.

Haney, A. F. & Hammond, M. G. (1983). Infertility in women exposed to diethylstilbestrol in utero. *Journal of Reproductive Medicine for the Obstetrician and Gynecologist*, Vol.28, No.12, pp. 851-856.

Hartmann, S.; Lacorn, M. & Steinhart, H. (1998). Natural occurrence of steroid hormones in food. *Food Chemistry*, Vol.62, No.1, pp. 7-20.

Hebestreit, M.; Flenker, U. ; Buisson, C.; Andre, F.; Le Bizec, B.; Fry, H.; Lang, M.; Weigert, A.P.; Heinrich, K.; Hird, S. & Schanzer, W. (2006). Application of stable carbon isotope analysis to the detection of testosterone administration to cattle. *Journal of Agricultural and Food Chemistry*, Vol.54, No.8, pp. 2850-2858.

Herman-Giddens, M. E.; Slora, E. J.; Wasserman, R. C.; Bourdony, C. J.; Bhapkar, M. V.; Koch, G. G. & Hasemeier, C.M. (1997). Secondary sexual characteristics and menses in young girls seen in office practice: A study from the pediatric research in office settings network. *Pediatrics*, Vol.99, No.4, pp. 505-512.

Hooijerink, H.; Lommen, A.; Mulder, P. P. J.; Van Rhijn, J. A. & Nielen, M. W. F. (2005). Liquid chromatography-electrospray ionisation-mass spectrometry based method for the determination of estradiol benzoate in hair of cattle. *Analytica Chimica Acta*, Vol.529, (1-2 SPEC. ISS.), pp. 167-172.

Hunter R.A. (2010). Hormonal growth promotant use in the australian beef industry. *Animal Production Science*, Vol.50, pp. 637-659.

Huyghe, E.; Matsuda, T. & Thonneau, P. (2003). Increasing incidence of testicular cancer worldwide: A review. *Journal of Urology*, Vol.170; No.1, pp. 5-11.

Igo, J. L.; Brooks, J. C.; Johnson, B. J.; Starkey, J.; Rathmann, R. J.; Garmyn, A. J.; Nichols, W.T.; Hutcheson, J.P. & Miller, M.F. (2011). Characterization of estrogen-trenbolone acetate implants on tenderness and consumer acceptability of beef under the effect of 2 aging times. *Journal of Animal Science*, Vol.89, No.3, pp. 792-797.

Kaaks, R.; Rinaldi, S.; Key, T. J.; Berrino, F.; Peeters, P. H. M.; Biessy, C. et al. (2005). Postmenopausal serum androgens, oestrogens and breast cancer risk: The european prospective investigation into cancer and nutrition. Endocrine-Related Cancer, 12(4), 1071-1082.

Kaklamanos, G.; Theodoridis, G.; Papadoyannis, I. N. & Dabalis, T. (2007). Determination of anabolic steroids in muscle tissue by liquid chromatography-tandem mass spectrometry. *Journal of Agricultural and Food Chemistry*, Vol.55, No.21, pp. 8325-8330.

Kaklamanos, G.; Theodoridis, G. A.; Dabalis, T. & Papadoyannis, I. (2011). Determination of anabolic steroids in bovine serum by liquid chromatography–tandem mass spectrometry. *Journal of Chromatography B*, Vol.879, No.2, pp. 225-229.

Kaklamanos, G.; Theodoridis, G. & Dabalis, T. (2009a). Determination of anabolic steroids in bovine urine by liquid chromatography–tandem mass spectrometry. *Journal of Chromatography B*, Vol.877, No.23, pp. 2330-2336.

Kaklamanos, G.; Theodoridis, G. & Dabalis, T. (2009b). Gel permeation chromatography clean-up for the determination of gestagens in kidney fat by liquid chromatography–tandem mass spectrometry and validation according to 2002/657/EC. *Journal of Chromatography A*, Vol.1216, No.46, pp. 8067-8071.

Kieken, F.; Pinel, G.; Antignac, J.P.; Paris, A.; Garcia, P.; Popot, M. G.; Grall, M.; Mercadier, V.; Toutain, P.L. ; Bonnaire, Y. & Le Bizec, B. (2011). Generation and processing of urinary and plasmatic metabolomic fingerprints to reveal an illegal administration of recombinant equine growth hormone from LC-HRMS measurements. *Metabolomics*, Vol.7, No.1, pp. 84-93.

Kushnir, M. M.; Blamires, T.; Rockwood, A. L.; Roberts, W. L.; Yue, B.; Erdogan, E.; Bunker, A.M. & Meikle, A.W. (2010). Liquid chromatography-tandem mass spectrometry assay for androstenedione, dehydroepiandrosterone, and testosterone with pediatric and adult reference intervals. *Clinical Chemistry*, Vol.56, No.7, pp. 1138-1147.

Kvarnryd, M.; Grabic, R.; Brandt, I. & Berg, C. (2011). Early life progestin exposure causes arrested oocyte development, oviductal agenesis and sterility in adult xenopus tropicalis frogs. *Aquatic Toxicology*, Vol.103, No.1-2, pp.18-24.

Lane, E. A.; Austin, E. J. & Crowe, M. A. (2008). Oestrous synchronisation in cattle-current options following the EU regulations restricting use of oestrogenic compounds in food-producing animals: A review. *Animal Reproduction Science*, Vol.109, No.1-4, pp. 1-16.

Le Bizec, B.; Pinel, G. & Antignac, J. (2009). Options for veterinary drug analysis using mass spectrometry. *Journal of Chromatography A*, Vol.1216. No.46, pp. 8016-8034.

Lee, C. Y.; Lee, H. P.; Jeong, J. H.; Baik, K. H.; Jin, S. K.L; Lee, J. H. & Sohnt, S.H. (2002). Effects of restricted feeding, low-energy diet, and implantation of trenbolone acetate plus estradiol on growth, carcass traits, and circulating concentrations of insulin-like growth factor (IGF)-I and IGF-binding protein-3 in finishing barrows. *Journal of Animal Science*, Vol.80, No.1, pp. 84-93.

Levy, L. S. (2010). CHAPTER 1. The use of hormomally active substances in veterinary and zootechnical uses – the continuing scientific and regulatory challenges. In : *Analyses for hormonal substances in food-producing animals*, J.F. Kay (Ed.), The Royal Society of Chemistry.

Lindon, J. C.; Holmes, E.; Bollard, M. E.; Stanley, E. G. & Nicholson, J. K. (2004). Metabonomics technologies and their applications in physiological monitoring, drug safety assessment and disease diagnosis. *Biomarkers*, Vol.9, No.1, pp. 1-31.

Loizzo, A. (1984). The case of diethylstilbestrol treated veal contained in homogenized baby-foods in italy. Methodological and toxicological aspects. *Annali Dell'Istituto Superiori Di Sanità*, Vol.20, No.2-3, pp. 215.

Loizzo, A.; Gatti, G. L.; Macri, A.; Moretti, G.; Ortolani, E. & Palazzesi, S. (1984). Italian baby food containing diethylstilbestrol - 3 years later. *Lancet*, Vol.1, No.8384, pp. 1014-1015.

Lukanova, A.; Lundin, E.; Micheli, A.; Arslan, A.; Ferrari, P.; Rinaldi, S.; Krogh, V.; Lenner, P.; Shore, R.E.; Biessy, C.; Muti, P.; Riboli, E.; Koenig, K.; Levitz, M.; Stattin, P.; Berrino, F.; Hallmans, G.; Kaaks, R.; Toniolo, P. & Zeleniuch-Jacquotte, A. (2004). Circulating levels of sex steroid hormones and risk of endometrial cancer in postmenopausal women. *International Journal of Cancer*, Vol.108, No.3, pp. 425-432.

Lusk, J. L. & Fox, J. A. (2002). Consumer demand for mandatory labeling of beef from cattle administered growth hormones or fed genetically modified corn. *Journal of Agricultural and Applied Economics*, Vol.34, No.1, pp. 27-38.

Lutz, U. ; Lutz, R. W. & Lutz, W. K. (2006). Metabolic profiling of glucuronides in human urine by LC-MS/MS and partial least-squares discriminant analysis for classification and prediction of gender. *Analytical Chemistry*, Vol.78, No.13, pp. 4564-4571.

Martínez, M. F.; Kastelic, J. P.; Adams, G. P.; Cook, B.; Olson, W. O. & Mapletoft, R. J. (2002). The use of progestins in regimens for fixed-time artificial insemination in beef cattle. *Theriogenology*, Vol.57, No.3, pp. 1049-1059.

Maruyama, K.; Oshima, T. & Ohyama, K. (2010). Exposure to exogenous estrogen through intake of commercial milk produced from pregnant cows. *Pediatrics International*, Vol.52, No.1, pp. 33-38.

Maume, D.; Deceuninck, Y.; Pouponneau, K.; Paris, A.; Le Bizec, B. & André, F. (2001). Assessment of estradiol and its metabolites in meat. *Acta Pathologica, Microbiologica et Immunologica Scandinavica*, Vol.109, No.1, pp. 32-38.

Maume, D.; Le Bizec, B.; Pouponneau, K.; Deceuninck, Y.; Solere, V.; Paris, A.; Antignac, J.P. & André, F. (2003). Modification of 17β-estradiol metabolite profile in steer edible tissues after estradiol implant adminsitration. *Analytica Chimica Acta*, Vol.483, pp. 289-297.

McPhee, M. J.; Oltjen, J. W.; Famula, T. R. & Sainz, R. D. (2006). Meta-analysis of factors affecting carcass characteristics of feedlot steers. *Journal of Animal Science*, Vol.84, pp. 3143-3154.

Meireles, S. I.; Esteves, G. H.; Hirata, R.; Peri, S.; Devarajan, K.; Slifker, M.; Mosier, S.L.; Peng, J.; Vadhanam, M.V.; Hurst, H.E.; Neves, E.J.; Reis, L.F.; Gairola, C.G.; Gupta, R.C. & Clapper, M.L. (2010). Early changes in gene expression induced by tobacco smoke: Evidence for the importance of estrogen within lung tissue. *Cancer Prevention Research*, Vol.3, No.6, pp. 707-717.

Melnik, B. C. (2009). Milk - the promoter of chronic western diseases. *Medical Hypotheses*, Vol.72, No.6, pp. 631-639.

Micheli, A.; Muti, P. ; Secreto, G.; Krogh, V.; Meneghini, E.; Venturelli, E.; Sieri, S.; Pala, V. & Berrino, F. (2004). Endogenous sex hormones and subsequent breast cancer in premenopausal women. *International Journal of Cancer*, Vol.112, No.2, pp. 312-318.

Moal, V.; Mathieu, E.; Reynier, P.; Malthièry, Y. & Gallois, Y. (2007). Low serum testosterone assayed by liquid chromatography-tandem mass spectrometry. Comparison with five immunoassay techniques. *Clinica Chimica Acta*, Vol.386, No.1-2, pp. 12-19.

Mooney, M. H.; Elliott, C. T. & Le Bizec, B. (2009). Combining biomarker screening and mass-spectrometric analysis to detect hormone abuse in cattle. *Trends in Analytical Chemistry*, Vol.28, No.6, pp. 665-675.

Mooney, M. H.; Situ, C.; Cacciatore, G.; Hutchinson, T.; Elliott, C. & Bergwerff, A. A. (2008). Plasma biomarker profiling in the detection of growth promoter use in calves. *Biomarkers*, Vol.13, No.3, pp. 246-256.

Nelson, R. E.; Grebe, S. K.; O'Kane, D. J. & Singh, R. J. (2004). Liquid chromatography - tandem mass spectrometry assay for simultaneous measurement of estradiol and estrone in human plasma. *Clinical Chemistry*, Vol.50, No.2, pp. 373-384.

Noppe, H.; Le Bizec, B.; Verheyden, K. & De Brabander, H. F. (2008). Novel analytical methods for the determination of steroid hormones in edible matrices. *Analytica Chimica Acta*, Vol.611, No.1, pp. 1-16.

Pape-Zambito, D. A.; Roberts, R. F. & Kensinger, R. S. (2010). Estrone and 17β-estradiol concentrations in pasteurized-homogenized milk and commercial dairy products. *Journal of Dairy Science*, Vol.93, No.6, pp. 2533-2540.

Parent, A.; Teilmann, G.; Juul, A.; Skakkebaek, N. E.; Toppari, J. & Bourguignon, J. (2003). The timing of normal puberty and the age limits of sexual precocity: Variations around the world, secular trends, and changes after migration. *Endocrine Reviews*, Vol.24, No.5, pp. 668-693.

Parr, S. L.; Chung, K. Y.; Hutcheson, J. P.; Nichols, W. T.; Yates, D. A.; Streeter, M. N.; Swingle, R.S.; Galyean, M.L. & Johnson, B.J. (2011). Dose and release pattern of anabolic implants affects growth of finishing beef steers across days on feed. *Journal of Animal Science*, Vol.89, No.3, pp. 863-873.

Partsch, C. & Sippell, W. G. (2001). Pathogenesis and epidemiology of precocious puberty. effects of exogenous oestrogens. *Human Reproduction Update*, Vol.7, No.3, pp. 292-302.

Pedreira, S.; Lolo, M.; Vázquez, B. I.; Franco, C. M.; Cepeda, A. & Fente, C. (2007). Liquid chromatography-electrospray ionization-mass spectrometry method in multiple reaction monitoring mode to determine 17α-ethynylestradiol residues in cattle hair without previous digestion. *Journal of Agricultural and Food Chemistry*, Vol.55, No.23, pp. 9325-9329.

Pinel, G.; Weigel, S.; Antignac, J.P.; Mooney, M. H.; Elliott, C.; Nielen, M. W. F. & Le Bizec, B. (2010). Targeted and untargeted profiling of biological fluids to screen for anabolic practices in cattle. *Trends in Analytical Chemistry*, Vol.29, No.11, pp. 1269-1280.

Pleadin, J.; Terzić, S.; Perši, N. & Vulić, A. (2011). Evaluation of steroid hormones anabolic use in cattle in croatia. *Biotechnology in Animal Husbandry*, Vol.7, No.2, pp. 7-2.

Poelmans, S.; De Wasch, K.; Noppe, H.; Van Hoof, N.; Van Cruchten, S.; Le Bizec, B.; Deceuninck, Y.; Sterk, S.; Van Rossum, H.J.; Hoffman, M.K. & De Brabander, H.F. (2005a). Endogenous occurrence of some anabolic steroids in swine matrices. *Food Additives and Contaminants*, Vol.22, No.9, pp. 808-815.

Poelmans, S.; De Wasch, K.; Noppe, H.; Van Hoof, N.; Van De Wiele, M.; Courtheyn, D.; Gillis, W.; Vanthemsche, P.; Janssen, C.R. & De Brabander, H.F. (2005b). Androstadienetrione, a boldenone-like component, detected in cattle faeces with GC-MSn and LC-MSn. *Food Additives and Contaminants*, Vol.22, No.9, pp. 798-807.

Pohjanen, E.; Thysell, E.; Jonsson, P.; Eklund, C.; Silfver, A.; Carlsson, I.B.; Lundgren, K.; Moritz, T.; Svensson, M.B. & Antti, H. (2007). A multivariate screening strategy for investigating metabolic effects of strenuous physical exercise in human serum. *Jounal of Proteome Research*, Vol.6, No.6, pp. 2113-2120.

Prins, G. S. (2008). Endocrine disruptors and prostate cancer risk. *Endocrine-Related Cancer*, Vol.15, No.3, pp. 649-656.

Rambaud, L.; Bichon, E.; Cesbron, N.; André, F. & Le Bizec, B. (2005). Study of 17β-estradiol-3-benzoate, 17α-methyltestosterone and medroxyprogesterone acetate fixation in bovine hair. *Analytica Chimica Acta*, Vol.532, No.2, pp. 165-176.

Regal, P.; Vázquez, B. I.; Franco, C. M.; Cepeda, A. & Fente, C. A. (2008). Development of a rapid and confirmatory procedure to detect 17β-estradiol 3-benzoate treatments in bovine hair. *Journal of Agricultural and Food Chemistry*, Vol.56, No.24, pp. 11607-11611.

Riedmaier, I.; Becker, C.; Pfaffl, M. W. & Meyer, H. H. D. (2009). The use of omic technologies for biomarker development to trace functions of anabolic agents. *Journal of Chromatography A*, Vol.1216, pp. 8192-8199.

Rijk, J. C. W.; Lommen, A.; Essers, M. L.; Groot, M. J.; Van Hende, J. M.; Doeswijk, T. G. & Nielen, M.W.F. (2009). Metabolomics approach to anabolic steroid urine profiling of bovines treated with prohormones. *Analytical Chemistry*, Vol.81, No.16, pp. 6879-6888.

Rosenfeld, D. L. & Bronson, R. A. (1980). Reproductive problems in the DES-exposed female. *Obstetrics and Gynecology*, Vol.55, No.4, pp. 453-456.

Salman, M.; New, J. C. J.; Bailey, M.; Brown, C.; Detwiler, L.; Galligan, D.; Hall, C.; Kennedy, M.; Lonergan, G.; Mann, L.; Renter, D.; Saeed, M.; White, B. & Zika, S. (2008).

Global food systems and public health: Production methods and animal husbandry, A national commission on industrial farm animal production report. pew commission on industrial farm animal production. In: *Comparative Medicine Publications and Other Works*, 01.09.2011, Available from http://trace.tennessee.edu/utk_compmedpubs/32

Scheffler, J. M.; Buskirk, D. D.; Rust, S. R.; Cowley, J. D. & Doumit, M. E. (2003). Effect of repeated administration of combination trenbolone acetate and estradiol implants on growth, carcass traits, and beef quality of long-fed holstein steers. *Journal of Animal Science*, Vol.81, pp. 2395-2400.

Scheltema, R. A., Kamleh, A.; Wildridge, D.; Ebikeme, C.; Watson, D. G.; Barrett, M. R.; Jansen, R.C. & Breitling, R. (2008). Increasing the mass accuracy of high-resolution LC-MS data using background ions - a case study on the LTQ-orbitrap. *Proteomics*, Vol.8, No.22, pp. 4647-4656.

Scippo, M.; Degand, G.; Duyckaerts, A.; Maghuin-Rogister, G. & Delahaut, P. (1994). Control of the illegal administration of natural steroid hormones in the plasma of bulls and heifers. *Analyst*, Vol.119, No.12, pp. 2639-2644.

Senger, P. L. (1994). The estrus detection problem: New concepts, technologies, and possibilities *Journal of Dairy Science*, Vol.77, No.9, pp. 2745-2753.

Shao, B.; Zhao, R.; Meng, J.; Xue, Y.; Wu, G.; Hu, J. & Tu, X. (2005). Simultaneous determination of residual hormonal chemicals in meat, kidney, liver tissues and milk by liquid chromatography-tandem mass spectrometry. *Analytica Chimica Acta*, Vol.548, No.1-2, pp. 41-50.

Shatalova, E. G.; Klein-Szanto, A. J. P.; Devarajan, K.; Cukierman, E. & Clapper, M. L. (2011). Estrogen and cytochrome P450 1B1 contribute to both early- and late-stage head and neck carcinogenesis. *Cancer Prevention Research*, Vol.4, No.1, pp. 107-115.

Sheridan, P. J.; Austin, F. H.; Bourke, S. & Roche, J. F. (1990). The effect of anabolic agents on growth rate and reproductive organs of pigs. *Livestock Production Science*, Vol.26, No.4, pp. 263-275.

Simontacchi, C.; Perez De Altamirano, T.; Marinelli, L.; Angeletti, R. & Gabai, G. (2004). Plasma steroid variations in bull calves repeatedly treated with testosterone, nortestosterone and oestradiol administered alone or in combination. *Veterinary Research Communications*, Vol.28, No.6, pp. 467-477.

Sorwell, K. G. & Urbanski, H. F. (2010). Dehydroepiandrosterone and age-related cognitive decline. *Age*, Vol.32, No.1, pp. 61-67.

Sottas, P.; Saudan, C.; Schweizer, C.; Baume, N.; Mangin, P. & Saugy, M. (2008). From population- to subject-based limits of T/E ratio to detect testosterone abuse in elite sports. *Forensic Science International*, Vol.174, No.2-3, pp. 166-172.

Stanczyk, F. Z.; Lee, J. S. & Santen, R. J. (2007). Standardization of steroid hormone assays: Why, how, and when? *Cancer Epidemiology Biomarkers and Prevention*, Vol.16, No.9, pp. 1713-1719.

Stephany, R. (2001). Hormones in meat: Different approaches in the EU and in the USA. *Acta Pathologica, Microbiologica et Immunologica Scandinavica*, Vol.109, pp. S357-S363

Stolker, A.; Groot, M.; Lasaroms, J.; Nijrolder, A.; Blokland, M. & Riedmaier, I. (2009). Detectability of testosterone esters and estradiol benzoate in bovine hair and plasma following pour-on treatment. *Analytical and Bioanalytical Chemistry*, Vol.395, No.4, pp. 1075-1087.

Stolker, A. A. M.; Rutgers, P.; Oosterink, E.; Lasaroms, J. J. P.; Peters, R. J. B.; Van Rhijn, J. A. & Nielen, M.W.F. (2008). Comprehensive screening and quantification of veterinary drugs in milk using UPLC-ToF-MS. *Analytical and Bioanalytical Chemistry*, Vol.391, No.6, pp. 2309-2322.

Strahm, E.; Baume, N.; Mangin, P.; Saugy, M.; Ayotte, C. & Saudan, C. (2009). Profiling of 19-norandrosterone sulfate and glucuronide in human urine: Implications in athlete's drug testing. *Steroids*, Vol.74, No.3, pp. 359-364.

Strous, R. D.; Maayan, R.; Lapidus, R.; Goredetsky, L.; Zeldich, E.; Kotler, M. & Weizman, A. (2004). Increased circulatory dehydroepiandrosterone and dehydroepiandrosterone sulphate in first-episode schizophrenia: Relationship to gender, aggression and symptomatology. *Schizophrenia Research*, Vol.71, No.2-3, pp. 427-434.

Swan, S. H.; Liu, F.; Overstreet, J. W.; Brazil, C. & Skakkebaek, N. E. (2007). Semen quality of fertile US males in relation to their mothers' beef consumption during pregnancy. *Human Reproduction*, Vol.22, No.6, pp. 1497-1502.

Torrado, S.; Roig, M.; Farré, M.; Segura, J. & Ventura, R. (2008). Urinary metabolic profile of 19-norsteroids in humans: Glucuronide and sulphate conjugates after oral administration of 19-nor-4-androstenediol. *Rapid Communications in Mass Spectrometry*, Vol.22, No.19, pp. 3035-3042.

U.S. Department of Health and Human Services (FDA) (2006). Guidance for industry: General principles for evaluating the safety of compounds used in food-producing animals.

U.S. FDA (2010). Animal drugs @ FDA. 13/07/2010, Available from http://www.accessdata.fda.gov/scripts/animaldrugsatfda/index.cfm

USDA (2009). Beef 2007–08, part III: Changes in the U.S. beef cow-calf industry, 1993–2008. In: USDA:APHIS:VS, CEAH. Fort Collins, CO(#N512-1008)

Vallejo, M.; Garcia, A.; Tunon, J.; García-Martínez, D.; Angulo, S.; Martín-Ventura, J. L.; Blanco-Colio, L.M.; Almeida, P.; Egido, J. & Barbas, C. (2009). Plasma fingerprinting with GC-MS in acute coronary syndrome. *Analytical and Bioanalytical Chemistry*, Vol.394, No.6, pp. 1517-1524.

Vanhaecke, L.; Bussche, J. V.; Wille, K.; Bekaert, K. & De Brabander, H. F. (2011). Ultra-high performance liquid chromatography–tandem mass spectrometry in high-throughput confirmation and quantification of 34 anabolic steroids in bovine muscle. *Analytica Chimica Acta*, Vol.700, No.1-2, pp. 70-77.

Vilariño, M.; Rubianes, E.; van Lier, E. & Menchaca, A. (2010). Serum progesterone concentrations, follicular development and time of ovulation using a new progesterone releasing device (DICO®) in sheep. *Small Ruminant Research*, Vol.91, No.2-3, pp. 219-224.

Wang, Y.; Tang, H.; Nicholson, J. K.; Hylands, P. J.; Sampson, J. & Holmes, E. (2005). A metabolic strategy for the detection of the metabolic effects of chamomile (matricaria recutita L.) ingestion. *Journal of Agricultural and Food Chemistry*, Vol.53, pp. 191-196.

Werner, E.; Croixmarie, V.; Umbdenstock, T.; Ezan, E.; Chaminade, P.; Tabet, J. C. & Junot, C. (2008). Mass spectrometry-based metabolomics: Accelerating the characterization of discriminating signals by combining statistical correlations and ultrahigh resolution. *Analytical Chemistry*, Vol.80, No.13, pp. 4918-4932.

Wigle, D. T.; Turner, M. C.; Gomes, J. & Parent, M. (2008). Role of hormonal and other factors in human prostate cancer. *Journal of Toxicology and Environmental Health - Part B: Critical Reviews*, Vol.11, No.3-4, pp. 242-259.

Wiley, A. S. (2011). Milk intake and total dairy consumption: Associations with early menarche in NHANES 1999-2004. *PLoS ONE*, Vol.6, No.2, pp. e14685.

Xu, C. L.; Chu, X. G.; Peng, C. F.; Jin, Z. Y. & Wang, L. Y. (2006). Development of a faster determination of 10 anabolic steroids residues in animal muscle tissues by liquid chromatography tandem mass spectrometry. *Journal of Pharmaceutical and Biomedical Analysis*, Vol.41, No.12, pp. 616-621.

Yager, J. D. & Davidson, N. E. (2006). Estrogen carcinogenesis in breast cancer. *New England Journal of Medicine*, Vol.354, No.3, pp. 270-282.

Yang, Y.; Shao, B.; Zhang, J.; Wu, Y. & Duan, H. (2009). Determination of the residues of 50 anabolic hormones in muscle, milk and liver by very-high-pressure liquid chromatography–electrospray ionization tandem mass spectrometry. *Journal of Chromatography B*, Vol.877, pp. 489-496.

Zhong, S.; Ye, W.; Feng, E.; Lin, S.; Liu, J.; Leong, J.; Ma, C. & Lin, Y.C. (2011). Serum derived from zeranol-implanted ACI rats promotes the growth of human breast cancer cells in vitro. *Anticancer Research*, Vol.31, No.2, pp. 481-486.

12

Milk Biodiversity: Future Perspectives of Milk and Dairy Products from Autochthonous Dairy Cows Reared in Northern Italy

Ricardo Communod[1], Massimo Faustini[1],
Luca Maria Chiesa[1], Maria Luisa Torre[2],
Mario Lazzati[3] and Daniele Vigo[1]
[1]Department of Veterinary Sciences and Technologies for Food Safety,
Faculty of Veterinary Medicine, University of Milan,
[2]Department of Drug Sciences,
Faculty of Pharmacy, University of Pavia
[3]Pavia Breeders Association Director,
Italy

1. Introduction

The United Nations is alerting worldwide population: our planet will undergo a drastic change in less than 40 years. Human population will increase to reach 9 billion inhabitants; consequently, cereal annual production should increase to be about 3 billion tons compared to current 2.1tons. According to a Food and Agriculture Organization's report FAO, "FAO 2050", human birthrate is expected to prevail in developing countries. Thus, food resources, such as wheat, corn, barley, and others, will be used to ensure people survival in these countries, which will reduce availability of silage and concentrates addressed to cattle breeding. This redistribution of primary resources will force bovine species to a diet based on forage biomass with low energy and water consumption, associated with the use of less productive areas such as foothills, forests and pastures, thus implementing availability of cereals for human consumption. Noticeably, climate changes will persist and modify any environment where plants grow. In this context, farmers and breeders will be faced with new challenges. Maintaining genetic biodiversity in plants including some wild peculiarities may help to solve some of current agricultural problems and protecting such biodiversity appear therefore crucial to increase a sustainable and efficient use of land. According to 'Europe 2020' protocol established by United Nations, greenhouse gas emissions must be reduced by 20%, energy efficiency increased by 20% with a reduction in consumption, and energy from renewable sources pushed up by 20%. Therefore, it is expected that in certain areas long-term and perennial crops will be implemented; energy consumption for sowing, harvesting, and drying products for livestock consumption will be reduced; availability of mature manure to improve soil fertilization, reducing easily washable nitrogen, will be increased. In the coming decades, an actually thorny scenario will affect future generations: processes such as steady population and consumption growth will cause a reduction in the response capacity of ecosystems, and a consequent decrease in food resources, water and energy. The agricultural system is strictly implied in the abovementioned issues, and plays its role as a multifaceted character. In fact, agriculture can give answers to the drawing demand for food while consuming huge quantities of water and fuels on the planet to achieve its goals. To date, care for biodiversity is pivotal to improve and increase a sustainable and efficient use of land.

2. Dairy cattle breeding: European and Italian scenarios

In parallel with vegetal selection aimed to maximize production, even farm animal biodiversity has been damaged to select animals with fast food conversion into milk and meat. This mono-aptitude selective criterion has caused a decline in several sectors, the main ones being reproductive performances and quality of products (Schennink et al., 2007), problems easily noticed in the dairy cattle scenario. In fact, mainly in Friesian breeds, reproductive performances have decreased worldwide, with negative consequences to cows robustness and longevity by increasing stress, udder health disturbances and locomotion disorders (Roxström et al., 2001a). Holstein, Brown Swiss and Jersey, subjected to a mono-aptitude selection in the last 40 years, almost aimed to a quantitative milk production, reach, from a physiological point of view, a very critical situation: they have missed good reproductive efficiency characters, (e.g. calving interval and conception rate (Sørensen et al., 2007), excellent longevity in farm, resistance to stress and diseases (metabolic syndrome, ketosis, mastitis and foot diseases)(Roxström et al., 2001a; Roxström et al., 2001b; Carlén, 2004)), whereas they have dramatically increased "energy and financial voracity" (diet based on starch and protein meals, great health and structural investments due to several high recurring diseases (Ingvartsen et al., 2003; Collard et al., 2000; Carlén et al., 2004).

Italian dairy breeding has followed the European trend, by selecting high yielding Holstein cows, influencing some important milk quality parameters, such as milk fat, protein and somatic cell count, all involved in cheese making processes. Recently, Italian researchers have demonstrated that milk with a somatic cell content greater than 400,000 cells/ml evidences a scarce aptitude to rennet coagulation and, in general, it does not seem to be suitable for cheese production, with particular reference to Grana cheese production (Sandri et al., 2010). Italy is the most important cheese producer and exporter in Europe, with its about 460,000 tons of products and almost 3 billion Euros (data from www.clal.it, www.ismea.it) derived from PDO (Protected Designation Origin) and PGI (Protected Geographical Indication). The most representative Italian products are certainly *Parmigiano Reggiano* and *Grana Padano*, which have recently increased by 9.8% their export trend to Germany, the United States, France, Switzerland and the United Kingdom. Therefore, both for the Italian breeding system and for the whole country it is very important to try to solve current problems related to health and welfare of their dairy cows, in order to maintain a unique cheese production and exportation.

An innovative idea lies in finding solutions to the current dairy breeding scenario by orienting our sight on the existing Italian bovine heritage, and diversifying from the current thought that has driven genetic selection of dairy cosmopolitan breeds in the last 40 years. Using very rustic and frugal cattle with double (milk and meat) or threefold (milk, meat and work) aptitude, such as Italian autochthonous cows, showing a good food conversion into milk and meat, would be highly advantageous, compared with the more cosmopolitan and selected Holstein, Brown Swiss and Jersey, unable to maintain high milk yield and reproduction standards, reared with a protein-deficient diet mainly based on forage and cereals.

3. Autochthonous dairy cattle of northern Italy

3.1 Breed: Origin, diffusion, traits, aptitude – Cabannina (Bigi and Zanon, 2008)

These cows are native to the province of Genoa and are reared in Liguria region and in Pavia province in Lombardy; currently, about 220 heads are enrolled in the population register. They have a dark brown coat with light-colored lines and reddish shades. The head is small, short, light; the muzzle is black and widely white-bordered. The reduced size (withers height of 122 cm for females and 134 cm for males, with a maximum adult live weight of 4.5 quintals), short and powerful limbs, and very hard claws, make these animals excellent to grazing, the only animals able to effectively use plant resources in the high slopes of the Ligurian Apennines. The Cabannina Breeding, based on local grazing for a large part of the year, determines the specific characteristics of flavor and authenticity of the final products, milk and cheese. For this reason, it can be said that Cabannina is the testimony for the province of Genoa to the indissoluble link between land and its products, and that it perfectly expresses the adaptation process of the characteristics of both breed and environment, where it has evolved and maintained over time. Milk production from this cow breed is kept on 20 to 30 quintals/year and its high longevity (it is common to find in farm cows aged over 12 years) confirms their excellent adaptation to their territory. Their cheese, entirely derived from raw milk, is called "U Cabanin" (The Cabanin). It was established in 2007 thanks to the intervention of the Breeders Association and the Chamber of Commerce of Genoa. From 2010, U Cabanin is one of Slow Food Presidia and according to Carlo Petrini, Slow Food Association founder and President, Cabannina cow breed represents a correct mix of pleasure for food and responsibility, sustainability and harmony with nature.

Fig. 3.1. Cabannina cow.

3.2 Breed: Origin, diffusion, traits, aptitude - Varzese-Ottonese-Tortonese (Bigi and Zanon, 2008)

As clearly shown by its long compound name, this breed's area of origin includes the northern Italian Apennines and 4 neighbouring regions: Lombardy, Emilia Romagna, Liguria and Piedmont. Currently there are about 240 heads enrolled in the population register, atomized in few farms in Pavia, Piacenza, Alessandria and Milano provinces. The cows in this breed show a uniform reddish-blond coat, more or less intense, with limited lighter shades around their muzzles, eyes, bellies, inner thighs and distal limbs. Their medium size (withers height of 135 cm for females and 145 cm for males with a maximum adult live weight of 5.5 quintals), and their distinct characteristics of rusticity, frugality, fertility and longevity (this breed can easily reach 10 births and the goal of one calf/year) make these animals be used as first choice in marginal areas like mountain, wood and foothill grazes. Born to be primarily used for work - in fact, bulls were famous for their strength, endurance, docility and for their resistant hoofs), and even cows were employed for rapid and light draft - this breed was then appreciated for the production of meat and milk, used to make excellent cheese today including some important cheese named "Nisso", "Robiola", "Montebore" and "Molana", all produced in Oltrepo regions, areas lying south of the River Po. The Varzese breed has been recently required by farms where tourists can be lodged and/or consume local products. In addition, it can be found in teaching farms and, as it recalls traditional aspects of the ancient rural world, it plays a role in folklore events.

Fig. 3.2. Varzese cow

4. Comparative study among breeds

To understand to what extent the regulatory mechanisms of production and reproduction of high yielding dairy cows can be altered, comparative physiological studies within Italian dairy breeds experiencing no genetic improvement were necessary. This investigation aimed to increase scientific information about these animals in the perspective of developing a more efficient and sustainable dairy breeding. Data now available, provided by ongoing research at our laboratory, are actually encouraging. In our research, two endangered local bovine breeds of Northern Italy, Varzese and Cabannina, were considered and compared with Friesian cows in relation to some aspects characterizing lactation and reproduction.

4.1 Milk study: Milk fat, the most variable component

Milk is one of the most complete foods in nature; its nutritional value is mainly attributable to proteins and fats, the latter being its most variable component. Milk fat is present as a suspension of defined globules showing a biological membrane giving each globules an identity as well as a precise structural and functional behavior. The globule dimensions range from 0.5 to 20 μm; they have been extensively described in mass milk of Holstein cows (Lopez, 2005), both as native and as subjected to treatments, i.e. homogenization and pasteurization (Michalski et al., 2003). The fat globules interact with milk caseins of curd and their size can influence the processes of lipolysis and ripening of cheese (Lopez, 2005; Michalski et al., 2004). Our research was oriented toward fat and fatty acid composition, important features in milk since they influence its physical, organoleptic and nutritional properties (Chilliard et al., 2000). Recent studies showed that milk composition is determined by several factors such as diet (Banks et al., 1983; Grummer, 1990; Perfield et al., 2006; Perfield et al., 2007), stage of lactation and season (Coulon, 1994), genetics and management (Schennink et al., 2007). Milk fat is characterized by high amounts of saturated fatty acids (SFA), especially myristic acid (C14:0) and palmitic acid (C16:0), and by a low amount of mono- and poly-unsaturated fatty acids (Soyeurt et al., 2006). Milk is the major contributor of SFA to human diet, a fact that led to a widespread conviction for which milk and dairy products can negatively affect human health. A recent meta-analysis conducted by Elwood et al. (2010) demonstrated a reduction in risk for several causes of death (ischaemic heart disease, stroke and incident diabetes) in subjects with the highest consumption of dairy products compared to those with the lowest intake. In order to understand lifestyle-related diseases, such as obesity, hyperlipidemia, arteriosclerosis, diabetes mellitus and hypertension, researchers oriented their attention to diet and in particular to dietary lipids (Vessby, 2003). The most studied milk components are ω-6 and ω-3 poli-unsaturated fatty acids (PUFA), conjugated linoleic acids (CLA) and SFA, but to date also mono-unsaturated fatty acids (MUFA) should be considered as confering important nutritional quality to milk. Not only could MUFA profile be interesting, but also Δ9-desaturase activity since it is the key enzyme to convert SFA in MUFA and to control conjugated linoleic acids concentration. Some researchers (Schennink et al., 2007; Soyeurt et al., 2007) estimated that a moderate heritability of this enzyme and the concentration of MUFA have another chance to improve unsaturated fatty acids (UFA) in milk composition. In mammary glands, Δ9-desaturase catalyzes the insertion of a double bond between carbon atoms 9 and 10 of fatty acid (Pereira et al., 2003). In mammary glands, fatty acids originating from blood or from *de novo* fatty acid synthesis can be desaturated and the degree of unsaturation is often calculated by a so-called "desaturase index".

4.2 Reproduction study: Uterine involution monitored by a new non invasive method

Fertility is a complex parameter undoubtedly influenced by genetics, environment and management. These components exist and act in synergy and simultaneously, making it extremely difficult for any strategies and technologies applied to set up a stabilization of reproductive efficiency. The post-partum period has a basic effect on the resumption of ovarian cyclicity and future reproductive efficiency, a parameter that unequivocally determines the career of a cow in the herd. By definition, the postpartum period is, , a physiological process between birth and complete uterine involution, essential to bring the female genital apparatus back to favorable conditions for optimal embryonic development and implantation (Badinand 1993; Kaidi et al., 1991b). Immediately after birth a cow uterus weighs about 10 kg and in a month it reaches a weight of 1 kg physiologically (Badinand 1993). Its involution can be appreciated clinically by transrectal palpation: under normal conditions a uterus can be completely taken in a hand after 2-3 weeks and never before 10-12 days after birth (Badinand 1993). Uterine involution in cattle can be affected by many factors, such as dystocia, uterine prolapse, abortion and bacterial infections (93% of uteri in cattle can be infected by a large amount of bacteria until 15 days post-partum) (Elliot et al., 1968). Uterine involution is characterized by a significant tissue remodeling: measuring cellular turnover markers, such as hydroxyproline (HYPRO), the most abundant amino acid present in uterine collagen, could provide useful information on involution progress. The serum concentration of this protein increases gradually in late gestation and it is related to the mechanism of placental separation and uterine involution (Kaidi et al., 1991a).

5. Materials and methods to describe milk production

A total number of 13 lactating cows were enrolled. Two local breeds were taken into account (Varzese, n=4 and Cabannina, n=4) and compared with a cosmopolite breed (Friesian, n=5). All animals were raised in the same center located in Northern Italy and fed with the same diet. Milk samples from the whole udder were collected during morning milking and from a lactometer. Samples, kept at 4°C, were forwarded to a laboratory and analyzed within 2 hours. The animals considered were in early and mid lactation, ranging from 40 to 180 days, and the intervals of collection were at 20 days. The starting point of milk collection was chosen in order to assure the full physiological uterine involution in all heads.

5.1 Milk fat globule granulometry

Distributions of fat globules diameters were performed using a granulometer laser scatter, according to the method proposed by Lopez, 2005. Size distributions were characterized by volume weighted diameter of the globules (d_{43}, in microns) and by specific surface area (SSA, in $cm^2 * ml^{-1}$). Data obtained were subjected to analysis of covariance (ANCOVA, GLM procedure, SPSS ver. 17.0 per Windows); breed was taken into account as a fixed factor, and day of lactation as a covariate. The differences between breeds were evaluated by LSD (least significant difference) for multiple comparisons. To assess the evolution of globular diameter (expressed as d_{43}) in relation to the days of lactation the following relationship was applied :

$$D_{43} = (A+B*(days\ of\ lactation))/(1+C*(days\ of\ lactation))$$

where A, B, C are constants.

5.2 Milk fat content, fatty acid extraction, derivatization and desaturase indices

The determination of milk fat content was performed by a UV Spectrophotometric method, according to Forcato et al. (2005). Lipids extraction was performed according to a modified Bligh & Dyer method (Manirakiza et al., 2001) and derivatization according to Molto-Puigmartì et al. (2007). Fatty acids were identified using external standards (Standard containing 37 fatty acids, FAME Mix 37, Supelco, USA) and two CLA standards (Matreya, USA), and quantified using 19:0 (nonadienoic acid) as internal standard. Peak areas were corrected according to the theoretical relative FID response correction factors (TRFs) published by Ackman (2002). Results are presented as g/100 g fatty acids (% by weight). Percentage of the single contribute of each fatty acid was calculated on the total of the area under known peaks. Percentages of myristoleic acid (C14:1-*cis-9*), palmitoleic acid (C16:1-*cis-9*) and oleic acid (C18:1-*cis-9*) were taken into account. Desaturase indices (Δ) were calculated according to Schennink et al. (2008) and total desaturase index (TDI) was calculated according to Mele et al. (2007) on C14, C16 and C18 fatty acids; briefly, the individual Δ was calculated as $Cx:1/(Cx+Cx:1)*100$, where x is the number of carbons of fatty acid. The total desaturase index was calculated as

$$(C14:1+C16:1+C18:1/(C14+C14:1+C16+C16:1+C18+C18:1)*100.$$

5.3 Statistical analysis

All data were analyzed by JMP® software ver. 7.0.2 (SAS Institute Inc.) for Windows platform.

6. Materials and methods to describe reproduction

A total number of 16 lactating cows were enrolled. All animals were multiparous, with eutocic stillbirth, normal post-partum period and raised in the same center located in Valle Salimbene (Pavia) in Northern Italy. All animals were followed for the first 30 days after birth and were divided into 3 treatment groups based on breed, specifically:

• 4 Varzese cows (VAR group);
• 5 Cabannina cows (CAB group);
• 7 Friesian cows (FRI group);

All cows did not undergo any gynecological clinic examination as manipulation (palpation and retraction of uterus in the pelvic cavity could produce stimuli influencing a physiological uterine involution) (Rosemberger, 1979). In this regard Hurtgen and Ganiam (1979), in a trial conducted on mares, showed that intracervical or intrauterine manipulation during the luteal phase of the estrous cycle may directly or indirectly stimulate release of endogenous prostaglandins that trigger regression of the corpus luteum, followed by oestrus onset and ovulation (Hurtgen and Ganiam (1979). In our trial animals that showed persistent hyperthermia (40.5 °C) for over 48 hours, or post-partum syndromes requiring drug administration that could affect uterine involution (e.g., administration of boron gluconate drugs, steroidal and non steroidal anti-inflammatory drugs) were excluded.

6.1 Serum hydroxyproline determination

From the coccygeal vein of each subject a single operator collected blood samples without anticoagulant additives in order to obtain serum. Sampling was performed on days 0 (day of birth),1,2,3,4,5,10,15,20,25,30 post-partum and hydroxyproline content was analyzed by a spectrophotometric method as reported by Huszar et al. (1980).

6.2 Statistical analysis

All data were analyzed by JMP® software ver. 7.0.2 (SAS Institute Inc.) for Windows platform.

7. Results

7.1 Milk fat globules

Dimensional analysis of milk fat globules highlighted deep differences among breeds, mainly between Friesian and the others. Friesian cows are characterized by larger milk fat globules (Fig. 7.1., p <0.05) and by a significantly lower specific surface area (Fig. 7.2., p <0.05). Cabannina breed shows the smallest fat globules (Fig. 7.1., p <0.05) and the highest specific surface area (Fig. 7.2., p <0.05). Varzese breed occupies an intermediate position (Figs. 7.1. and 7.2., p<0.05). In all breeds, globule diameter trend decreases during lactation. This reduction is more evident in the first weeks of lactation and slows down until it reaches *a plateau* in the following months. These features can elucidate some physiological functions of fat globules on mammalian offspring and on industrial transformation of milk from *niche* breeds. Key data from our research are collected and presented in the following figures:

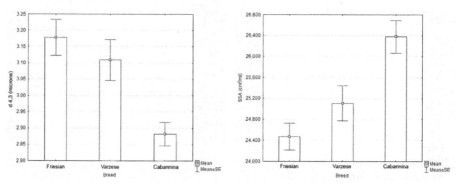

Fig. 7.1. and 7.2. Mean+SE for d_{43} (volume weighted diameter) and for SSA (Specific Surface Area) in breeds during survey. Differences among groups were statistically significant (p<0.05).

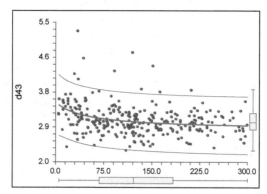

Fig. 7.3. Trend of mean globular diameter d43 during lactation. Regression curve and limits are given after months.

7.2 Milk fatty acids profile

Mean milk production of the three groups of cows during observation period was 14.24±4.36 kg, 24.17±6.84 kg, and 11.84±4.59 kg for the Cabannina, Holstein and Varzese, respectively. Mean productions were significantly different (ANOVA, p<0.001), and the three groups were all significantly different when compared pairwise (Tukey test, p<0.05). The three breeds showed an overall fat percentage of 4.05±1.14% (Cabannina, n=36), 3.53±1.01% (Holstein, n=39) and 4.27±0.87 (Varzese, n=30). The univariate descriptive statistics for the variables considered are summed up in Table 7.1. Results also report ANCOVA significances for breed and time. At first glance, great differences among groups can be noticed. In particular, the Varzese breed shows higher percentages of C16:1c, C18:1c, ΣMUFA, and higher levels in Δ16, Δ18 and ΣΔ; Cabannina milk reveals significantly higher levels in C14:1c and Δ14, whereas it is in an intermediate position about Δ16 and Δ18. Holstein cows maintain the lowest percentages for all the MUFAs determined. Positive significant temporal trends (p<0.05) were observed for C14:1c, C16:1c and for Δ14, Δ16, Δ18 and ΣΔ. A significant trend for the other variables was not evidenced.

Fatty acid (%)	Breed			ANCOVA		
	Cabannina	Friesian	Varzese	Breed	Day of lactation	Time trend
C14:1-*cis*	1.14±0.25[a]	0.96±0.32[b]	1.03±0.40[ab]	*	***	↑
C16:1-*cis*	1.93±0.75	1.80±0.73	2.13±0.80	n.s.	***	↑
C18:1-*cis*	19.60±3.64[b]	16.78±3.89[c]	22.30±3.81[a]	***	n.s.	=
Σ*cis*-MUFA	22.40±5.35[b]	19.91±4.11[c]	25.31±5.27[a]	***	n.s.	=
Desaturase						
Δ14	6.95±1.44[a]	6.06±2.08[b]	6.77±2.63[ab]	0.07	***	↑
Δ16	5.15±1.80[ab]	4.20±1.72[b]	5.71±2.21[a]	*	*	↑
Δ18	67.32±3.70[ab]	65.06±5.14[b]	69.85±5.90[a]	**	*	↑
ΣΔ	26.58±4.18[b]	23.00±4.41[c]	29.98±4.05[a]	***	n.s.	=

Table 7.1. Univariate descriptive statistics and ANCOVA results for *cis*-MUFA profiles and desaturase indices in different dairy cow breeds. *cis*-MUFA = *cis*-monounsaturated fatty acids - [a-c]Different superscripts indicate a p<0.05 difference (correct t-test for multiple comparisons) - *=p<0.05; **=p<0.01; ***=p<0.001 - Δ = Desaturase Index; ΣΔ = Total Desaturase Indices - n.s. = not significant - ↑ = increasing time trend - ↓ = decreasing time trend - = = constant time trend.

7.3 Hydroxyproline

Preliminary results show some differences from what has been described in Friesian cows: in fact, autochthonous cows show more outstanding curves inclination than Friesian and a shorter timeframe for uterine involution, as they achieve optimal levels (about 12 μg/ml) already 20 days after birth (Fig. 7.4.).

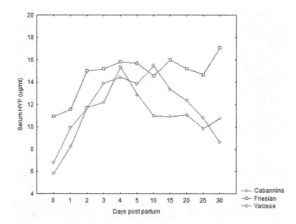

Fig. 7.4. Comparison of serum hydroxyproline content trend in Varzese cows (green), Friesian cows (red) and Cabannina cows (blue) during post partum period.

8. Discussion and conclusions

The present investigation highlights important among-breed variations. First of all, variations in milk fat dimension and composition, evidence of milk biodiversity deriving from breed biodiversity. In fact, fat globules are very different in the three breeds we considered; in particular, Cabannina breed has the smallest ones. As described by Lopez et al., 2011 both size and surface area of fat globules are important parameters that can influence the mechanisms of hydrolysis of lipids by digestive enzymes. Smaller globules are better as they present a bigger surface area to enzymes, which increases digestive processes. According to Fauquant et al. (2005), a different composition and size of fat globules could change functional and sensory properties of dairy products. Michalsky et al. in 2007 showed that cheese with small-fat globules exhibited greater stretching and elasticity and improved sensory characteristics. The content in *cis*-MUFA found in the present research is quite similar to data reported in recent literature (Michalsky et al., 2007; Lopez et al., 2011). Enhancement of *cis*-MUFA in milk is desirable for human consumption, as reported by Givens (2008): higher intakes of *cis*-MUFA and a reduction in short fatty acids decrease plasma insulin levels, total plasma cholesterol and LDL-cholesterol concentrations, therefore reducing risks for coronary heart diseases (CHD). A higher intake of *cis*-MUFA is also well-considered in type-2 diabetes, as reported by Ros (2003), that underlines as *cis*-MUFAs are an alternative to low-fat diets in the management of diabetes. Moreover, Lauszus et al. (2001) indicate that intake of MUFAs may prevent blood pressure from rising in gestational diabetes mellitus, with no influence on lipid and lipoprotein concentrations. In a survey conducted in 11 EU member states, intake of *cis*-MUFA from dairy products ranges from 8.3 to 28.6% of total intake of *cis*-MUFA (Givens, 2008), confirming the huge contribution of milk and dairy products in *cis*-MUFA intake. The report by Givens suggests that modification of fatty acid content in milk obtained by replacing short fatty acids with *cis*-MUFA may reduce risk for CHD in the population, which implies that European agricultural policies should be deeply changed. Positive effects of *cis*-MUFA can also be exerted on udder health: during mastitis, for example, an enhancement in lipase activity can be appreciated, with the increase of free fatty acids, mainly short chain ones (Randolph and Erwin, 1974); several fatty acids are endowed of a good antibiotic power, that can be expressed via inhibition of enzyme/fatty acid synthesis/nutrient uptake, cell lysis, metabolites leakage, disruption of electron transport chain, interference with oxidative phosphorylation and lipid peroxidation (Desbois and Smith, 2010; Clément et al., 2007). Furthermore, by enhancing the activity of stearoyl CoA desaturase (SCD), the nutritional value of milk would be ameliorated, but the simple up-regulation of its activity seems to be limited, as reported in a comprehensive milk lipid synthesis model (Shorten et al., 2004). In the present research, the local breeds considered show either higher levels in cis-MUFAs or in desaturase indices: features that are likely to be linked to genetics, as evidenced by Schennink et al. (2008), by a complex interaction in gene/allele expressions, and that could be used to improve the nutritional value of milk.

About reproduction physiology, the results obtained indicate that the reproductive physiology of Varzese and Cabannina is characterized by an early resumption of ovarian activity and by an early fecundation opportunity: in fact, the onset of first estrus can be observed 20 days after birth and the opportunity to impregnate can occur in the following cycle, i.e. approximately 40 from birth. That would allow farmers to achieve the goal of a calf/year, as the primary indicator of welfare, reproductive efficiency and good mammary function. According to unpublished data, obtained during trials, it could be said that autochthonous breeds have peculiar features to solve current problems of the scenario of high yielding dairy cows. As previously said, in the current system of cattle breeding, cows have dramatically increased "energy and financial voracity "(diet based on starch and protein meals, great health and structural investments due to several high recurring diseases (Ingvartsen et al., 2003; Collard et al., 2000; Carlén et al., 2004). In post partum period, energy needs required by high-yielding Holstein cows has increased by 25% compared to thirty years ago, despite the considerably limited growth in muscle masses (Agnew et al., 2003). All experts know about mobilization of various constituents from adipose tissue to support breast functions in producing milk (Veerkamp, 1998), but few know that the muscle is an important structure for reserves of amino acids. In highly selected cows this phenomenon is much more marked than in cows genetically less selected (Pryce 2004). A cow's energy balance decreases even a couple of weeks before parturition, as a result of the animal's reduced ability of food ingestion. In the first weeks after birth, food ingestion cannot compensate the wide adipose tissue mobilization. Therefore, cows maintain this status of negative energy balance (NEB) for 5-7 weeks from birth (Grummer 2007). At the beginning of lactation, mobilization of adipose tissue and low blood glucose bioavailability are key events to induce metabolic syndromes (Ingvartsen et al., 2003), ketosis, liver diseases, paretic-spastic syndromes and foot diseases (Collard et al., 2000). In autochthonous dairy farms ketosis and other metabolic syndromes are hardly ever present: in fact, these cows can keep up their double aptitude for maintaining a good milk production and creating a favorable muscle mass. A feature giving Cabannina and Varzese cows an interesting physiological ability to solve imbalances during NEB status through abundant energy reserves (consisting of subcutaneous and inframisial adipose tissue and muscle itself) immediately available to provide the animals with glucose and amino acids.

In conclusion, restoration of endangered niche breeds can undoubtedly give a boost to local products and to conservation of livestock biodiversity; FAO sustains livestock biodiversity as a "safety net for the future", mainly in developing countries, as reported in a recent document, FAO, 2010. These principles can also be extended to developed countries with the aim to better exploit local resources and preserve relic breeds from an impending extinction which would mean the loss of a priceless legacy. In the forthcoming years, the peculiar nutritional and nutraceutical aspects present in milk and in dairy products deriving from biodiversity farms will hopefully show up.

9. Acknowledgment

The authors are grateful for animal and sampling supply to Mr Luigi Antonio Chierico, a precursory breeder in Valle Salimbene, (Pavia, Lombardy), who runs the only and unique bovine biodiversity farm existing in the world.

10. References

Ackman, R.G. (2002). The gas chromatograph in practical analyses of common and uncommon fatty acids for the 21st century. *Analytica Chimica Acta*, Vol.465, No.1-2, (August 2002), pp. 175-192, ISSN 0003-2670

Agnew, R.E.; Yan, T.; Murphy, J.J.; Ferris, C.P. & Gordon, F.J. .(2003). Development of maintenance requirement and energetic efficiency for lactation from production data of dairy cows. *Livestock Production Science*, Vol.82, No.(1-2), pp. 151-162, ISSN 1871-1413

Badinand, F. (1993). Involution uterine-physiologie-pathologie. *Atti della Società Italiana di Buiatria*, Vol XXV, pp. 41-57, ISSN

Banks, W., Clapperton, J. L. & W. Steele. 1983. Dietary manipulation of the content and fatty acid composition of milk fat. *Proceedings of the Nutrition Society*. Vol.42, No.3, pp. 399-406, ISSN 0029-6651

Bigi, D. & Zanon, A. (2008). *Atlante delle razze autoctone*, Edagricole, ISBN 978-88-506-5259-4, Milano, Italia.

Carlén, E.; Strandberg, E. & Roth, A. 2004. Genetic Parameters for Clinical Mastitis, Somatic Cell Score, and Production in the First Three Lactations of Swedish Holstein Cows. *Journal of Dairy Science*, Vol.87, No.9, (September 2004), pp. 3062-3070, ISSN 0022-0302

Chilliard, Y., Ferlay, A.; Mansbridge, R.M. & Doreau, M. (2000). Ruminant milk fat plasticity: nutritional control of saturated, polyunsaturated, trans and conjugated fatty acids. *Annales de Zootechnie*, Vol.49, No.3, (November 2000), pp. 181-205, ISSN 0003-424X

Clément, M.; Tremblay, J.; Lange, M.; Thibodeau, J. & Belhumeur, P. (2007). Whey-derived free fatty acids suppress the germination of Candida albicans in vitro. *Fems Yeast Research*, Vol.7, No.2, (March 2007), pp. 276-285, ISSN 1567-1356

Collard, B.L.; Boettcher, P.J.; Dekkers, J.C.M.; Petitclerc, D. & Schaeffer, L.R. (2000). Relationships between energy balance and health traits of dairy cattle in early lactation. *Journal of Dairy Science*, Vol.83, (May 2000), pp.2683- 2690, ISSN 0022-0302

Desbois, A.P. & Smith, V.J.. (2010). Antibacterial free fatty acids: activities, mechanisms of action and biotechnological potential. *Applied Microbiology and Biotechnology*, Vol.85, No.6, pp. 1629-1642, ISSN 1432-0614

Elliot, K.; McMahon, K.J.; Gier, H.T. & Marion G.B. (1968). Uterus of the cow after parturition: bacterial content. *American Journal of Veterinary Research*, Vol.29, pp. 77-81, ISSN 0002-9645

Elwood, P.C.; Pickering, J.E.; Givens, D.I. & Gallacher, J.E. (2010). The consumption of milk and dairy foods and the incidence of vascular disease and diabetes: an overview of the evidence. *Lipids*, Vol.5, No.10, (March 2010), pp. 925-939, ISSN 1558-9307

Fauquant, C.; Briard, V.; Leconte, N. & Michalski, MC. (2005). Differently sized native milk fat globules separated by microfiltration: fatty acid composition of the milk fat globule membrane and triglyceride core. You have full text access to this content, *European Journal of Lipid Science and Technology*, Vol.107, (February 2005), pp.80-86, ISSN 1438-9312

Forcato, D.O.; Carmine, M.P. ; Echeverria, G.E.; Pecora, R.P. & Kivatinitz, S.C. (2005). Milk fat content measurement by a simple UV spectrophotometric method: An alternative screening method. *Journal of Dairy Science*, Vol.88, No.2, (February 2005), pp.478-481, ISSN 0022-0302

Givens, D.I. (2008). Session 4: Challenges facing the food industry in innovating for health: impact on CVD risk of modifying milk fat to decrease intake of SFA and increase intake of cis-MUFA, *Proceedings of the Nutrition Society*, ISSN 0029-6651, Symposium, Dublin, IRAN, REPUBLIQUE ISLAMIQUE (18/06/2008), pp. 419-427

Grummer, R.R. (1990). Effect of feed on the composition of milk fat. *Journal of Dairy Science*, Vol.73, pp. 88, ISSN 0022-0302

Grummer, R.R. (2007). Strategies to improve fertility of high yielding dairy farms: Management of the dry period. *Theriogenology*, Vol.S68, pp. S281-S288, ISSN 0093-691X

Hurtgen, J.P. & Ganiam, V.K. (1979). The effect of intrauterine and cervical manipulation on the equine oestrous cycle and hormone profiles. *Journal of Reproduction and Fertility*, Vol.27, pp. 191-197, ISSN 0022-4251

Huszar, G.; Maiocco, J. & Naftolin F. (1980). Monitoring of collagen and collagen fragments in chromatography of protein misture. *Analytical Biochemistry*, Vol.105, No.1, (June 1980), pp. 424-429, ISSN 0003-2697

Ingvartsen, K.L.; Dewhurst, R.J. & Friggens, N.C. (2003). On the relationship between lactational performance and health: is it yield or metabolic imbalance that causes diseases in dairy cattle? A position paper. *Livestock Production Science*, Vol.83, pp. 277-308, ISSN 1871-1413

Kaidi, R.; Brown, P.J.; David, J.S.E.; Etherington, D.J. & Robins S.P. (1991a). Uterine collagen during involution in cattle. *Matrix*, Vol.11, pp. 101-107, ISSN 1936-2994

Kaidi, R.; Brown, P.J. & David, J.S.E. (1991b). Uterine involution in cattle. *Veterinary Annual*, Vol.31, pp. 39-50, ISSN 0083-5870

Lauszus, F.F.; Rasmussen, O.W.; Henriksen, J.E.; Klebe, J.G.; Jensen, L.; Lauszus, K.S. & Hermansen, K. (2001). Effect of a high monounsaturated fatty acid diet on blood pressure and glucose metabolism in women with gestational diabetes mellitus. *European Journal of Clinical Nutrition*, Vol.55, No.6, (June 2001), pp. 436-443, ISSN 0954-3007

Lopez, C. (2005). Focus on the supramolecular structure of milk fat in dairy products. *Reproduction Nutrition Development*, Vol.45, No.4, (July-August 2005), pp. 497-511, ISSN 0926-5287

Lopez, C.; Briard-Bion, V.; Ménard, O.; Beaucher, E.; Rousseau, F.; Fauquant, J.; Leconte, N. & Robert, B. (2011). Fat globules selected from whole milk according to their size: different compositions and structure of the biomembrane, revealing sphingomyelin-rich domains. *Food Chemistry*, Vol.125, No.2, (March 2011), pp. 355-368, ISSN 0308-8146

Manirakiza, P.; Covaci, A. & Schepens, P. (2001). Comparative study on total lipid determination using Soxhlet, Roese-Gottlieb, bligh & dyer, and modified bligh & dyer extraction methods. *Journal of Food Composition and Analysis*, Vol.14, No.1, (February 2001), pp. 93-100, ISSN 1096-0481

Mele, M., Conte, G.; Castiglioni, B.; Chessa, S.; Macciotta, N.P.P.; Serra A.; Buccioni, A.; Pagnacco, G. & Secchiari, P. (2007). Stearoyl-Coenzyme A desaturase gene polymorphism and milk fatty acid composition in Italian Holsteins. *Journal of Dairy Science*, Vol.90,No.9, (September 2007), pp. 4458-4465, ISSN 0022-0302

Michalski, M.; Gassi, J.Y.; Famelart, M.H.; Leconte, N.; Camier, B.; Michel, F. & Briard, V. (2003). The size of native milk fat globules affects physical-chemical and sensory properties of Camembert cheese. *Dairy Science and Technology. Le Lait*, Vol.83, No.2 (June 2003), pp.131-143, ISSN 1958-5594

Michalski, M.C.; Ollivon, M.; Briard, V.; Leconte, N. & Lopez, C. (2004). Native fat globules of different sizes selected from raw milk: thermal and structural behaviour. *Chemistry and Physics of Lipids*, Vol.132, No.2, (December 2004), pp. 247-261, ISSN 0009-3084

Michalski, M.C.; Camier, B.; Gassi, J.Y.; Briard-Bion, V.; Leconte, N.; Famelart, M.H. & Lopez, C. (2007). Functionality of smaller vs control native milk fat globules in Emmental cheeses manufactured with adapted technologies. *Food Research International*, Vol.40, No.1, (January 2007), pp. 191-202, ISSN 0963-9969

Molto-Puigmarti, C.; Castellote, A.I. & Lopez-Sabater, M.C. (2007). Conjugated linoleic acid determination in human milk by fast-gas chromatography. *Analytica Chimica Acta*, Vol.602, No.1, pp. 122-130, ISSN 0003-2670

Pereira, S.L.;. Leonard, A.E. & Mukerji, P. (2003). Recent advances in the study of fatty acid desaturases from animals and lower eukaryotes. *Prostaglandins Leukotrienes and Essential Fatty Acids*, Vol.68, No.2, (February 2003), pp. 97-106, ISSN 1098-8823

Perfield, J.W. II; Delmonte, P.; Lock, A.L.; Yurawecz, M.P. & Bauman, D.E. (2006). Trans-10, trans-12 conjugated linoleic acid does not affect milk fat yield but reduces Δ9-desaturase index in dairy cows. *Journal of Dairy Science*, Vol.89, No.7, (July 2006), pp. 2559-2566, ISSN 0022-0302

Perfield, J.W. II; Lock, A.L.; Griinari, J.M.; Sab, A.; Delmonte, P.; Dwyer, D.A. & Bauman, D.E. (2007). Trans-9, Cis-11 conjugated linoleic acid reduces milk fat synthesis in lactating dairy cows. *Journal of Dairy Science*, Vol.90, No.5, (May 2007), pp. 2211-2218, ISSN 0022-0302

Pryce, J.E., Royal, M.D.; Garnsworthy, P.C. & Mao, I.L. (2004). Fertility in the high-producing dairy cow. *Livestock Production Science*, Vol.86, No.1-3, (March 2004), pp. 125-135, ISSN 1871-1413

Randolph, H.E. & Erwin, R.E. (1974). Influence of mastitis on properties of milk.10. Fatty-acid composition. *Journal of Dairy Science*, Vol.57, No.8, (August 1974), pp. 865-868, ISSN 0022-0302

Ros, E. (2003). Dietary cis-monounsaturated fatty acids and metabolic control in type 2 diabetes. *The American Journal of Clinical Nutrition*, Vol.78, No.3suppl, (September 2003), pp. 617S-625S, ISSN 1938-3207

Rosenberger, G. (1979). *L'esame clinico del bovino*, Ed. Essegivi, Edagricole, ISBN 88-206-3713-8, Piacenza, Italy

Roxström, A.; Strandberg, E.; Berglund, B.; Emanuelson, U. & Philipsson, J. (2001a). Genetic and environmental correlations among female fertility traits and milk production in different parities of Swedish Red and White dairy cattle. *Acta Agriculturae Scandinavica, Section A - Animal Science*, Vol.51, No.1, pp.7-14, ISSN 0906-4702

Roxström, A.; Strandberg, E.; Berglund, B.; Emanuelson, U. & Philipsson, J. (2001b). Genetic and environmental correlations among female fertility traits and the ability to show oestrus, and milk production. *Acta Agriculturae Scandinavica, Section A - Animal Science*, Vol.51, No.3, pp. 192-199, ISSN 0906-4702

Sandri, S.; Summer, A.; Tosi, F.; Mariani, M.S.; Pecorari, M.; Franceschi, P.; Formaggioni, P.; Pederzani, D. & Malacarne, M. (2010). Influence of somatic cell content on dairy aptitude of milk. *Scienza e Tecnica Lattiero-Casearia*, Vol.61, No.1, pp. 5-18, ISSN 0390-637X

Schennink, A.; Stoop, W.M.; Visker, M.H.P.W.; Heck, J.M.L.; Bovenhuis, H.; Poel, J.J.v.d.; Valenberg, H.J.F.v. & Arendonk, J.A.M.v. (2007). DGAT1 underlies large genetic variation in milk-fat composition of dairy cows. *Animal Genetics*, Vol.38, No.5, (October 2007), pp. 467–473, ISSN: 1365-2052

Schennink, A.; Heck, J.M.L.; Bovenhuis, H.; Visker, M.H.P.W.; Valenberg, H.J.F.v. & Arendonk, J.A.M.v. (2008). Milk fatty acid unsaturation: genetic parameters and effects of Stearoyl-CoA desaturase (SCD1) and Acyl CoA: diacylglycerol acyltransferase 1 (DGAT1). *Journal of Dairy Science*, Vol.91, No.5, (May 2008), pp. 2135-2143, ISSN 0022-0302

Shorten, P.R.; Pleasants, T.B. & Upreti, G.C. (2004). A mathematical model for mammary fatty acid synthesis and triglyceride assembly: the role of stearoyl CoA desaturase (SCD). *Journal of Dairy Research*, Vol.71, No.4, pp. 385-397, ISSN 0022-0299

Sørensen, A.C.; Lawlor, T. & Ruiz, F. (2007). A survey on fertility in the Holstein populations of the world. In: *Proceedings of the Int Conf on Fertility in dairy cows "EAAP Satellite Meeting"*, pp. 1:17, ISBN 91-576-5678-9, Liverpool Hope University, UK, 30-31 August 2007

Soyeurt, H.; Dardenne, P.; Gillon, A.; Croquet, Vanderick, C.S.; Mayeres, P.; Bertozzi, C. & Gengler, N. (2006). Variation in fatty acid contents of milk and milk fat within and across breeds. *Journal of Dairy Science*, Vol.89, No.12, (December 2006), pp. 4858-4865, ISSN 0022-0302

Soyeurt, H.; Gillon, A.; Vanderick, S., Mayeres, P.; Bertozzi, C. & Gengler, N. (2007). Estimation of heritability and genetic correlations for the major fatty acids in bovine milk. *Journal of Dairy Science*, Vol.90, No.9, (September 2007), pp. 4435-4442, ISSN 0022-0302

Soyeurt, H.; Dehareng, F.; Mayeres, P.; Bertozzi, C. & Gengler, N. (2008). Variation of Δ9-desaturase activity in dairy cattle. *Journal of Dairy Science*, Vol.91, No.8, (August 2008), pp. 3211-3224, ISSN 0022-0302

Veerkamp, R.F. (1998). Selection for economic efficiency of dairy cows using information on live weight and feed intake: A review. Journal of Dairy Science, Vol.81, No.(4), pp. 1109-1119, ISSN 0022-0302

Vessby, B. (2003). Dietary fat, fatty acid composition in plasma and the metabolic syndrome. *Current Opinion in Lipidology*, Vol.14, No.1, (February 2003), pp.15-19, ISSN 1473-6535

13

Aluminium in Acid Soils: Chemistry, Toxicity and Impact on Maize Plants

Dragana Krstic[1], Ivica Djalovic[2],
Dragoslav Nikezic[1] and Dragana Bjelic[3]
[1]*University of Kragujevac, Faculty of Science*
[2]*Institute of Field and Vegetable Crops*
[3]*University of Kragujevac, Faculty of Agronomy*
Serbia

1. Introduction

Soil acidity is a limiting factor affecting the growth and yield of many crops all over the world. The basic problems concerning chemical properties of more acid soils are, besides acidity itself, the presence of toxic compounds and elements, such as soluble forms of Al, Fe and Mn, nitrites and various toxic organic acids. Aluminium (Al) toxicity is one of the major constraints on crop productivity on acid soils, which occur on up to 40% of the arable lands of the world. Al is the third most abundant element in the earth's crust and is toxic to plants when solubilised into soil solution at acidic pH values (Kochian, 1995). A total of 3950 million hectares of land is classed as having acidic soil, of which 15% is used for planting of annual and perennial crops (von Uexküll & Mutert, 1995).

Northern belt of acid soils occurring in the humid northern temperate zone is comprised of predominantly organic acid soils and supports coniferous forests. A southern belt of mineral acid soils occurs in the humid tropics. Currently approximately 12% of land in crop production is acidic (Uexküll & Mutert, 1995), however, the nutriextent of acid soil is increasing world-wide. Mineral acid soils result from parent materials that are acidic and naturally low in the basic cations (Ca, Mg, K and Na), or because these elements are leached from the soil, reducing pH and the buffering capacity of the soil. As soil pH decreases, aluminum (Al) is solubilized and the proportion of phytotoxic aluminium ions increases in the soil solution. In most mineral soils there is sufficient Al present to buffer the soil to around pH 4. Organic acid soils, consisting of large amounts of humic acids and partially decomposed plant matter, typically have little Al buffering and the pH of these soils can fall rather below pH 4 (Kidd, 2001).

A number of factors contribute to acid soil toxicity transdepending on soil composition. In acid soils with a high mineral content, the primary factor limiting plant growth is Al toxicity. The Al released from soil minerals under acid conditions occurs as $Al(OH)_2^+$, $Al(OH)$ and $Al(H_2O)^{3+}$, the latter commonly referred to as Al (Kinraide, 1991). For most agriculturally important plants, Al ions rapidly inhibit root growth at micromolar concentrations.

The primary target of Al toxicity is the root apex. Aluminium affects a host of different cellular functions, frustrating attempts to identify the principal effect(s) of Al toxicity. Exposure to Al causes stunting of the primary root and inhibition of lateral root formation. Affected root tips are stubby due to inhibition of cell elongation and cell division. The resulting restricted root system is impaired in nutrient and water uptake, making the plant more susceptible to drought stress. Plants sensitive to Al toxicity have greatly reduced yield and crop quality (Samac & Tesfaye, 2003; Jovanovic et al., 2006; 2007).

The influence of physical-chemical characteristics of soil on distribution of some elements and availability for plants in vertisols of Serbia are confirmed (Krstic et al., 2004; Krstic et al., 2007; Dugalic et al., 2010; Jelic et al., 2010; Milivojevic et al., 2011) Aim of this study was testing of soil pH, exchangeable acidity and mobile aluminium (Al) status in profiles of pseudogley soils of Čačak–Kraljevo basin.

1.1 Aluminium chemistry in the soil

Aluminium, bound as oxides and complex aluminosilicates, ranks third in abundance among the elements in the Earth's crust. Despite much research since Hartwell and Pember first postulated, nearly 90 years ago, that soluble aluminium is a major inhibitor of plant growth in acid soils, the mechanism of aluminium phytotoxicity is not yet fully understood. Aluminium can inhibit root growth at the organ, tissue, and cellular levels at micromolar concentrations (Ciamporová, 2002). Acid soils, present mostly in humid tropical and subtropical areas of the world, are characterized by having excess H^+, Mn^{2+}, and Al_3^+, with deficiencies of Ca^{2+}, Mg^{2+}, and PO_4^{3-}. Additionally, sulfur dioxide and other air pollutants cause acid soil stress in areas other than the tropics (Foy, 1984). In acidic soils, hydroxyl-rich aluminium compounds solubilize to an extent in the soil solution. Forty percent of the arable land globally is acidic because of solubilization of the abundantly present aluminium, greatly limiting crop productivity.

Aluminium chemistry is quite complex. It has a high ionic charge and a small crystalline radius, which gives it a level of reactivity that is unmatched by other soluble metals. When the pH of a solution is raised above 4.0, Al^{3+} forms the mononuclear species $AlOH_2^+$, $Al(OH)_2^+$, $Al(OH)_3$, and $Al(OH)_4^+$, and soluble complexes with inorganic ligands such as sulfate and fluoride, AlF_2^+, AlF_3^+, $Al(SO)_4^+$, and also with many organic compounds. Larger polynuclear hydroxyl aluminium species also form as metastable intermediates during $Al(OH)_3$ precipitation. The mononuclear Al^{3+} species seems to be most toxic at low pH, at which it exists as an octahedral hexahydrate. With increasing pH, $Al(H_2O)^{3+}_6$ undergoes repeated deprotonations to form insoluble $Al(OH)_3$ at pH 7.0. At cytosolic pH, 7.4 aluminate ion, $Al(OH)_4^-$, is formed. In near neutral solutions, polynuclear forms of aluminium, which contain more than one aluminium atom, occurs. One of the most important polymer triskaidekaaluminium, $AlO_4Al_{12}(OH)_{24}(H_2O)_{127}^+$ refered as Al_{13} (Parker & Bertsch, 1992), seems to be the most toxic Al specie.

2. Material and methods

2.1 General characteristics of the Čačak–Kraljevo Basin

Čačak-Kraljevo basin is part of western Serbia (Morava river area). It is narrow belt longitude approximately 70 km in NW–SE direction and width from 5 to 18 km. Kablar,

Ovčar, Troglav, Stolovi, Goč, Suvobor, Vujno and Kotlenik mountains are border toward SW and NE directions. Pseudogley soils of this area (approximately 32.000 hectares situated mainly in latitudes between 180 and 200 m above sea level) have been developed on diluvial–holocene terrace of Western Morava and its tributaries. Climate of this area is moderate continental characterizing mean annual air–temperature 11.2°C (winter 1.4°C, summer 20.5°C) and precipitation 715.8 mm (Kraljevo Weather Bureau; means 1961–1990).

2.2 Sampling and chemical analysis

Total 102 soil profiles were opened during 2008 at certain sites of the Čačak–Kraljevo basin. The tests encompassed 54 field, 28 meadow, and 20 forest profiles. From the opened profiles, samples of soil in the disturbed state were taken from the humus and Eg horizons (102 profiles); then from the B_1tg horizon of 39 fields, 24 meadows and 15 forest profiles (total 78) and from the B_2tg horizon of 14 fields, 11 meadows, and 4 forest profiles (total 29). Laboratory determination of exchangeable acidity was conducted in a suspension of soil with a 1.0 M KCl solution (pH 6.0) using a potentiometer with a glass electrode, as well as by Sokolov's method, where the content of Al ions in the extract is determined in addition to total exchangeable acidity (H^+ + Al^{3+} ions) (Jakovljevic et al., 1995).

3. Results and discussion

This mean pH (1M KCl) of tested soil profiles were 4.28, 3.90 and 3.80, for Ah, Eg and B_1tg horizons, respectively. Also, soil pH of forest profiles was lower in comparison with meadows and arable lands (means: 4.06, 3.97 and 3.85, for arable lands, meadows and forest, respectively). Soil acidification is especially intensive in deeper horizons because 27% (Ah), 77% (Eg) and 87% (B_1tg) soil profiles have pH lower than 4.0 (Table 1).

Distribution of pH (1M KCl) in soil profiles (a=arable land; m=meadow; f=forest)								
Horizons	n	pH (1M KCl)					pH values	
		< 4.0	4.1–4.5	4.6–5.1	> 5.1	Sum	Mean	Range
		pH (1M KCl) in % of total (n) tested profiles						
Ah (a)	54	18.5	57.5	22.2	1.8	100	4.33	3.7–5.2
Ah (m)	24	20.8	75.0	4.2	0.0	100	4.25	3.9–4.8
Ah (f)	20	55.0	30.0	5.0	10.0	100	4.18	3.6–5.3
Ah (total)	98	26.5	56.1	14.3	3.1	100	4.28	3.6–5.3
Eg (a)	54	64.8	27.8	7.4	0.0	100	3.99	3.6–5.1
Eg (m)	24	91.7	8.3	0.0	0.0	100	3.89	3.6–4.5
Eg (f)	20	90.0	10.0	0.0	0.0	100	3.69	3.4–4.1
Eg (total)	98	76.5	19.4	4.1	0.0	100	3.90	3.4–5.1
B_1tg (a)	39	76.9	20.5	2.6	0.0	100	3.86	3.5–4.6
B_1tg (m)	20	95.0	5.0	0.0	0.0	100	3.78	3.6–4.4
B_1tg (f)	17	100.0	0.0	0.0	0.0	100	3.69	3.5–4.0
B_1tg (total)	76	86.9	11.8	1.3	0.0	100	3.80	3.5–4.6
B_2tg (total)	31	90.2	6.6	3.2	0.0	100	3.83	3.6–4.8

Table 1. Distribution of pH (1M KCl) in soil profiles

Mean total exchangeable acidity (TEA) of tested soil profiles were 1.55, 2.33 and 3.40 meq 100g^{-1}, for Ah, Eg and B$_1$tg horizons, respectively. However, it is considerably higher in forest soils (mean 3.39 meq 100g^{-1}) than in arable soils and meadows (means 1.96 and 1.93, respectively).

The deeper horizons (Eg and B$_1$tg) of meadows and forest soil profiles have especially high TEA values. Especially high frequencies of the high TEA values (over 3.0 meq 100g^{-1}) were found in forest soil profiles (Table 2).

		TEA (meq 100g^{-1})					TRA (meq 100g^{-1})	
Horizons	n	<1.0	1–2	2–3	>3.0	Sum	Mean	Range
		TEA in % of total (n) tested profiles						
Ah (a)	53	86.8	13.2	0.0	0.0	100	0.96	0.07–1.84
Ah (m)	27	85.2	14.8	0.0	0.0	100	0.90	0.22–1.58
Ah (f)	20	55.0	20.0	5.0	20.0	100	2.79	0.09–5.49
Ah (total)	100	80.0	15.0	1.0	4.0	100	1.55	0.07–5.49
Eg (a)	53	35.8	37.8	22.6	3.8	100	1.80	0.16–3.44
Eg (m)	27	18.5	63.0	11.1	7.4	100	1.85	0.37–3.33
Eg (f)	20	10.0	20.0	30.0	40.0	100	3.34	0.58–6.09
Eg (total)	100	26.0	41.0	21.0	12.0	100	2.33	0.16–6.09
B$_1$tg (a)	37	24.3	18.9	24.3	32.5	100	3.12	0.23–6.01
B$_1$tg (m)	23	8.7	21.7	39.2	30.4	100	3.05	0.60–5.49
B$_1$tg (f)	14	0.0	21.4	28.6	50.0	100	4.03	1.36–6.69
B$_1$tg (total)	74	14.8	20.3	29.6	35.3	100	3.40	0.23–6.69
B$_2$tg (total)	29	14.8	27.6	36.5	21.1	100	2.62	0.70–5.54

Total exchangeable acidity (TEA) in soil profiles (a=arable land; m=meadow; f=forest)

Table 2. Distribution of total exchangeable acidity (sum of H^+ and Al^{3+}) in soil profiles

Mean mobile Al contents of tested soil profiles were 11.02, 19.58 and 28.33 mg Al 100 g^{-1}, for Ah, Eg and B$_1$tg horizons, respectively. Soil pH and TEA in forest soils are considerably higher (mean 26.08 meq Al 100 g^{-1}) than in arable soils and meadows (means 16.85 and 16.00 Al 100 g^{-1}, respectively). The Eg and B$_1$tg horizons of forest soil profiles have especially high mobile Al contents (means 28.50 and 32.95 mg Al 100 g^{-1}, respectively). Frequency of high levels of mobile Al is especially high in forest soils because 35% (Ah), 85.0% (Eg) and 93.3% (B$_1$tg) of tested profiles were in range above 10 mg Al 100 g^{-1} (Table 3).

Increased TEA is characteristics of soils in which acidification processes are rather for advanced, the reaction of their soil solutions being fairly acidic, which pH values are lower than 5.0. This is typical for pseudogley which is the most widely disseminated type of soil in the Čačak-Kraljevo basin. Due to the fact that Al ions in an increased concentration are much more dangerous for plants than H^+ ions in the same concentration at the same value of TEA, plants increasingly suffer if a higher share of Al ions is present in it. Already at the content of 6–10 mg 100 g^{-1} of readily mobile Al in the soil, plant growth is retarded to a greater or lesser extent depending on the species (Rengel, 2004). High TEA, created predominantly by Al ions, is among the most important causes of the low productive capacity of pseudoglay in the indicated basin where, despite of fertilizer use and application of different agrotechnical measures, average yields of cultivated plants are low and vary fairly greatly depending on weather conditions of the year.

Mobile aluminum contents in soil profiles (a = arable land; m = meadow; f = forest)								
Horizons	n	Mobile aluminum (mg Al 100 g^{-1})					mg Al 100 g^{-1}	
		<3.0	3.1–6.0	6.1–10	>10	Sum	Mean	Range
		Mobile aluminum in % of total						
Ah (a)	54	63.0	18.5	11.1	7.4	100	8.15	0.2 – 16.1
Ah (m)	28	64.4	17.8	7.1	10.7	100	8.10	1.0 – 15.2
Ah (f)	20	40.0	10.0	15.0	35.0	100	16.80	0.4 – 33.2
Ah (total)	102	58.8	16.3	10.8	14.1	100	11.02	0.2 – 33.2
Eg (a)	54	20.4	13.0	12.9	53.7	100	15.40	0.5 – 30.3
Eg (m)	28	10.7	7.1	21.4	60.8	100	14.85	0.3 – 29.4
Eg (f)	20	0.0	15.0	0.0	85.0	100	28.50	3.5 – 53.5
Eg (total)	102	13.7	11.8	12.8	61.7	100	19.58	0.3 – 53.5
B_1tg (a)	39	12.8	12.8	7.7	66.7	100	27.00	1.0 – 53.0
B_1tg (m)	24	0.0	8.3	12.5	79.2	100	25.05	3.2 – 46.9
B_1tg (f)	15	0.0	0.0	6.7	93.3	100	32.95	7.9 – 58.0
B_1tg (total)	78	6.4	9.0	8.9	75.7	100	28.33	1.0 – 58.0
B_2tg (total)	29	0.0	14.2	13.5	72.3	100	20.50	3.6 – 37.4

Table 3. Distribution of mobile aluminium in soil profiles

3.1 Aluminium Influence on maize plants

Al ions translocate very slowly to the upper parts of plants (Ma et al., 1997). Most plants contain no more than 0.2 mg Al g^{-1} dry mass. However, some plants, known as Al accumulators, may contain over 10 times more Al without any injury. Tea plants are typical Al accumulators: the Al content in these plants can reach as high as 30 mg g^{-1} dry mass in old leaves (Matsumoto et al., 1976). Approximately 400 species of terrestrial plants, belonging to 45 families, have so far been identified as hyperaccumulators of various toxic metals (Baker et al., 2000).

The main aluminum toxicity symptom is inhibition of root elongation with simultaneous induction of β-1,3-glucan (callose) synthesis, which is apparent alter even a short exposure time. Aluminium causes extensive root injury, leading to poor ion and water uptake (Barcelo & Poschenrieder, 2002). One of hypothesis is that the sequence of toxicity starts with perception of aluminum by the root cap cells, followed by signal transduction and a physiological response within the root meristem. However, recent work has ruled out a role of the root cap and emphasizes that the root meristem is the sensitive site. Root tips have been found to be the primary site of aluminum injury, and the distal part of the transition zone has been identified as the target site in maize (*Zea mays*) (Sivaguru & Horst, 1998). Root cells division results in root elongation. Aluminum is known to induce a decrease in mitotic activity in many plants, and the aluminum-induced reduction in the number of proliferating cells is accompanied by the shortening of the region of cell division in maize (Panda, 2007).

Blancaflor et al. (1998) have studied Al-induced effects on microtubules and actin microfilaments in elongating cells of maize root apices, and related the Al-induced growth inhibition to stabilization of microtubules in the central elongation zone. With respect to growth determinants (auxin, gibberelic acid and ethylene), Al apparently interacts directly and/or indirectly with the factors that influence organization of the cytoskeleton, such as cytosolic levels of Ca^{2+} (Jones et al., 2006), Mg^{2+} and calmodulin (Grabski et al., 1998), cell–surface electrical potential (Takabatake & Shimmen, 1997), callose formation (Horst et al., 1997) and lipid composition of the plasma membrane.

Genetic variability for Al resistance in maize has been reported (Jorge & Arruda, 1997; Pintro et al., 1996 and Al-resistant maize cultivars have been selected for acidic soils (Pandey & Gardner, 1992). Maize grain-yield increase has been obtained on acid soils through selection for tolerant cultivars in tropical maize populations. Most breeding work designed at increasing productivity on acid soil, focused on tolerance to Al toxicity (Garvin & Carver, 2003).

Al resistance mechanisms can be grouped into two categories, exclusion of Al from the roots, and detoxification of Al ions in the plant (Taylor, 1991; Heim et al., 1999; Kochian et al., 2005; Zhou et al., 2007). Exclusion mechanisms include binding of Al in the cell wall, a plant-induced rhizosphere pH barrier, and root exudation of Al–chelating compounds. Organic acids have been reported to play a role both in Al exclusion, via release from the root and Al detoxification in the symplasm, where organic acids such as citric acid and malic acid could chelate Al and reduce or prevent its toxic effects at the cellular level, in particular protecting enzyme activity internally in the plant from the deleterious effect of Al (Delhaize et al., 1993). Genetic adaptation of plants to Al toxicity may provide a sustainable strategy to increase crop yield in the tropics at relatively low costs and low environmental impacts. This approach is particularly interesting for maize, where Al tolerant germplasm is available for selection and for genetic studies. A number of studies have been carried out to elucidate the genetic control of Al tolerance in maize, resulting in controversial results. However, a consensus among the authors has shown that the trait is quantitatively inherited under the control of few genes (Lima et al., 1995). Most of the genetics studies on aluminum tolerance in maize have evaluated the seminal root growth under nutrient solution as screening technique. Nutrient solutions with high concentration of aluminum have proven to be an effective way to discriminate tolerant and susceptible maize genotypes (Martins et al., 1999; Cancado et al., 1999). Although a large number of studies have been conducted, the genetic basis and the molecular mechanisms responsible for the genetic variability in maize Al tolerance are still poorly understood.

3.2 Al toxicity and root growth

High Al concentrations are particularly difficult to interpret in terms of physiological responses. A high proportion of Al in the nutrient growth medium might become inert by precipitation (e.g., with phosphate) or by polymerisation and complexation. Thus, the concentration of free Al promoting toxicity in plant metabolism can be much lower than that existing in the growth medium (Mengel & Kirkby, 1987). Low concentrations of Al can also lead to a stimulation of root growth in tolerant genotypes of *Zea mays* L.

In non-accumulators plant species the negative effects of Al on plant growth prevail in soils with low pH (Marschner, 1995), the reduction in root growth being the most serious consequence (Tabuchi & Matsumoto, 2001). This symptom of Al toxicity has been related to the linkage of Al to carboxylic groups of pectins in root cells (Klimashevsky & Dedov, 1975) or to the switching of cellulose synthesis into callose accumulation (Teraoka et al., 2002), to Al inhibition of mitosis in the root apex (Rengel, 1992; Delhaize & Ryan, 1995) implicating blockage of DNA synthesis, aberration of chromosomal morphology and structure occurrence of anaphase bridges and chromosome stickness and to Al-induced programmed cell death in the root-tip triggered by reactive oxygen species (Pan et al., 2001).

According to Comin et al. (1997) tolerant cultivars of *Zea mays* L. have different toxicity mechanisms, following monomeric or polymeric forms of Al supplied to the growth medium. Aluminum can easily polymerise, transforming the monomeric form (Al^{3+}) into a polymeric form (Al_{13}), which is much more phytotoxic in maize. Yet, although Bashir et al. (1996) had noticed Al nucleotypic effects on maize, a lack of nuclear DNA content variability was found among wheat isolines differing in Al response as well as four genes that ameliorate Al toxicity (Ezaki et al., 2001). Indeed, the general responses to Al excess by tolerant genotypes deal with the varying ability of plants to modify the pH of the soil-root interface (Mengel & Kirkby, 1987; El-Shatnawi & Makhadmeh, 2001).

4. Conclusion

Soil acidity and aluminium toxicity is certain one of the most damaging soil conditions which affecting the growth of most crops. In this paper soil pH, exchangeable acidity and mobile aluminium (Al) status in profiles of pseudogley soils of Western Serbia region were studied. Total 102 soil profiles were opened during 2008 in the Western Serbia. The tests encompassed 54 field, 28 meadow, and 20 forest profiles. From the opened profiles, samples of soil in the disturbed state were taken from the humus and Eg horizons (102 profiles); then from the B_1tg horizon of 39 fields, 24 meadows and 15 forest profiles (total 78) and from the B_2tg horizon of 14 fields, 11 meadows, and 4 forest profiles (total 29). Laboratory determination of exchangeable acidity was conducted in a suspension of soil with a 1.0 M

KCl solution (pH 6.0) using a potentiometer with a glass electrode, as well as by Sokolov's method, where the content of Al ions in the extract is determined in addition to total exchangeable acidity ($H^+ + Al^{3+}$ ions). Mean pH (1M KCl) of tested soil profiles were 4.28, 3.90 and 3.80, for Ah, Eg and B_1tg horizons, respectively. Also, soil pH of forest profiles was lower in comparison with meadows and arable lands (means: 4.06, 3.97 and 3.85, for arable lands, meadows and forest, respectively). Soil acidification is especially intensive in deeper horizons because 27% (Ah), 77% (Eg) and 87% (B_1tg) soil profiles have pH lower than 4.0. Mean total exchangeable acidity (TEA) of tested soil profiles were 1.55, 2.33 and 3.40 meq $100g^{-1}$, for Ah, Eg and B_1tg horizons, respectively. However, it is considerably higher in forest soils (mean 3.39 meq $100g^{-1}$) than in arable soils and meadows (means 1.96 and 1.93, respectively). Mean mobile Al contents of tested soil profiles were 11.02, 19.58 and 28.33 mg Al 100 g^{-1}, for Ah, Eg and B_1tg horizons, respectively. Soil pH and TEA in forest soils are considerably higher (mean 26.08 meq Al $100g^{-1}$) than in arable soils and meadows (means 16.85 and 16.00 Al 100 g^{-1}, respectively). The Eg and B_1tg horizons of forest soil profiles have especially high mobile Al contents (means 28.50 and 32.95 mg Al 100 g^{-1}, respectively). Frequency of high levels of mobile Al is especially high in forest soils because 35% (Ah), 85.0 % (Eg) and 93.3% (B_1tg) of tested profiles were in range above 10 mg Al 100 g^{-1}.

Al ions translocate very slowly to the upper parts of plants. Most plants contain no more than 0.2 mg Al g^{-1} dry mass. However, some plants, known as Al accumulators, may contain over 10 times more Al without any injury. Tea plants are typical Al accumulators: the Al content in these plants can reach as high as 30 mg g^{-1} dry mass in old leaves. Approximately 400 species of terrestrial plants, belonging to 45 families, have so far been identified as hyperaccumulators of various toxic metals.

The main aluminum toxicity symptom is inhibition of root elongation with simultaneous induction of glucan (β-1,3-callose) synthesis, which is apparent alter even a short exposure time. Aluminium causes extensive root injury, leading to poor ion and water uptake. Aluminum is known to induce a decrease in mitotic activity in many plants, and the aluminum-induced reduction in the number of proliferating cells is accompanied by the shortening of the region of cell division in maize.

Genetic adaptation of plants to Al toxicity may provide a sustainable strategy to increase crop yield in the tropics at relatively low costs and low environmental impacts. This approach is particularly interesting for maize, where Al tolerant germplasm is available for selection and for genetic studies.

High Al concentrations are particularly difficult to interpret in terms of physiological responses. A high proportion of Al in the nutrient growth medium might become inert by precipitation (e.g., with phosphate) or by polymerisation and complexation. Thus, the concentration of free Al promoting toxicity in plant metabolism can be much lower than that existing in the growth medium.

5. Acknowledgment

This research was supported by a grant from the Ministry of Science of the Republic of Serbia (Projects TR 31073 III 41011 and ON 171021)

6. References

Baker, A. J. M.; McGrath, S. P.; Reeves, R. D. & Smith, J. A. C. (2000). Metal hyperaccumulator plants: A review of the ecology and physiology of a biological resource for phytoremediation of metal–polluted soils. In: *Phytoremediation of Contaminated Soil and Water.* N. Terry & G. Banuelos (Eds.), 85–107, Lewis Publisher, Boca Raton

Barcelo, J. & Poschenrieder, C. (2002). Fast root growth responses, root exudates and internal detoxification as clues to the mechanisms of aluminium toxicity and resistance: A review. *Env. Exp. Bot.,* 48, 75–92

Bashir, A.; Biradar, D.P.& Rayburn, A.L. (2006). Determining relative abundance of specific DNA sequences in flow cytometrically sorted maize nuclei. *J. Exper. Botany,* 46, 451–457

Blancaflor, E. B.; Jones, D. L. & Gilroy S. (1998). Alterations in the cytoskeleton accompany aluminum–induced growth inhibition and morphological changes in primary roots of maize. *Plant Physiol.,* 118, 159–172

Ciamporová, M. (2002). Morphological and structure responces of plant roots to aluminium at organ, tissue, and cellular levels. *Biol. Pl.,* 45, 161-171

Cançado, G. M. A.; Loguercio, L. L.; Martins, P. R.; Parentoni, S. N.; Borém, A.; Paiva, E. & Lopes, M. A. (1999). Hematoxylin staining as a phenotypic index for aluminum tolerance selection in tropical maize (*Zea mays* L.). *Theor. Appl. Genet.,* 99, 747–754

Comin-Chiaramonti, P.; Cundari, A.; Piccirillo, E.M.; Gomes, C.B.; Castorina, F.; Censi , P.; Demin A.; Marzoli, A.; Speziale, S. & Velázquez, V.F. (1997). Potassic and sodic igneous rocks from Eastern Paraguay: their origin from the lithospheric mantle and genetic relationships with the associated Paraná flood tholeiites. *J. Petrology,* 38, 495-528

Delhaize, E.; Craig, S.; Beaton, C. D,.; Bennet, R. J.; Jagadish, V. C. & Randall, P. J. (1993). Aluminum tolerance in wheat (*Triticum aestivum* L.) I. Uptake and distribution of aluminum in root apices. *Plant Physiol.,* 103, 685–693

Delhaize, E. & Ryan, P. R. (1995). Aluminium toxicity and tolerance in plants. *Plant Physiol.,* 107, 315–321

Dugalic, G.; Krstic, D.; Jelic, M.; Nikezic, D.; Milenkovic, B.; Pucarevic, M. & Zeremski-Skoric, T. (2004). Heavy metals, organics and radioactivity in soil of western Serbia . *J. Hazard. Mat.,* 177, 697-702

El-Shatnawi, M. K. & Makhadmeh, I. M. (2001). A Review- Ecophysiology of the plant-rhizosphere system. *J. Agronomy & Crop Science,* 187, 1-9

Ezaki, B.; Katsuhara, M.; Kawamura, M. & Matsumoto, H. (2001). Different mechanisms of four aluminium (Al)-resistant transgenes for Al toxicity in Arabidopsis. *Plant Physiol.,* 127, 918–927

Foy, C. D. (1984). Physiological effects of hydrogen, Al and manganese toxicities in acid soil. In: *Soil acidity and liming.* F. Adams, (Ed.), 57-97, American Society of Agronomy, Madison, Wisconsin

Garvin, D. F. & Carver B. F. (2003), The Role of the Genotype in Tolerance to Acidity and Aluminium Toxicity. In: *Handbook of Soil Acidity.* Z. Rengel (Ed.), 387–406, Marcel Dekker, New York

Grabski, S.; Arnoys, E.; Busch, B. & Schindler, M. (1998). Regulation of actin tension in plant cells by kinases and phosphatases. *Plant Physiol.*, 116, 279–290

Heim, A.; Luster, J.; Brunner, I.; Frey, B. & Frossard, E. (1999). Effects of aluminium treatment on Norway spruce roots: aluminium bindings forms, element distribution, and release of organic substances. *Plant and Soil*, 216, 103-116

Horst, W. J.; Püschel, A. K. & Schmohl, N. (1997). Induction of callose formation is a sensitive marker for genotypic aluminium sensitivity in maize. *Plant Soil*, 192, 23–30

Jakovljevic, M.; Pantovic, M. & Blagojevic, S. (1995). *Laboratory Manual of Soil and Water Chemistry* (in Serbian), Faculty of Agriculture, Belgrade

Jelic, M.; Djalovic, I.; Milivojevic, J. & Krstic, D. (2010). Mobile aluminium content of vertisols as dependent upon fertilization system and small grains genotypes, *Proceedings of 3nd International Scientific/Professional Conference Agriculture in Nature and Environment Protection*, pp. 137-142, ISBN 978-953-7693-008, Vukovar, Croatia, May 31- June 2, 2010

Jones, D. L.; Blancaflor, E. B.; Kochian, L. V. & Gilroy S. (2006). Spatial coordination of aluminium uptake, production of reactive oxygen species, callose production and wall rigidification in maize roots. *Plant Cell Environ.*, 29, 1309–1318

Jorge, R. A. & Arruda, P. (1997). Aluminum–induced organic acid exudation by roots of aluminum-tolerant tropical maize. *Phytochemistry*, 45, 675–681

Jovanovic,., Z.; Djalovic, I.; Komljenovic, I.; Kovacevic, V. & Cvijovic, M. (2006). Influences of liming on vertisol properties and yields of the field crops. *Cereal Res. Commun.*, 34, 517-520

Jovanovic, Z.; Djalovic, I.; Tolimir, M. & Cvijovic, M. (2007). Influence of growing sistem and NPK fertilization on maize yield on pseudogley of Central Serbia. *Cereal Res. Commun.*, 35, 1325-1329

Kidd, P. S. & Proctor, J. (2001). Why plants grow poorly on very acid soils: are ecologists missing the obvious? *J. Exp. Bot.*, 52, 791-799

Kinraide, T. B. (1991). Identity of rhizotoxic aluminium species. *Plant Soil*, 134, 167-178

Kochian, K. V. (1995). Cellular mechanisms of aluminium toxicity and resistance in plant. *Annu. Rev. Plant Physiol. Mol. Biol.*, 46, 237-260

Klimashevskii, E. L. & Dedov, V. M. (1975). Localization of growth inhibiting action of aluminium ions in alongating cell walls. *Fiziologiia Rastenii*, 22, 1183-1190

Kochian, K. V. (1995). Cellular mechanisms of aluminium toxicity and resistance in plant. *Annu. Rev. Plant Physiol. Mol. Biol.*, 46, 237-260

Kochian, L. V.; Piñeros, M. A. & Hoekenga O. A. (2005). The physiology, genetics and molecular biology of plant aluminum resistance and toxicity. *Plant and Soil*, 274, 175–195

Krstic, D.; Nikezic, D.; Stevanovic, N. & Jelic, M. (2004). Vertical profile of ^{137}Cs in soil. *Appl. Radiat. Issotopes*, 61, 1487-1492

Krstic, D.; Stevanovic, N.; Milivojevic, J. & Nikezic, D. (2007). Determination of the soil-to-grass transfer of ^{137}Cs and its relation to several soil properties at various locations in Serbia. Isotopes Environ. Health St., 43, 65-73

Lima, M.; Miranda, Filho, J. B. & Furlani, P. R. (1995). Diallel cross among inbred lines of maize differing in aluminum tolerance. *Braz. J. Genet.*, 4, 579–584

Ma, Q.; Hiradate, J. F.; Nomoto, K.; Iwashita, T. & Matsumoto, H. (1997). Internal detoxification mechanism of Al in hydrangea: Identification of Al form in the leaves. *Plant Physiol.*, 113, 1033–1039

Marschner, H. (1995). *Mineral nutrition of higher plants* (2nd ed.), Academic Press, London

Martins, P. R.; Parentoni, S. N.; Lopes, M. A. & Paiva, E. (1999). Eficiěncia de indices fenotĭpicos de comprimento de raiz seminal na avaliaĉăo de plantas individuais de milho quanto ă tolerăncia ao aluminio. *Pesquisa Agropecuăria Brasileira*, 34, 1897–1904

Matsumoto, H.; Hirasawa, E.; Torikai, H. & Takahashi, E. (1976). Localization of absorbed aluminum in pea root and its binding to nucleic acids. *Plant Cell. Physiol.*, 17, 127–137

Mengel, K. & Kirkby, E.A. (1987). *Principles of Plant Nutrition* (4th ed.), International Potash Institute, IPI, Bern, Switzerland, pp. 685.

Milivojevic, J.; Nikezic, D.; Krstic, D.; Jelic, M. & Djalovic, I. (2011). Influence of Physical-Chemical Characteristics of Soil on Zinc Distribution and Availability for Plants in Vertisols of Serbia. *Pol. J. Environ. Stud.*, 20, 993-1000

Pan, J. M.; Zhu, M. & Chen, H. (2001). Aluminium-induced cell death in root tip cells of barley. *Environm. Exp. Bot.*, 46, 71-79

Panda, S. K. & Matsumoto, H. (2007). Molecular physiology of aluminium toxicity and tolerance in plants. *The Botanical Revew*, 73, 326-347

Pandey, S. & Gardner, C. O. (1992). Recurrent selection for population, variety and hybrid improvement in tropical maize. *Adv. Agron.*, 48, 1–87

Parker, D. R. & Bertsch, E. M. (1992). Formation of the „Al₁₃" tridecameric polycation under diverse synthesis conditions. *Environm. Sci. Technol.*, 26, 914-921

Pintro, J.; Barloy, J. & Fallavier, P. (1996). Aluminium effects on the growth and mineral composition of corn plants cultivated in nutrient solution at low aluminum activity. *J. Plant Nutr.*, 19, 729–741

Rengel, Z. (1992). Role of calcium in aluminium toxicity. *New Phytol.*, 121, 499-513

Rengel, Z. (2004). Aluminium cycling in the soil-plant-animal-human continuum. *Biometals*, 17, 669-689

Samac, D. A. & Tesfaye, M. (2003). Plant improvement for tolerance to aluminium in acid soils. *Plant Cell, Tissue and Organ Culture*, 75, 189-207

Sivaguru, M. & Horst, W. J. (1998). Differential impacts of aluminum on microtubule organization depend on growth phase in suspension-cultured tobacco cells. *Physiol. Plant*, 107, 110–119

Tabuchi, H. & Matsumoto, H. (2001). Changes in cell wall properties on wheat (*Triticum aestivum*) roots during aluminium-induced growth inhibition. *Physiol. Plant*, 112, 353-358

Takabatake, R. & Shimmen, T. (1997). Inhibition of electrogenesis by aluminum in characean cells. *Plant Cell Physiol.*, 38, 1264–1271

Taylor, G. J. (1991). Current views of the aluminum stress response: the physiological basis of tolerance. *Curr Top Plant Biochem Physiol.*, 10, 57–93

Teraoka, T.; Kaneko, M.; Mori, S. & Yoshimura, E. (2002). Aluminium rapidly inhibits cellulose synthesis in roots of barley and wheat seedings. *J. Plant Physiol.*, 123, 987-996 von Uexküll, H. R. & Mutert, E. (1995). Global extent, development and economic impact of acid soils. *Plant Soil*, 171, 1-15

Zhou L. L., Bai G. H., Carver B., Zhang D. D. (2007): Identification of new sources of aluminum resistance in wheat. Plant Soil, 297: 105–118

14

Genetic Characterization of Global Rice Germplasm for Sustainable Agriculture

Wengui Yan

United States Department of Agriculture
Agricultural Research Service (USDA-ARS),
Dale Bumpers National Rice Research Center,
USA

1. Introduction

Crop genebanks or germplasm collections store thousands of crop varieties. Each variety has unique genetic traits to be used in fighting diseases and insects, increasing yield and nutritional value and adjusting to environmental changes such as drought, soil salinity, etc. The Germplasm Resources Information Network (GRIN, 2011) of the United States (US) manages germplasm of plants, animals, microbes and invertebrates. Currently, there are 540,935 accessions of plant germplasm for 95,800 taxonomic names, 13,388 species of 2,208 genera along with 1,866,764 inventory records, 1,628,283 germination records, 7,291,757 characteristic records and 201,156 images in the GRIN (GRIN, 2011).

Rice is one of the most important food crops because it feeds more than half of the world's population (Yang and Hwa, 2008). There are some 4,500,000 accessions in plant germplasm collections worldwide (FAO, 1996), about 9% or 400,000 accessions are rice (Hamilton and Raymond, 2005). The United States Department of Agriculture (USDA) has started collecting rice germplasm from all over the world since the 1800s (Bockelman et al., 2002). At present, the global collection has 18,729 accessions of rice germplasm originated from 116 countries, stored in the National Small Grains Collection (NSGC, 2011) and managed by the GRIN. Great majority of these accessions (18,476 or 98.7%) belong to Asian cultivated species *Oryza sativa* in the US Department of Agriculture (USDA) rice germplasm collection. Africa cultivated species *Oryza glaberrima* has 175 accessions, and other nine species of *Oryza* have very few accessions ranging from 1 for *O. grandiglumis* to 19 for *O. glumipatula*. Some 94% of the accessions in the USDA rice germplasm collection were obtained internationally, and the remainder domestically (Yan et al., 2007). All public cultivars registered in the US can be entered in the collection. Foreign germplasm accessions must be grown for one generation in a plant quarantine greenhouse isolated from commercial rice growing areas to prevent accidental introduction of new disease and insect pests.

Evaluation of germplasm collections is essential for maintenance of the diversity and identification of valuable genes. The USDA-Agricultural Research Service (ARS) coordinates the National Plant Germplasm System (NPGS) and its related germplasm activities in the US, including germplasm acquisition, rejuvenation, storage, distribution, evaluation, and enhancement (Bretting, 2007). The NPGS is a cooperative effort by public and private organizations to preserve the genetic diversity of plants. Crop Germplasm Committees (CGC), representing the federal, state, and private sectors in various scientific disciplines, determine the set of descriptors to be managed by GRIN for most crops. Rice CGC has requested 42 descriptors plus panicle and kernel images to characterize the collection (Rice Descriptors, 2011). The USDA-ARS Dale Bumpers National Rice Research Center (DBNRRC) coordinates germplasm activities of rice including evaluation of the collection for the 42 descriptors and constantly updating the GRIN database. Furthermore, the DBNRRC manages the Genetic Stocks – Oryza collection including more than 30,000 accessions of genetic materials donated from national and international research programs (GSOR, 2011).

Comprehensive evaluation of the collection for such a large number of descriptors has been hindered by the sheer number of accessions, particularly those involving grain quality and resistances to biotic and abiotic stresses which require sophisticated instruments and significant resources. It also is difficult to characterize such a large collection using molecular means. For practical evaluation and effective management of large collections in crops, the core collection concept was proposed in the 1980s (Brown, 1989).

2. USDA rice core collection

A core collection is a subset of a large germplasm collection that contains chosen accessions capturing most of the genetic variability within the entire gene bank (Brown, 1989). With the strategy of comprehensive evaluation and accurate analysis of the core collection, the genetic diversity of the collection can be assessed, genetic distances among the accessions can be estimated for identification of special divergent subpopulations, genetic gaps of the existing collection can be identified for planning acquisition strategies and joint analysis of phenotype and genotype can be conducted for molecular understanding of the collection (Steiner et al., 2001). These analyses can help users effectively find the traits in which they are interested along with molecular information. The information is useful for determining strategies for transferring desirable traits found in the collection into new commercial cultivars.

2.1 Establishment of the core collection

The USDA rice core subset (RCS) or collection was assembled by sampling from over 18,000 accessions in the working collection of the NSGC in 1998 and 2002, respectively (Yan et al., 2007). A method of stratification by country and then random sampling was adapted by: 1) recording the number of accessions from each country or region of origin; 2) calculating the logarithm (log) of the number of accessions from each country or region of origin; 3) randomly choosing the accessions within each country or region based on the relative log numbers, with a minimum of one accession per country or region; and 4) removing obvious duplications by plant introduction (PI) number and cultivar name. In addition to the stratified sampling, additional emphasis was placed on some newly introduced Chinese germplasm (Yan et al., 2002) and newly released accessions from quarantine programs (Yan et al., 2003). The resultant RCS consists of 1,794 entries from 112 countries and represents approximately 10% of the rice whole collection (RWC).

2.2 Evaluation of the core collection

The RCS was evaluated at Stuttgart, Arkansas in 2002. Seeds of each accession were visually purified by seed shape and hull color as described in the GRIN before planting in a plot consisting of two rows, 0.3 m apart and 1.4 m long using a Hege 500 planter. Plots were separated by 0.9 m to avoid biological and mechanical contamination. A permanent flood was established after 67 kg ha⁻¹ of nitrogen as urea was applied at about 5-leaf stage.

Agronomic descriptors were recorded in the field using standard criteria described in the GRIN. Rough or paddy rice is the mature rice grain as harvested, and becomes brown rice when the hulls are removed. Rough and brown rice samples were analyzed on an automated grain image analyzer (GrainCheck 2312; Foss Tecator AB, Hoganas, Sweden) to determine rice kernel dimensions (length, width and length/width ratio), hull and seed pericarp (bran) colorations, and 1000 grain weight. Samples were milled for determination of apparent amylose content (Pérez and Juliano, 1978; Webb, 1972) and alkali spreading value (ASV) (Little et al., 1958). Fourteen important traits were selected for comparison with the whole collection.

2.3 Comparative study of the RCS with RWC

Statistical analysis was conducted using the univariate and correlation procedures of SAS statistical software, Version 9.1.3 (SAS Institute, 2004). Frequency distributions for each of 14 traits were determined using Microsoft Office Excel software. Frequency refers to how often data values occur within a range of values in an Excel bins-array that is an array of data intervals into which the data values are grouped. For example, days to flower had a bins-array of 40, 50, 60, 70, 80, 90, 100, 110, 120, 130, 140, 150, 160, 170, 180 and 190 (Fig. 1), e.g., all accessions ranging from 36 to 45 days were grouped in bin 40. Frequencies (%) of the respective bins were 0.02, 0.05, 1.15, 2.91, 7.54, 16.01, 20.33, 21.16, 14.91, 6.65, 4.07, 2.29, 1.83, 0.48, 0.52 and 0.10 among 15,097 accessions in RWC, and 0, 0.24, 1.26, 4.56, 10.43, 23.38, 27.40, 13.73, 9.53, 3.54, 2.82, 1.50, 0.96, 0.48, 0.18 and 0 among 1,668 RCS entries that headed in the field (others failed to head). Paired frequencies of the RWC and the RCS on each bin were used for correlation analysis, which measures the correspondence between the two collections. The RCS data of 1,794 accessions were from above field evaluation the RWC data of ~15,000 accessions were extracted from the GRIN.

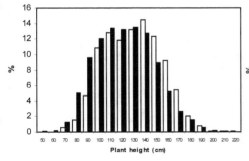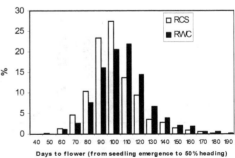

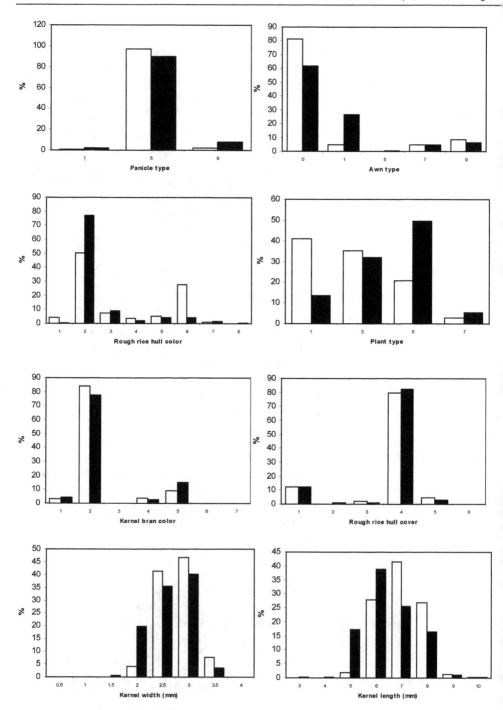

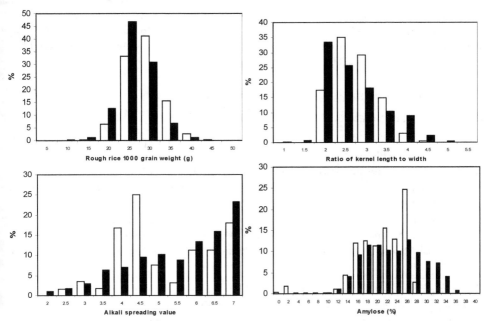

Fig. 1. Comparative distributions of frequency (%) for 14 traits of 1,794 core accessions field-evaluated in 2002 with ~15,000 accessions which data were extracted from the GRIN. Those with no unit are categorical traits, and their category classifications are explained in the GRIN, i.e. Awn type: 0-absent; 1-short and partly awned; 5- short and fully awned; 7-long and partly awned and 9-long and fully awned (Rice Descriptors, 2011).

2.4 Frequency analysis of 14 traits proves that the RCS well represents the RWC

As displayed in Fig. 1, the correlation coefficient (r) of the RCS distribution frequency with RWC was 0.90 for Days to flower, 0.93 for Plant height, 0.93 for Awn type, 0.99 for Panicle type, 0.69 for Plant type, 0.88 for Hull color, 0.99 for Hull cover, 0.99 for Bran color, 0.83 for Kernel length, 0.94 for Kernel width, 0.85 for Kernel length/width ratio, 0.91 for Grain weight, 0.82 for Amylose content and 0.65 for Alkali spreading value (Yan et al., 2007). Taken together, the 14 traits had a high correlation of distribution frequency (r=0.94, P<0.0001) between the RCS and RWC, resulting a determination coefficient (r²) of 0.88. The high correlation of the RCS with the RWC demonstrates that a stratified set of 10% of the accessions can be effectively used to assess the variability in the whole rice collection with 88% certainty. The correlation analysis validates the RCS to be well representative of the RWC for genetic assessment of global rice germplasm.

2.5 The RCS improves genetic characterization of germplasm collection

In an effort to better characterize genetic diversity of the rice collection, the RCS with 10% of over 18,000 accessions in the whole collection is a reasonable size for replicated evaluations. As a result, this core subset has been evaluated for agronomic descriptors (Yan et al., 2005a), kernel dimension traits that impact milling yield and market class (Yan et al., 2005b), resistance to physiological disease 'straighthead' (Agrama and Yan, 2010) and fungal disease 'sheath blight' (*Rhizoctonia solani*) (Jia et al., 2011) and 'blast' (*Magnaporthe oryzae*) (Agrama et al., 2009),   and DNA markers associated with cooking quality and blast resistance (McClung et al., 2004, 2006; Fjellstrom et al., 2006).

3. Geographic analysis of global rice germplasm

3.1 Genotyping and statistical analysis

Total genomic DNA was extracted using a rapid alkali extraction procedure (Xin et al., 2003) from a bulk of five plants derived from a single plant selected to represent each accession in the core collection. Seventy-two (71 SSR and an indel) molecular markers, covering the entire rice genome, approximately with an average of one marker per 30 cM, were used to genotype the 1,794 accessions. PCR amplification of the markers followed the procedure that was described by Agrama et al. (2009). DNA samples were separated on an ABI Prism 3730 DNA analyzer according to the manufacturer's instructions (Applied Biosystems, Foster City, CA, USA). Fragments were sized and binned into alleles using GeneMapper v. 3.7 software.

The 112 countries or districts from which the 1,794 accessions originated were classified into 14 geographic regions according to groupings of the United Nations Statistic Division (UNSD, 2009). Each accession was plotted on the global map using its latitude and longitude coordinates according to the GRIN passport database. The map was built using the 'prcomp' procedure in the statistics module (version 2.8.1) of the R statistical package including 'spatial', 'maps' and 'fields' (Venables and Ripley, 1998, Venables et al., 2008).

PowerMarker software (Liu and Muse, 2005) was used to calculate allele frequencies and polymorphism information content (PIC) values (Botstein et al., 1980) for each marker, region and country. Analysis of molecular variance (AMOVA; Excoffier et al., 1992) was conducted for variance components within and among regions and countries of origin, respectively, using ARLEQUIN 3.0 software (Schneider et al., 2000). Significance of variance components was tested using a non-parametric procedure based on 1,000 random permutations of individuals using the software ARLEQUIN 3.0 (Schneider et al., 2000). Genetic diversity was estimated using Nei diversity index for each accession according to Lynch and Milligan (1994). Geographical distribution of diversity index represented by Kriging methods was globally mapped using the R-script (François et al., 2008).

Genetic relationships among accessions represented by regions and countries were determined by the unweighted pair-group method with an arithmetic mean (UPGMA) analysis based on Nei (Nei, 1973) genetic similarity estimated using the 72 markers. The UPGMA trees were constructed from 1,000 bootstrap replicates using the software PowerMarker (Liu and Muse, 2005) and drawn with MEGA v. 3.1 (Kumar et al., 2004). The number of alleles, which are private to a population and do not exist in other populations, is especially informative when populations are studied with highly variable multi-allelic markers, such as SSRs (Szpiech et al., 2008). The average number of private alleles per locus for core accessions originating in each of 14 geographic regions was estimated using ADZE (Allelic Diversity AnalyZEr) software (Szpiech et al., 2008) with the 72 molecular markers.

3.2 Allelic diversity among 14 geographic regions

A total of 1,005 alleles were revealed by 72 molecular markers, averaging 14 alleles per locus and ranging from 2 to 36. Polymorphic information content (PIC) values averaged 0.66 ± 0.02 ranging from 0.17 to 0.92 with the majority distributed between 0.50 and 0.90. Sixty markers (83%) were highly informative (PIC>0.50), 10 (14%) reasonably informative (0.50>PIC>0.25) and 2 (3%) slightly informative (PIC<0.25), demonstrating a high discriminatory power of these selected markers (Yan et al., 2010).

The 1,794 core accessions were introduced from 112 countries and distributed to 14 worldwide geographic regions with the most countries in Africa and the least in North America (Table 1). Accession number ranged from 57 in Oceania to 224 in South America. China had the most accessions (135), while 34 countries had less than five accessions each

Geographic region	Countries	Accessions	Alleles/ locus	PIC
Africa	26	198	9.32	0.64
Central America	12	116	8.01	0.59
Central Asia	7	61	6.71	0.59
China	4	212	8.58	0.58
Eastern Europe	7	102	6.96	0.45
Middle East	6	91	7.47	0.62
North America	2	75	6.06	0.46
North Pacific	3	108	7.50	0.52
Oceania	6	57	6.79	0.61
South America	12	224	8.44	0.62
South Pacific	4	120	8.42	0.64
Southeast Asia	6	114	8.86	0.66
Southern Asia	7	215	10.06	0.64
West Europe	10	101	6.00	0.39
Total	112	1794		
Mean			7.80	0.57

Table 1. Allelic analysis of 1,794 accessions in the USDA rice core collection genotyped with 72 DNA markers among 14 geographic regions.

AMOVA showed that the majority (89%) of total genetic variance attributed to differences within regions and the rest (11%) was due to variance among regions (Table 2). Likewise, when countries were taken into account, 82 % of the total variation was due to the differences within countries, and the remaining portion of the variance was equally shared by both among regions and among countries. Genetic variations were significantly differentiated among regions (Φ_{st} =0.10, P<0.001) and among countries (Φ_{st} =0.12, P<0.001), and very highly and significantly differentiated within countries (Φ_{st} =0.85, P<0.001).

Source	df	Sum of squares	Mean squares	Φ_{st}	P-value	Estimated variance	Percentage of total variance
Among regions	13	18956.7	1458.2	0.11	<0.001	10.8	11%
Within regions	1780	164044.8	92.2	0.89	<0.001	92.2	89%
Total	1793	183001.5				103.0	100%
Among region	13	19674.3	1513.4	0.10	<0.001	8.9	8.6%
Among countries	65	17106.2	263.2	0.12	<0.001	9.2	8.9%
Within countries	1672	142286.8	84.8	0.85	<0.001	84.8	82.5%
Total	1750	179067.2				102.9	

Table 2. Analysis of molecular variance (AMOVA) in 14 regions for 1,794 accessions in the USDA rice core collection genotyped with 72 DNA markers.

3.3 Genetic diversity and genetic relationships among geographic regions

Rice accessions collected from Southern Asia had the most number of alleles per locus, followed by Africa, Southeast Asia, China, South America, South Pacific and Central America, while those in Western and Eastern Europe, North America and Central Asia had the least (Table 1). As demonstrated by the PIC value, the accessions derived from Southeast Asia had the greatest diversity, followed by Southern Asia, South Pacific, Africa, Middle East, South America and Oceania, while those in Western and Eastern Europe and North America had the lowest diversity. Visualized by Nei Genetic Diversity index on the world map using the Kriging method, germplasm accessions collected from Southern Asia, Southeast Asia, Central America and Africa were mostly diversified, while those from North Pacific, Oceania, Western and Eastern Europe and North America had the lowest diversity (Fig. 2).

Germplasm accessions that were introduced from Southern Asia had the most private alleles per locus, followed by Africa, Central America, Southeast Asia, South Pacific, China, Oceania and Middle East, while those in Eastern Europe, Central Asia, North and South America and Western Europe had the least private alleles per locus (Fig. 3).

Three main clusters were resulted from the UPGMA analysis based on Nei (Nei, 1973) genetic similarity (Fig. 4). In cluster 1, germplasm accessions from South America were mostly related to Central America, and then to Africa, Oceania and North America. Two sub-groups of the originating region among rice accessions obviously existed in cluster 2, while Eastern Europe and Western Europe were in sub-group 1 and Central Asia, Middle East and North Pacific in sub-group 2. In cluster 3, germplasm accessions originating in Southeast Asia were closest to those in the South Pacific, and then to China and the Southern Asia. Cluster 1 was closer to cluster 2 than to cluster 3.

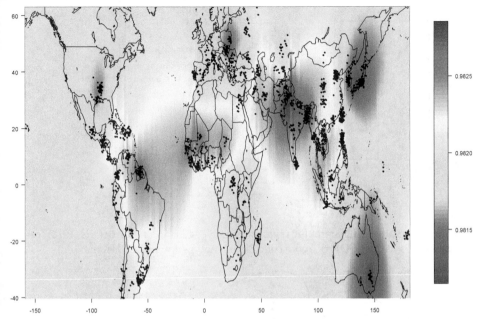

Fig. 2. Geographic diversity of rice germplasm demonstrated by Nei Genetic Diversity Index in the USDA rice core collection genotyped with 72 DNA markers. The deeper the red color is, the greater the genetic diversity is for the area. The deeper the blue color is, the smaller the genetic diversity is for the area. Each dot represents an accession placed on the world map according to its latitude and longitude.

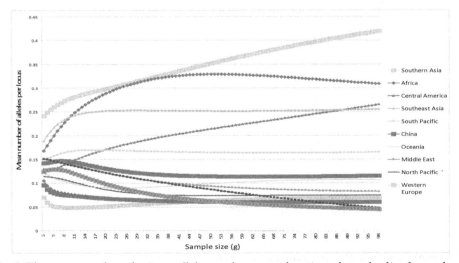

Fig. 3. The mean number of private alleles per locus as a function of standardized sample size (g) for 14 geographic regions arranged from high on the top to low on the bottom for 1,794 accessions in the USDA rice core collection.

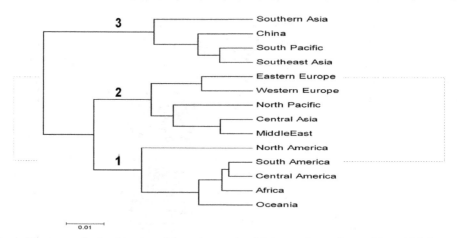

Fig. 4. Cluster analysis of geographic regions using Nei genetic similarity (Nei, 1973) for 1,794 accessions in the USDA rice core collection genotyped with 72 DNA markers

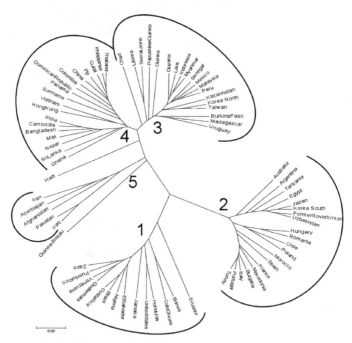

Fig. 5. Cluster analysis of countries having five or more accessions in the USDA rice core collection genotyped with 72 DNA markers

3.4 Genetic diversity and genetic relationships among countries

Among the 78 countries from which 5 or more accessions were introduced, Myanmar had the most diversification indicated by the highest PIC (0.65). The PICs measuring genetic diversities ranged 0.60-0.63 in 13 countries: four in Africa, three in Southeast Asia and two each in South America, South Pacific, and Southern Asia; and 0.50-0.60 in 27 countries: four each in Africa and Central America, three each in South America, and Southern Asia, two each in Central Asia, China, Middle East, and North Pacific, and one each in Eastern Europe, North America, Oceania, Southeast Asia, and South Pacific. There were 22 countries with the PICs ranging 0.40-0.50: five in South America, four each in Africa and Central America, two each in Middle East and Oceania, and one each in Central Asia, China, North Pacific, Southern Asia, and Western Europe. France and Spain in Western Europe and Romania in Eastern Europe had the lowest PIC value.

Cluster analysis of 78 countries from which 5 or more accessions were present in the core collection formed five distinctive groups (Fig. 5). Fourteen countries were placed in Cluster 1, six in Central America, four in South America, three in Africa, and one in North America which is the United States. Cluster 2 contained 20 countries, six in Eastern Europe, four in Western Europe, three in Middle East, two each in North Pacific and South America and one each in Africa, Central Asia and Oceania. Cluster 3 included 19 countries, seven in Africa, three in South America, two each in South Pacific and Southeast Asia, and one each in Central Asia, China, North America, North Pacific and Oceania. Cluster 4 had 18 countries, four in Southern Asia, three each in Central America and Southeast Asia, two each in Africa, China and South America, and one each in Oceania and South Pacific. Cluster 5 was the smallest, including five countries, two each in Middle East and Southern Asia, and one in Central Asia. Two countries each with five accessions were independent of these clusters. Haiti in Central America was between Cluster 4 and 5, while Guinea-Bissau in Africa was between Cluster 1 and 5. The vast diversity found in the USDA global rice collection is an important genetic resource that can effectively support breeding programs in the U.S. and worldwide.

4. Genetic differentiation of global rice germplasm

Cultivated rice (*Oryza sativa* L.) is structured into five genetic groups, *indica* (IND), *aus* (AUS), *tropical japonica* (TRJ), *temperate japonica* (TEJ) and *aromatic* (ARO) (Izawa, 2008; Caicedo et al., 2007; Garris et al., 2005). Genetic characterization of rice germplasm collections will enhance their utilization by the global research community for improvement of rice.

4.1 Statistical analysis

Genotypic data of 71 SSR plus an indel markers for the core collection plus 23 reference cultivars were used to decide putative number of structures at first. Genetic structure was inferred using the admixture analysis model-based clustering algorithms implemented in TESS v. 2.1 (Chen et al., 2007). TESS implements a Bayesian clustering algorithm for spatial population genetics. Multi-locus genotypes were analyzed with TESS using the Markov Chain Monte Carlo (MCMC) method, with the F-model and a ψ value of 0.6 which assumes 0.0 as non-informative spatial prior. To estimate the K number of ancestral-genetic populations and the ancestry membership proportions of each individual in the cluster analysis, the algorithm was run 100 times, each run with a total of 70.000 sweeps and 50.000 burn-in sweeps for each K value from 2 to 15. For each run we computed the Deviance Information Criterion (DIC) (Spiegelhalter et al., 2002), a model-complexity penalized measure to show how well the model fits the data. The putative number of clusters was obtained when the DIC values were the smallest and estimates of data likelihood were the highest in 10% of the runs. Similarity coefficients between runs and the average matrix of ancestry membership were calculated using CLUMPP v. 1.1 (Jakobsson and Rosenberg, 2007).

Each accession in the core collection was grouped to a specific cluster or population by its K value resulted from cluster analysis using TESS. The sub-species ancestry of each K was inferred by the reference cultivars for *indica, AUS, aromatic, temperate japonica,* and *tropical japonica* rices. Analysis of molecular variance (AMOVA; Excoffier et al., 1992) was used to calculate variance components within and among the populations obtained from TESS in the collection. Estimation of variance components was performed using the software ARLEQUIN 3.0 (Schneider et al., 2000). The AMOVA-derived Φ_{ST} (Weir and Cockerham, 1984) is analogous to Wright's F statistics differing only in their assumption of heterozygosity (Paun et al., 2006). Φ_{ST} provides an effective estimate of the amount of genetic divergence or structuring among populations (Excoffier et al., 1992). Significance of variance components was tested using a non-parametric procedure based on 1,000 random permutations of individuals. The computer package ARLEQUIN was used to estimate pair-wise F_{ST} (Goudet, 1995) for the populations obtained from TESS.

Multivariate analysis such as principle component analysis (PCA) provides techniques for classifying the inter-relationship of measured variables. Multivariate geo-statistical methods combine the advantages of geo-statistical techniques and multivariate analysis while incorporating spatial or temporal correlations and multivariate relationships to detect and map different sources of spatial variation on different scales (Goovaerts, 1992; Wackernagel, 1994). Geographical spatial interpolation of principal coordinates of latitude and longitude and admixture ancestry matrix coefficients (Ks) calculated in TESS for each accession were represented by kriging method (François et al., 2008) as implemented in the R statistical packages 'spatial', 'maps' and 'fields' (Venables and Ripley, 1998; Venables et al., 2008) for visualizing distribution in the world map.

Principal components analysis (PCA) was conducted using GenAlex 6.1 (Peakall and Smouse, 2006) software to structure the core collection genotyped by 72 molecular markers, and generate a PC-matrix. Geo-statistical and geographic analysis was based on CNT coordinates of latitude and longitude where a core accession originated using the R statistical packages. Polymorphism information content (PIC) and number of alleles per locus in each sub-species population were estimated using PowerMarker software (Liu and Muse, 2005). Number of distinct alleles in each population and number of alleles private to each population, that is not found in other populations, were calculated using ADZE program (Allelic Diversity AnalyZEr, Szpiech et al., 2008). ADZE uses the rarefaction method to trim unequal accessions to the same standardized sample size, a number equal to the smallest accessions across the populations.

4.2 Number of populations and ancestry determination

Structural analysis resulted in the lowest Deviance Information Criterion (DIC) or highest log likelihood scores when the putative number (K) of populations was set at five, and the ancestry coefficient of each accession in each K was estimated accordingly (Fig. 6) (Agrama et al., 2010). Similarly, principle coordinate (PC) analysis of Nei's genetic distance (Nei, 1973; 1978) classified the core accessions into five clusters by PC1 and PC2 including 71% of total variances (Fig. 7). Both structure and PC analyses indicated that five populations sufficiently explained the genetic diversity in the core collection. Analysis of molecular variance (AMOVA) showed that 38% of the variance was due to genetic differentiation among the populations (Table 3). The remaining 62% of the variance was due to the differences within the populations. The variances among and within the populations were highly significant ($P<0.001$).

Source	df	SS	MS	Est. Var.	%	Φ_{ST}	P-value[a]
Among Pops	4	57383	14346	43	38	0.38	<0.001
Within Pops	1781	124086	70	70	62	0.62	<0.001
Total	1785	181470		112	100		

[a] Probability of obtaining a more extreme random value computed from non-parametric procedures (1,000 permutations).

Table 3. Analysis of molecular variance (AMOVA) for the 1,763 core accessions and 23 reference cultivars for five populations (Pops) of ARO, AUS, IND, TEJ and TRJ based on 72 DNA markers.

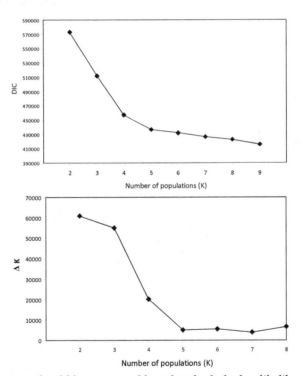

Fig. 6. Five populations should be structured based on both the log-likelihood values (Deviance Information Criterion, DIC) and the change rate of log-likelihood values (ΔK) for estimated number of populations over 50 structure replicated runs using TESS program. Where relatively flat change of both DIC and ΔK occurs indicates the most likely number of populations.

Among 40 reference cultivars, 20 that are known *tropical japonica* (TRJ) were classified in K1, four known *temperate japonica* (TEJ) in K2, eight known *indica* (IND) in K3, three known *AUS* (AUS) in K4 and five known *aromatic* (ARO) in K5, indicating the correspondent ancestry of each population. Based on the references, each accession was clearly assigned to a single population when its inferred ancestry estimate was 0.6 or larger and admixture between populations when its estimate was less than 0.6. Admixture was based on proportion of the estimate, i.e. GSOR 310002 was assigned TEJ-TRJ because of its estimate 0.5227 in K2 and 0.4770 in K1.

K1 or TRJ population included 353 (19.8%) absolute accessions, 41 (2.3%) admixtures with K2 or TEJ population, 26 (1.5%) admixtures with K3 or IND and one admixture with K4 or AUS. In K2, 420 (23.5%) accessions had absolute ancestry, 52 (2.9%) admixed with K1 and seven admixed with other populations. K3 or IND population had 625 (35.0%) accessions among which 595 were clearly assigned, twelve admixed with K4 or AUS, and 18 admixed with other populations. One hundred sixty-five (9.8%) accessions were clearly grouped in K4, 13 were admixed with K3 and two admixed with K5 or ARO population. Seventy-two (4.0%) accessions were clearly structured in K5, five were admixed with K2 and three admixed with other population.

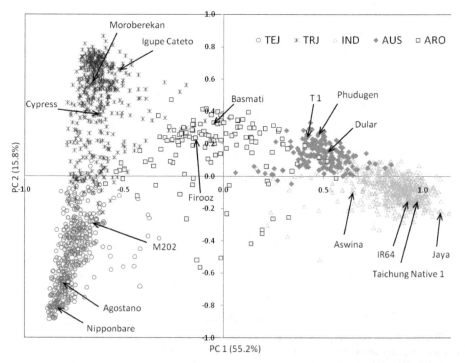

Fig. 7. Principle coodinates analysis of five populations inferred by highlighted reference cultivars (*temperate japonica* – TEJ, *tropical japonica* – TRJ, *indica* - IND, *aus* - AUS *and aromatic* - ARO) for the core accessions genotyped with 72 DNA markers.

4.3 Genetic relationship and global distribution of ancestry populations

All pair-wise estimates of F_{ST} using AMOVA for the populations were highly significant ranging from 0.240 to 0.517 (Table 4). IND was equally distant from ARO and AUS, but more distant from TEJ and TRJ. AUS and IND were mostly differentiated from TEJ. However, TEJ, TRJ and ARO were close to each other in comparison with others. These relationships were consistent with structure analysis revealed by the PCA (Fig. 7).

	ARO	AUS	IND	TEJ	TRJ
ARO		0.001	0.001	0.001	0.001
AUS	0.253		0.001	0.001	0.001
IND	0.284	0.308		0.001	0.001
TEJ	0.317	0.517	0.500		0.001
TRJ	0.240	0.475	0.462	0.273	

Table 4. Pairwise estimates of F_{ST} (lower diagonal) and their corresponding probability values (upper diagonal) for five rice populations, K5 - *aromatic* (ARO), K4 - *aus* (AUS), K3 - *indica* (IND), K2 - *temperate japonica* (TEJ) and K1 - *tropical japonica* (TRJ) for 1,763 core accessions genotyped with 72 DNA markers based on 999 permutations.

Among 421 accessions of TRJ rice in the core collection, the majority is collected from Africa (23%) and South America (21%), followed by Central America (15%), North America (13%), South Pacific (6%), Southeast Asia and Oceania (5% each) (Fig. 8A). North America had 75 accessions in total and 55 were grouped in TRJ, which was the highest percentage (73%) among 14 regions, followed by Central America (56%), Africa (49%) and South America (41%). Among 112 countries, the U.S. in North America had the highest percentage (92%) of accessions, followed by Cote d'Ivoire and Zaire (91%) in Africa and Puerto Rico (72%) in Central America.

Most TEJ rice is collected from Western and Eastern Europe (20% each), followed by North Pacific (14%), South America (10%), Central Asia (7%) and North China (7%) (Fig. 8B). Similarly, Western and Eastern Europe had the highest percentage (85% each) of TEJ, followed by North Pacific (55%) and South America (20%). Hungary accessions had the highest percentage (97%), followed by Italy (89%), Russian Federation and Portugal (83% each).

Based on United Nations' classification, region China includes Mongolia, Hong Kong, Taiwan and China itself. Most IND rice (25%) is collected from region China, followed by the South Asia (14%), South America (13%), Southeast Asia and Africa (10% each) (Fig. 8C). Region China had the highest percentage (72%) of IND, followed by South Pacific (57%), Southeast Asia (53%), Southern Asia (38%) and Africa (29%). Also, country China had the highest percentage (84%) of IND, followed by Columbia (81%), Sri Lanka (80%) and Philippines (68%).

About half of the AUS rice in the collection was sampled from the South Asia (48%), followed by Africa (16%), Middle East (11%), South America and Southeast Asia (7% each) (Fig. 8D). South Asia had the highest percentage (40%) of AUS, followed by

Middle East (21%), Africa (14%) and Southeast Asia (10%). Bangladesh had the highest percentage (63%) of AUS, followed by Iraq (64%), Pakistan (49%) and India (40%).

Aromatic rice in the collection originated mainly from Pakistan (20%) and Afghanistan (13%) in the South Asia and Azerbaijan (15%) in Central Asia, representing 37%, 44% and 57% of total core accessions from these countries, respectively (Fig. 8E).

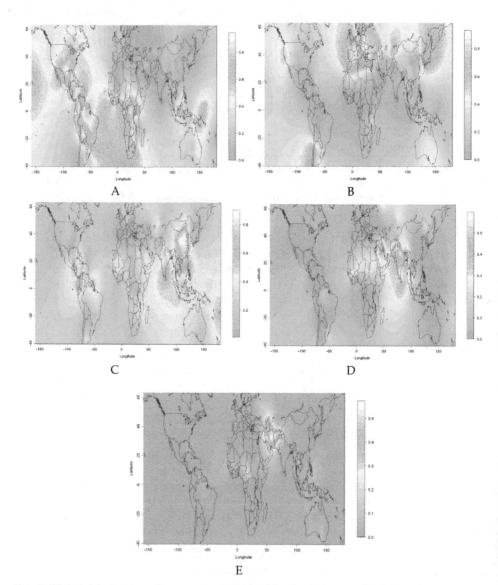

Fig. 8. Global distribution of core accessions in each population resulted from cluster analysis and inferred by reference cultivars based on geographical coordinates of latitude and longitude in K1 (*tropical japonica* – TRJ), A; K2 (*temperate japonica* – TEJ), B; K3 (*indica* – IND), C; K4 (*aus* – AUS), D and K5 (*aromatic* – ARO), E.

4.4 Genetic diversity of the populations

Average alleles per locus were the highest in IND, followed by AUS, ARO, TRJ and TEJ (Fig. 9). IND had 45% more alleles per locus than TEJ. ARO had the highest polymorphic information content (PIC), followed by AUS, IND, TRJ and TEJ. The PIC value of TEJ was 72% less than that of ARO. AUS had the most alleles per locus corrected for difference in sample size distinctly (Fig. 10A) and privately (Fig. 10B) from others. Although IND and ARO had same distinct alleles per locus, which was next to AUS, there were much more private alleles per locus in IND than in ARO. TEJ had either the lowest distinct alleles or private alleles per locus among the populations.

Genetic characterization of the USDA rice world collection for genetic structure, diversity, and differentiation will help design cross strategy to avoid sterility for gene transfer and exchange in breeding program and genetic studies, thus better serve the global rice community for improvement of cultivars and hybrids because this collection is internationally available, free of charge and without restrictions for research purposes. Seed may be requested from GRIN (GRIN, 2011) for the whole collection, and from GSOR (GSOR, 2011) for the core collection.

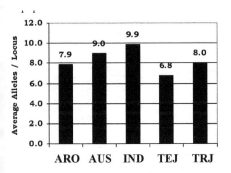 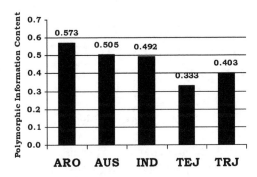

Fig. 9. Average alleles per locus and polymorphic information content for five populations resulted from cluster analysis and inferred by reference cultivars K1 (*tropical japonica* – TRJ), K2 (*temperate japonica* – TEJ), K3 (*indica* – IND), K4 (*aus* – AUS) and K5 (*aromatic* – ARO).

5. USDA rice mini-core collection

Development of core collections is an effective tool to extensively characterize large germplasm collections, and the utilization of a mini-core sub-sampling strategy further increases the effectiveness of genetic diversity analysis at detailed phenotype and molecular levels (Agrama et al., 2009). Using the advanced M strategy, Kim et al. (2007) presented PowerCore software that possesses the power to represent all the alleles identified by molecular markers and classes of the phenotypic observations in the development of core collections.

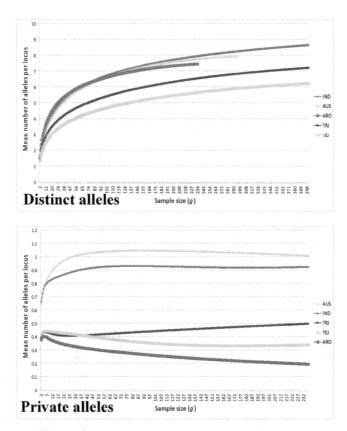

Fig. 10. The mean number of (A) distinct alleles per locus and (B) private alleles per locus to each of five populations, K1 (*tropical japonica* – TRJ), K2 (*temperate japonica* – TEJ), K3 (*indica* – IND), K4 (*aus* – AUS) and K5 (*aromatic* – ARO), as functions of standardized sample size *g*.

5.1 Phenotypic and genotypic data used to develop the USDA rice mini-core collection

Data of 26 phenotypic traits, 69 SSRs and one indel marker generated from 1,794 accessions in the USDA rice core collection at Stuttgart, Arkansas, USA were used to develop the mini-core. The phenotypic traits included 13 for morphology, two for cooking quality, 10 for rice blast disease resistance ratings from individual races of *Magnaporthe oryzae* Cav., and one for physiological disease, straighthead. Field evaluations of blast were conducted at the University of Arkansas Experiment Station, Pine Tree, AR following inoculation using a mixture of the most prevalent races (IB-1, IB-49, IC-17, IE-1, IE-1K, IG-1 and IH-1) found in the southern US rice production region using the method described by Lee et al. (2003). In greenhouse, seven blast races, IB-1, IB-33, IB-49, IC-17, IE-1K, IG-1, and IH-1 were individually inoculated and rated in a scale from 0 (no lesions) to 9 (dead).

5.2 Sampling strategy and representation analysis

Sampling the core collection was performed by the PowerCore software with an effort to maximize both the number of observed alleles at SSR loci and the number of phenotypic trait classes using the advanced M (maximization) strategy implemented through a modified heuristic algorithm (Agrama et al., 2009). The phenotypic traits were automatically classified into different categories or classes by the PowerCore program based on Sturges' rule = $1 + Log_2(n)$, where n is the number of observed accessions (Kim et al., 2007).

The resulting mini-core was compared with the original core collection to assess its homogeneity. Nei genetic diversity index (Nei, 1973) was estimated for each molecular marker in both the core and mini-core collections. Chi-squared (χ^2) tests were used to test the similarity for number of marker alleles and frequency distribution of accessions. Homogeneity was further evaluated for the 26 phenotypic traits using the Newman-Keuls test for means, the Levene test (Levene, 1960) for variances, and the mean difference (MD%), variance difference (VD%), coincidence rate of range (CR%) and variable rate of coefficient of variance (VR%) according to Hu et al. (2000). Coverage of all the phenotypic traits in the original core collection was estimated in the mini-core as proposed by Kim et al. (2007):

$$\text{Coverage (\%)} = \frac{1}{m}\sum_{j=1}^{m}\frac{Dc}{De}\times 100$$

Where Dc is the number of classes occupied in the mini-core and De is the number of classes occupied in the original core collection for each trait and m is the number of traits which is 26 in this case.

5.3 Distribution frequency of accessions in the core and mini-core collections

The heuristic search based on the 26 phenotypic traits and the 70 markers sampled 217 accessions (12.1%) out of 1,794 accessions in the core collection. The 217 mini-core entries originated from 76 countries covering all the 15 geographic regions (Table 5). Five regions, Subcontinent, South Pacific, Southeast Asia, Africa and China accounted for the majority, 63.6% of the mini-core entries, while the fewest entries came from three regions, Australia, Mideast and North America, accounting for 5.5%. Two accessions in the mini-core are of unknown origin.

The similarity of distribution frequencies between the core and mini-core collections for each of the 15 regions was tested using χ^2 with one degree of freedom (Table 5). All 15 regions had non-significant χ^2 values ranging from 0.095 to 0.996 with probability (P) from 0.303 to 0.758, which proved a homogeneous distribution between the two collections.

Among the 217 mini-core *Oryza* entries, eight belong to *O. glaberrima*; two each of *O. nivara* and *rufipogon*; one each of *O. glumaepatula*, *latifolia*, and the remaining 203 entries belong to *O. sativa*.

Region	USDA Rice Core collection		Mini-core		χ^2	P
	Number	%	Number	%		
Africa	198	11.0	24	11.1	0.996	0.318
Australia	24	1.3	1	0.5	0.513	0.474
Balkans	61	3.4	9	4.2	0.786	0.375
Central America	116	6.5	12	5.5	0.787	0.375
China	208	11.6	20	9.2	0.602	0.438
Eastern Europe	102	5.7	9	4.2	0.624	0.430
Mideast	91	5.1	5	2.3	0.308	0.579
North America	71	4.0	6	2.8	0.646	0.422
North Pacific	108	6.0	11	5.1	0.775	0.379
South America	224	12.5	15	6.9	0.206	0.650
South Pacific	152	8.5	24	11.1	0.558	0.455
Southeast Asia	114	6.4	23	10.6	0.303	0.303
Subcontinent	215	12.0	47	21.7	0.095	0.758
Western Europe	101	5.6	9	4.2	0.635	0.425
Unknown	9	0.5	2	0.9	0.725	0.725
Total	1794	100	217	100		

*χ^2 values with one degree of freedom and the corresponding probability (P).

Table 5. Distribution frequency comparison of origin of accessions between the USDA rice core and mini-core collections among 15 geographical regions.

5.4 Phenotypic diversity in the core and mini-core collections

Comparative analysis of the ranges, means and variances for 26 phenotypic traits demonstrated that the mini-core covered full range of variation for each trait. The Newman-Keuls test results indicate the presence of homogeneity of means between the core collection and mini-core for 22 traits (85%). Sixteen (62%) of the traits had homogeneous variances revealed by the Levene's test. Among the 10 traits having heterogeneous variances, five morphological traits and amylose content had greater variances in the mini-core than in the core collection. However, hull cover and color, and two disease traits had smaller variances.

The mean difference percentage (MD%), the variance difference percentage (VD%), the coincidence rate (CR%) and the variable rate (VR%) are designed to comparably evaluate the property of core collection with its initial collection. Over the entire 26 phenotypic traits, the MD% was 6.3%, far less than the significance level of 20%. The VD% was 16.5%, less than the significance level of 20%, and six traits had much greater variances in the mini-core than in the core collection (Table 6). The VR% compares the coefficient of variation values and determines how well the variance is being represented in the mini-core. More than 100% of VR is required for a core collection to be representative of its original collection (Hu et al., 2000). The mini-core had 102.7% VR over its originating core, indicating good representation.

	USDA Rice Core Collection			Mini-core			Test[1]	
	Range	Mean	Variance	Range	Mean	Variance	N-K	Lev
Morphology								
Days to flower	42 - 174	95.8	355.5	46 - 166	96.2	469.6	n.s.	*
Plant height cm	60 - 212	125.8	627.3	70 - 202	135.7	646.6	**	n.s.
Plant type[2]	1 - 9	2.7	2.82	1 - 9	2.7	3.01	n.s.	n.s.
Lodging[2]	0 - 9	2.3	4.98	0 - 9	3.1	7.71	**	**
Panicle type[2]	1 - 9	4.9	1.20	1 - 9	4.8	2.31	n.s.	*
Awn type[2]	0 - 9	1.2	8.27	0 - 9	2.0	12.56	**	**
Hull cover[2]	1 - 6	3.6	1.20	1 - 6	3.7	0.78	n.s.	*.
Hull color[2]	1 - 8	3.5	3.55	1 - 8	3.7	1.94	n.s.	*
Bran color[2]	1 - 7	2.3	1.09	1 - 7	2.5	1.75	n.s.	*
Kernel length mm	4.2 -10.0	6.5	0.63	4.2 - 10.1	6.5	0.95	n.s.	n.s.
Kernel width mm	1.5 - 3.5	2.6	0.11	1.5 - 3.5	2.6	0.10	n.s.	n.s.
Kernel Length/Width	2.0 - 5.0	2.6	0.35	2.0 - 5.0	2.6	0.45	n.s.	n.s.
1000 kernel weight g	6.72 - 37.4	21.2	14.76	10 .0 - 37.4	21.0	18.45	n.s.	n.s.
Quality								
Amylose %	0 - 26.9	19.9	25.57	0.10 – 26.5	10.5	38.46	n.s.	*
ASV[2]	2.1 - 7	5.1	1.59	2.3 – 7.0	4.9	1.47	n.s.	n.s.
Disease								
Leaf blast	0 - 9	4.5	7.50	0.3 - 9	4.9	7.88	n.s.	n.s.
Early panicle blast	0 - 9	4.1	8.63	0 - 9	4.1	8.26	n.s.	n.s.
Final panicle	0 - 9	5.0	8.00	0 - 9	4.9	8.40	n.s.	n.s.

	USDA Rice Core Collection			Mini-core			Test[1]	
blast								
Blast IB-1	0 - 8	4.0	9.24	0 - 8	3.9	8.60	n.s.	n.s.
Blast IB-33	0 - 8	6.1	1.7	0 - 8	6.1	1.74	n.s.	n.s.
Blast IB-49	0 - 8	5.0	9.27	0 - 8	5.0	8.60	n.s.	n.s.
Blast IC-17	0 - 8	4.0	10.58	0 - 8	3.4	9.93	*	n.s.
Blast IG-1	0 - 8	4.0	10.68	0 - 8	4.0	9.73	n.s.	*
Blast IE-1K	0 - 8	4.3	8.74	0 - 8	4.6	7.75	n.s.	*
Blast IH-1	0 - 8	1.8	5.78	0 - 8	2.0	5.45	n.s.	n.s.
Straighthead[2]	1 - 9	7.3	1.90	1.3 - 9	7.-5	1.83	n.s.	n.s.

[1] Means were tested using Newman-Keuls test (N-K) and variances were tested by Levene's test (Lev) for homogeneity between the USDA rice core collection and mini-core, * and ** significant at 0.05 and 0.01 probability, respectively.

[2] Categorical data as described in the GRIN (GRIN, 2011).

Table 6. Comparison of range, mean and variance between the USDA rice core collection and the mini-core for 26 phenotypic traits.

The coincidence rate (CR%) indicates whether the distribution ranges of each trait in the mini-core are well represented when compared to the core collection. The resulting CR over the 26 traits was 97.5%, indicating homogeneous distribution ranges of the phenotypic traits because it was larger than the recommended 80% (Kim et al., 2007). The calculated Coverage value for the resulting mini-core was 100%, suggesting there is full coverage of all the diversity present in each class of phenotypic traits in the USDA rice core collection.

5.5 Molecular diversity in the core and mini-core collections

Both the USDA rice core collection and mini-core contained the same total number of polymorphic alleles (= 962 alleles) produced by the 70 markers, with an average of 14 alleles per locus, ranging from two for RM338 to 37 for RM11229 (Fig. 7A). Total alleles per locus ranged from 2 to 9 for 24 markers, from 10 to 19 for 32 markers and from 20 to 37 for 14 markers. The Nei genetic diversity index values reveal the allelic richness and evenness in the population. Distributions of the Nei indices among the 70 markers were very similar between the core and mini-core collections (Fig. 7B). The core collection had an average Nei diversity index of 0.72 with a minimum of 0.24 for AP5625-1 and maximum of 0.94 for RM11229 and RM302, while the average was 0.76 with a minimum of 0.37 for RM338 and AP5625-1 and maximum of 0.95 for RM11229 and RM302 in the mini-core. The minor difference of the molecular diversity was not statistically significant. Similarly, none of the 70 markers had significantly different Nei diversity index between the core and mini-core collections, indicated by the χ^2 test with values ranging from 0.000 to 0.022 and probabilities ranging from 0.882 to 0.999. More than 60% of the markers have a diversity index higher than 0.60 indicating high diversity across the markers (Fig. 7).

A

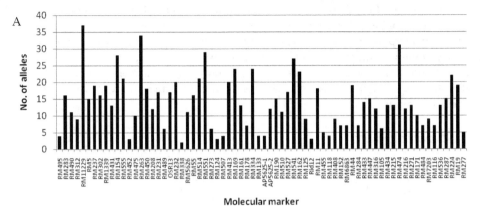

B

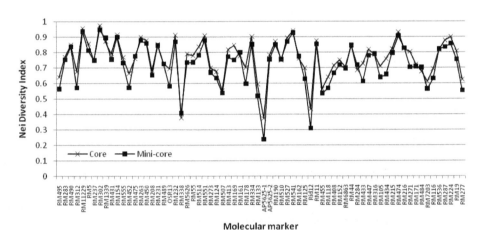

Fig. 7. Distribution of number of alleles per locus and Nei diversity index among the 70 DNA markers in the USDA rice core collection (Core) and mini-core (Mini-core). The markers were placed according to their potion within the rice genome.

6. Use the USDA rice mini-core collection for mining valuable genes

Demonstrated both phenotypically and genotypically, the USDA rice mini-core collection of 217 entries is a good representative of the core of 1,794 entries as well as the entire rice global genebank of more than 18,000 accessions in the US (Yan et al., 2007; Agrama et al., 2009). The vast genetic diversity means the richness of valuable genes that could be extracted for cultivar improvement (Li et al., 2010). The reasonable number of entries in the mini-core allows extensively phenotyping and genotyping for mining valuable genes. The phenotyping could be performed in replicated tests and in multi-locations for the traits that are largely affected by environments such as yield (Li et al., 2011) and that require large amount of resources such as biotic and abiotic stresses. The genotyping could be done genome-wide with high density of molecular markers such as simple sequence repeat (SSR) or single nucleotide polymorphism (SNP), or with sequencing the entire genome. The reliably phenotyping and densely genotyping genome-wide will improve the efficiency and accuracy of mining valuable genes for a globally sustainable agriculture. The core and mini-core collections are managed by the Genetic Stock *Oryza* Collection (GSOR, 2011) at the USDA-ARS Dale Bumpers National Rice Research Center and are available to the global research community.

7. Acknowledgement

The author thanks J.N. Rutger, R.J. Bryant, H.E. Bockelman, R.G. Fjellstrom, M.H. Chen, T.H. Tai, A.M. McClung, H.A. Agrama, F.N. Lee, M. Jia, T. Sookaserm, T. Beaty, A. Jackson, L. Bernhardt and Y. Zhou for their assistance to the project

8. References

Agrama H.A., W.G. Yan, F.N. Lee, R. Fjellstrom, M-H. Chen, M. Jia and A. McClung. 2009. Genetic assessment of a mini-core developed from the USDA rice genebank. Crop Sci. 49:1336-1346.

Agrama H.A., W.G. Yan, M. Jia, R. Fjellstrom and A.M. McClung. 2010. Genetic structure associated with diversity and geographic distribution in the USDA rice world collection. Natural Science 2:247-291.

Agrama, H.A. and W.G. Yan. 2010. Genetic diversity and relatedness of rice cultivars resistant to straighthead disorder. Plant Breeding 129:304-312.

Bockelman, H.E., R.H. Dilday, W. Yan, and D.M. Wesenberg. 2002. Germplasm collection, preservation, and utilization. C.W. Smith and R. H. Dilday (eds.) *Rice: Origin, History, Technology, and Production. Crop Production,* pp. 597-625. Ser. #6149. John Wiley & Sons, Inc. New York, NY.

Botstein, D., R.L. White, M. Skolnick, and M.R. Davis. 1980. Construction of a genetic linkage map in man using restriction fragment length polymorphisms. American J. of Human Genet. 32:314–331.

Bretting, P.K. 2007. The U.S. national plant germplasm system in an era of shifting international norms for germplasm exchange. Proc. XXVII IHC-S1 Plant Gen. Reso. Acta Hort. 760:55-60.

Brown, A.H.D. 1989. Core collections: a practical approach to genetic resources management. Genome, 31: 818-824.

Caicedo, A.L., S.H. Williamson, R.D. Hernandez, A. Boyko, et al. 2007. Genome-wide patterns of nucleotide polymorphism in domecticated rice. PloS Genetics 3:1745-1756.

Chen, C., E. Durand, F. Forbes, and O. François. 2007. Bayesian clustering algorithms ascertaining spatial population structure: A new computer program and a comparison study. Molecular Ecology Notes 7:747–756.

Excoffier. L., P.E. Smouse, and J.M. Quattro. 1992. Analysis of molecular variance inferred from metric distances among DNA haplotypes: Application to human mitocondrial DNA restriction sites. Genetics 131:479-491.

FAO. 1996. Report on the state of the world's plant genetic resources for food and agriculture, prepared for the FAO International Technical Conference on Plant Genetic Resources. FAO Leipzig, Germany, pp 17–23.

Fjellstrom, R.G., W. Yan, M.H. Chen, R.J. Bryant, H. Bockelman, and A.M. McClung. 2006. Genotypic and phenotypic assessment of the NSGC rice core collection for amylose content and alkali spreading value. Proc. 31st Rice Technical Working Group Conference. Feb. 26-Mar. 1, 2006, Houston, TX.

François, O., M.G.B. Blum, M. Jakobsson, and N.A. Rosenberg. 2008. Demographic history of European populations of *Arabidopsis thaliana*. PLoS Genetics 4(5): e1000075.

Garris, A.J., T.H. Tai, J. Coburn, S. Kresovich, and S. McCouch. 2005. Genetic structure and diversity in *Oryza sativa* L. Genetics 169:1631–1638

Goovaerts, P. 1992. Factorial kriging analysis: a useful tool for exploring the structure of multivariate spatial soil information. J Soil Sci 43:597–619.

Goudet, J. 1995. Fstat version 1.2: a computer program to calculate F statistics. J of Heredity 86: 485–486.

GRIN, 2011. Germplasm Resources Information Network.
    http://www.ars-grin.gov/ (Verified on September 12, 2011).

GSOR. 2011. Genetic Stock Oryza Cllection.
    http://www.ars.usda.gov/Main/docs.htm?docid=8318 (Verified on September 12, 2011).

Hamilton, R.S., and R. Raymond. 2005. Toward a global strategy for the conservation of rice genetic resources. In: Toriyama K, Heong KL, Hardy B (ed) Rice is life: scientific perspectives for the 21st century. Proceedings of the World Rice Research Conference held in Tsukuba, Japan, CD-ROM, pp 47–49.

Hu, J., J. Zhu, and H.M. Xu. 2000. Methods of constructing core collections by stepwise clustering with three sampling strategies based on the genotypic values of crops. Theor. Appl. Genet. 101:264–268.

Izawa, T. 2008. The process of rice domestication: a new model based on recent data. Rice 1:127–134.

Jakobsson, M., and N. A. Rosenberg. 2007. CLUMPP: a cluster matching and permutation program for dealing with label switching and multimodality in analysis of population structure. Bioinformatics 23:1801–1806.

Jia, L.M., W.G. Yan, H.A. Agrama, K. Yeater, X.B. Li, B.L. Hu, K. Moldenhauer, A. McClung and D.X. Wu. 2011. Searching for germplasm resistant to sheath blight from the USDA rice core collection. Crop Sci. 51:1507-1517.

Kim, K.W., H.K. Chung, G.T. Cho, K.H. Ma, D. Chandrabalan, J.G. Gwag, T.S. Kim, E.G. Cho, and Y.J. Park. 2007. PowerCore: a program applying the advanced M strategy with a heuristic search for establishing core sets. Bioinformatics 23:2155-2162.

Kumar, S., K. Tamura, and M. Nei. 2004. MEGA3: Integrated software for Molecular Evolutionary Genetics Analysis and sequence alignment. Brief Bioinform. 5: 150–163.

Lee, F.N., W.G. Yan, J.W. Gibbons, M.J. Emerson, and S.D. Clark. 2003. Rice blast and sheath blight evaluation results for newly introduced rice germplasm. *In* R.J. Norman, J-F.

Meullenet (eds.) BR Wells Rice Research Studies 2002. Univ. of Arkansas, Agric. Exp. Stn., Res. Ser. 504:85–92.

Levene, H. 1960. Robust test for equality of variances. *In* I. Olkin (ed). Contributions to probability and statistics: essays in honor of Harold Hotelling. Stanford University Press, Stanford, pp 278–292.

Li, X.B., W.G. Yan, H. Agrama, B.L. Hu, L.M. Jia, M. Jia, A. Jackson, K. Moldenhauer, A. McClung and D.X. Wu. 2010. Genotypic and phenotypic characterization of genetic differentiation and diversity in the USDA rice mini-core collection. Genetica 138:1221-1230.

Li, X.B., W.G. Yan, L.M. Jia, X. Sheng, A. Jackson, K. Moldenhauer, K. Yeater, A. McClung and D.X. Wu. 2011. Mapping QTLs for improving grain yield using the USDA rice mini-core collection. Planta, DOI 10.1007/s00425-011-1405-0.

Little, R. R., G. B. Hilder and E. H. Dawson. 1958. Differential effect of dilute alkali on 25 varieties of milled white rice. Cereal Chem. 35: 111-126.

Liu, K., and S.V. Muse. 2005. PowerMarker: integrated analysis environment for genetic marker data. Bioinformatics 21:2128–2129.

Lynch, M., and B.G. Milligan. 1994. Analysis of population genetic structure with RAPD markers. Mol. Ecol. 3: 91–99.

McClung, A., M. Chen, H.E. Bockelman, R.J. Bryant, W. Yan, and R. Fjellstrom. 2004. Characterization of a core collection of rice germplasm and elite breeding lines in the US with genetic markers associated with cooking quality. *Proc. 2nd Int'l Sym. Rice Functional Genomics*, Poster 127, Tucson, AZ, Nov. 15-17, 2004.

McClung, A.M., W. Yan, Y. Jia, F.N. Lee, M.A. Marchetti, and R.G. Fjellstrom. 2006. Genotypic and phenotypic assessment of the NSGC core collection of rice for resistance to *Pyricularia grisea*. Proc. 31st Rice Technical Working Group Conference. Feb. 26-Mar. 1, 2006, Houston, TX.

Nei, M. 1973. The theory and estimation of genetic distance. In: Morton NE (ed) Genetic structure of populations, Uni. Press Hawaii, Honolulu, HI, USA: 45–54.

Nei, M. 1978. Estimation of average heterozygosity and genetic distance from a small number of individuals. Genetics 89:583–590.

NSGC. 2011. National Small Grains Collection. http://www.ars-grin.gov/cgi-bin/npgs/html/site_holding.pl?NSGC (verified on August 29, 2011)

Paun, O., J. Greilhuber, E.M. Temsch, and E. Hörandl. .2006. Patterns, sources and ecological implications of clonal diversity in apomictic *Ranunculus carpaticola* (*Ranunculus auricomus* complex, Ranunculaceae). Mol Ecol 15:897–910.

Peakall, R., and P.E. Smouse. 2006. GenAlEx 6: Genetic analysis in Excel. Population genetic software for teaching and research. Mol Ecol Notes 6:288–295.

Pérez, C. M., and B.O., Juliano. 1978. Modification of the simplified amylose test for milled rice. Starch/Stärke, 30: 424-426.

Rice Descripotors. 2011. Multiple Descriptors Query for Rice. http://www.ars-grin.gov/cgi-bin/npgs/html/desc_form.pl?75 (Verified on September 12, 2011.

SAS Institute. 2004. SAS/STAT user's guide. Version 9.1.3 ed. SAS Inst., Cary, NC.

Schneider, S., D. Roessli., and L. Excoffier. 2000. Arlequin: a software for population genetic data. Genetics and Biometry Laboratory, University of Geneva, Switzerland.

Spiegelhalter, D.J., N.G. Best, B.P. Carlin, and van der Linde A. 2002. Bayesian measures of model complexity and fit (with discussion). J R Stat Soc B 64:583–639.

Steiner, J.J., P.R. Beuselinck, S.L. Greene, J.A. Kamn, J.H. Kirkbride, and C.A. Roberts. 2001. A description and interpretation of the NPGS birdsfoot trefoil core subset collection. Crop Sci. 41: 1968-1980.

Szpiech, Z.A., M. Jakobsson, and N.A. Rosenberg. 2008. ADZE: a rarefaction approach for counting alleles private to combinations of populations. Bioinfor. 24:2498-2504.

UNSD. 2009. United Nations Statistics Division, Geographic regions and composition. http://unstats.un.org/unsd/methods/m49/m49regin.htm

Venables, W.N. and B.D. Ripley. 1998. Modern Applied Statistics with S+, 2nd Edition. New York: Springer.

Venables, W.N., D.M. Smith and Development Core Team. 2008. Notes on R: A Programming Environment for Data Analysis and Graphics Version 2.8.1. Vienna: R Foundation for Statistical Computing.

Wackernagel, H. 1994. Cokriging versus kriging in regionalized multivariate data analysis. Geoderma 62:83–92.

Webb, B.D. 1972. An automated system of amylose analysis in whole-kernel rice. Cereal Sci. Today, 30: 284.

Weir, B., and C.C. Cockerham. 1984. Estimating F statistics for the analysis of population structure. Evolution 38:1358–1370.

Xin, Z., J.P. Velten, M.J. Oliver, and J.J. Burke. 2003. High-throughput DNA extraction method suitable for PCR. Bio. Techniques 34:820-826.

Yan, W., F.N. Lee, J.Neil Rutger, K.A.K. Moldenhauer, and J.W. Gibbons. 2002. Chinese germplasm evaluation for yield and disease resistance. R.J. Norman and J.-F. Meullenet, (eds.). B.R. Wells Rice Research Studies 2001. pp. 349-358, Univ. of AR, Agri. Exp. Sta., Res. Ser. 495.

Yan, W., J.Neil Rutger, R.J. Bryant, F.N. Lee, and J.W. Gibbons. 2003. Characteristics of newly introduced accessions in the USDA-ARS rice quarantine program. R.J. Norman and J.-F. Meullenet, (eds.). B.R. Wells Rice Research Studies 2002. pp. 112-124, Univ. of AR, Agri. Exp. Sta., Res. Ser. 504.

Yan, W., J. Neil Rutger, H.E. Bockelman, and T.H. Tai. 2005a. Agronomic evaluation and seed stock establishment of the USDA rice core collection. R.J. Norman, J.-F. Meullenet and K.A.K. Moldenhauer, (eds.), B.R. Wells Rice Research Studies 2004. pp. 63-68, Univ. of AR, Agri. Exp. Sta., Res. Ser. 529.

Yan, W., J. Neil Rutger, H.E. Bockelman, and T.H. Tai. 2005b. Evaluation of kernel characteristics of the USDA rice core collection. R.J. Norman, J.-F. Meullenet and K.A.K. Moldenhauer, (eds.), B.R. Wells Rice Research Studies 2004. pp. 69-74, Univ. of AR, Agri. Exp. Sta., Res. Ser. 529.

Yan, W., J. N. Rutger, R.J. Bryant, H.E. Bockelman, R.G. Fjellstrom, M.H. Chen, T.H.Tai, and A.M. McClung. 2007. Development and evaluation of a core subset of the USDA rice germplasm collection. Crop Sci. 47:869-878.

Yan, W.G., H. Agrama, M. Jia, R. Fjellstrom and A. M. McClung. 2010. Geographic description of genetic diversity and relationships in the USDA rice world collection. Crop Science 50:2406-2417.

Yang, X.C., and C.N. Hwa. 2008. Genetic modification of plant architecture and variety improvement in rice. Heredity, 101:396-404.

Permissions

The contributors of this book come from diverse backgrounds, making this book a truly international effort. This book will bring forth new frontiers with its revolutionizing research information and detailed analysis of the nascent developments around the world.

We would like to thank Dr. Anna Aladjadjiyan, for lending her expertise to make the book truly unique. She has played a crucial role in the development of this book. Without her invaluable contribution this book wouldn't have been possible. She has made vital efforts to compile up to date information on the varied aspects of this subject to make this book a valuable addition to the collection of many professionals and students.

This book was conceptualized with the vision of imparting up-to-date information and advanced data in this field. To ensure the same, a matchless editorial board was set up. Every individual on the board went through rigorous rounds of assessment to prove their worth. After which they invested a large part of their time researching and compiling the most relevant data for our readers. Conferences and sessions were held from time to time between the editorial board and the contributing authors to present the data in the most comprehensible form. The editorial team has worked tirelessly to provide valuable and valid information to help people across the globe.

Every chapter published in this book has been scrutinized by our experts. Their significance has been extensively debated. The topics covered herein carry significant findings which will fuel the growth of the discipline. They may even be implemented as practical applications or may be referred to as a beginning point for another development. Chapters in this book were first published by InTech; hereby published with permission under the Creative Commons Attribution License or equivalent.

The editorial board has been involved in producing this book since its inception. They have spent rigorous hours researching and exploring the diverse topics which have resulted in the successful publishing of this book. They have passed on their knowledge of decades through this book. To expedite this challenging task, the publisher supported the team at every step. A small team of assistant editors was also appointed to further simplify the editing procedure and attain best results for the readers.

Our editorial team has been hand-picked from every corner of the world. Their multi-ethnicity adds dynamic inputs to the discussions which result in innovative outcomes. These outcomes are then further discussed with the researchers and contributors who give their valuable feedback and opinion regarding the same. The feedback is then collaborated with the researches and they are edited in a comprehensive manner to aid the understanding of the subject.

Apart from the editorial board, the designing team has also invested a significant amount of their time in understanding the subject and creating the most relevant covers. They scrutinized every image to scout for the most suitable representation of the subject and create an appropriate cover for the book.

The publishing team has been involved in this book since its early stages. They were actively engaged in every process, be it collecting the data, connecting with the contributors or procuring relevant information. The team has been an ardent support to the editorial, designing and production team. Their endless efforts to recruit the best for this project, has resulted in the accomplishment of this book. They are a veteran in the field of academics and their pool of knowledge is as vast as their experience in printing. Their expertise and guidance has proved useful at every step. Their uncompromising quality standards have made this book an exceptional effort. Their encouragement from time to time has been an inspiration for everyone.

The publisher and the editorial board hope that this book will prove to be a valuable piece of knowledge for researchers, students, practitioners and scholars across the globe.

List of Contributors

Henning Høgh-Jensen
Department of Environmental Science, Aarhus University, Denmark

Clinton Beckford
Faculty of Education, University of Windsor, Ontario, Canada

R.A. Olawepo
Department of Geography and Environmental Management, University of Ilorin, Ilorin, Nigeria

Morgan Wairiu, Murari Lal and Viliamu Iese
Pacific Centre for Environment and Sustainable Development, University of the South Pacific, Fiji

Issa Ouedraogo, Korodjouma Ouattara, Séraphine Kaboré/Sawadogo and Souleymane Paré
Institut de l'Environnement et de Recherches Agricoles (INERA), Ouagadougou, Burkina Faso

Jennie Barron
Stockholm Environmental Institute (SEI), Stockholm, Sweden

Peter R.W. Gerritsen
Departamento de Ecología y Recursos Naturales - Imecbio, Centro Universitario de la Costa Sur,
Universidad de Guadalajara, Mexico

P. Ralevic and G.W. vanLoon
School of Environmental Studies, Queen's University, Kingston, Ontario, Canada

S.G. Patil
University of Agricultural Sciences (Raichur), India

P. Ralevic
Faculty of Forestry, University of Toronto, Ontario, Canada

Hans Morten Haugen
Diakonhjemmet University College, Oslo, Norway

Anna Aladjadjiyan
Agricultural University, Plovdiv, Bulgaria

Federica Cheli, Luciano Pinotti and Vittorio Dell'Orto
Dipartimento di Scienze e Tecnologie, Veterinarie per la Sicurezza Alimentare, Università degli Studi di Milano, Milano, Italy

Anna Campagnoli
Università Telematica San Raffaele Roma, Roma, Italy

Patricia Regal, Alberto Cepeda and Cristina A. Fente
University of Santiago de Compostela, Spain

Ricardo Communod, Massimo Faustini, Luca Maria Chiesa and Daniele Vigo
Department of Veterinary Sciences and Technologies for Food Safety, Faculty of Veterinary Medicine, University of Milan, Italy

Maria Luisa Torre
Department of Drug Sciences, Faculty of Pharmacy, University of Pavia, Italy

Mario Lazzati
Pavia Breeders Association Director, Italy

Dragana Krstic and Dragoslav Nikezic
University of Kragujevac, Faculty of Science, Serbia

Ivica Djalovic
Institute of Field and Vegetable Crops, Serbia

Dragana Bjelic
University of Kragujevac, Faculty of Agronomy, Serbia

Wengui Yan
United States Department of Agriculture, Agricultural Research Service (USDA-ARS), Dale Bumpers National Rice Research Center, USA